WITHDRAWN

Mycotoxins and N-Nitroso Compounds: Environmental Risks

Volume I

Editor

Ronald C. Shank
Associate Professor of Toxicology
Department of Community and Environmental Medicine and
Department of Medical Pharmacology and Therapeutics
College of Medicine
University of California
Irvine, California

CRC Press, Inc.
Boca Raton, Florida

Library of Congress Cataloging in Publication Data
Main entry under title:

Mycotoxins and N-nitroso compounds: environmental risks.

Bibliography: p.
Includes index.
1. Mycotoxicoses. 2. Mycotoxins--Environmental aspects. 3. Nitroso-compounds--Toxicology. 4. Nitroso-compounds--Environmental aspects. I. Shank, Ronald C. [DNLM: 1. Mycotoxins--Poisoning. 2. Carcinogens, Environmental--Poisoning. 3. Nitroso compounds--Poisoning. QW 630 M994]
RA1242.M94M94   615.9'529   81-623
ISBN 0-8493-5307-6 (v. 1)   AACR1
ISBN 0-8493-5308-4 (v. 2)

This book represents information obtained from authentic and highly regarded sources. Reprinted material is quoted with permission, and sources are indicated. A wide variety of references are listed. Every reasonable effort has been made to give reliable data and information, but the author and the publisher cannot assume responsibility for the validity of all materials or for the consequences of their use.

All rights reserved. This book, or any parts thereof, may not be reproduced in any form without written consent from the publisher.

Direct all inquiries to CRC Press, Inc., 2000 N.W. 24th Street, Boca Raton, Florida 33431.

© 1981 by CRC Press, Inc.

International Standard Book Number 0-8493-5307-6 (Volume I)
International Standard Book Number 0-8493-5308-4 (Volume II)

Library of Congress Card Number 81-623
Printed in the United States

PREFACE

The reader not familiar with the development of the field of environmental toxicology may find the combination of mycotoxins and N-nitroso compounds strange for the subject of one book. In fact, these two classes of toxic agents have been largely responsible for demonstrating the relevance of the recently formed subdiscipline of toxicology termed "food toxicology".

Recognition of human health problems associated with toxic mold metabolites (mycotoxins) is centuries old, but the whole impact of mycotoxin hazards was not realized until the problems of aflatoxins in the food supply were recognized. The aflatoxin story began as "the moldy peanut problem" and at first seemed to represent no more than an agricultural accident. Astute researchers, however, quickly determined that this group of compounds, and several other mycotoxins as well, represented a considerable potential for a serious threat to human health. Prof. Gerald N. Wogan and Prof. Paul M. Newberne, and their colleagues at the Massachusetts Institute of Technology, and Dr. Alex Ciegler and co-workers at the U.S. Department of Agriculture pioneered much of the work on the aflatoxins and other mycotoxins of environmental importance.

Shortly after the discovery that the highly potent carcinogen, aflatoxin B_1, was a contaminant of the human food supply on a world-wide scale, another "agricultural accident" pointed to the nitrosamines as another class of important food contaminants. Until dimethylnitrosamine was shown to be a contaminant in a sheep dietary, the N-nitroso compounds were considered by toxicologists only as excellent tools in the experimental study of chemical carcinogenesis. Dimethylnitrosamine had been used earlier in industry, but that use was greatly limited once the compound had been shown to be responsible for human liver poisoning. Prof. Peter N. Magee and Dr. John M. Barnes pioneered much of the basic research on these compounds. Today the associations between the use of sodium nitrite as a food additive and the formation of carcinogenic N-nitroso compounds are having considerable impact on the food industry.

Both groups of toxicants, the mycotoxins and the N-nitroso compounds, represent potentials for serious risks to human health as environmental carcinogens, especially as, but not limited to, contaminants of the food supply.

A great deal has been published recently on proposed mechanisms by which the N-nitroso compounds induce toxic injury and cancer, and this subject is reviewed in a compact chapter in Volume I of this book; in addition, Volume I is concerned with the occurrence, hazards, and risks of both mycotoxins and N-nitroso compounds. On the other hand, the time has come to distill from the vast mycotoxin literature a comprehensive and detailed review of mechanisms by which these natural environmental contaminants bring about toxicity and carcinogenicity. To that end it was decided not to restrain the contributors but rather encourage a rigorous analysis of mechanisms in mycotoxicoses and publish that treatise as the second volume of the book.

Ronald C. Shank
Editor

EDITOR

Dr. Ronald C. Shank has been active as a researcher and teacher in environmental toxicology for 15 years. He earned his Ph.D. in 1965 at the Massachusetts Institute of Technology while working on identifying biochemical lesions in animals during acute aflatoxin poisoning. His postdoctoral work was done at the Medical Research Council Laboratories Toxicology Unit at Carshalton, Surrey, England where Dr. Shank worked for 2 years on the biochemical actions of cycasin and dimethylnitrosamine. As an Assistant Professor, Dr. Shank set up a field laboratory for the Massachusetts Institute of Technology in Bangkok, Thailand at Mahidol University and spent 3 years conducting a study on the relationship between environmental aflatoxins and human liver cancer. He then returned to Cambridge, Massachusetts to conduct research on sodium nitrite as a precursor to carcinogenic nitrosamines and on the biochemical relationships between the toxicity and carcinogenicity of dimethylnitrosamine. In 1974 Dr. Shank moved to the University of California at Irvine to establish a graduate teaching and research program in environmental toxicology.

DEDICATION

This book is dedicated to Professor Gerald N. Wogan and Professor Peter N. Magee, without whose patience, instruction, guidance, and complete support this editor would be in no position to contribute to the field of environmental toxicology.

ACKNOWLEDGMENTS

The editor is most grateful to Mrs. Stella Nutting and Ms. Pamela Swick who contributed so much of their time and effort in preparing the manuscripts.

CONTRIBUTORS

Harland R. Burmeister
Research Microbiologist, Fermentation
  Laboratory
Northern Regional Research Center
U.S. Department of Agriculture
Peoria, Illinois

William F. Busby, Jr.
Research Scientist
Department of Nutrition and Food
  Science
Massachusetts Institute of Technology
Cambridge, Massachusetts

Alex Ciegler
Research Leader, Oilseed and Food
  Laboratory
Southern Regional Research Center
U.S. Department of Agriculture
New Orleans, Louisiana

Clifford W. Hesseltine
Laboratory Chief, Fermentation
  Laboratory
Northern Regional Research Center
U.S. Department of Agriculture
Peoria, Illinois

Peter N. Magee
Director
Fels Research Institute
Temple University School of Medicine
Philadelphia, Pennsylvania

Paul M. Newberne
Professor, Nutritional Pathology
Department of Nutrition and Food
  Science
Massachusetts Institute of Technology
Cambridge, Massachusetts

Adrianne E. Rogers
Senior Research Scientist
Department of Nutrition and Food
  Science
Massachusetts Institute of Technology
Cambridge, Massachusetts

Ronald F. Vesonder
Research Chemist, Fermentation
  Laboratory
Northern Regional Research Center
U.S. Department of Agriculture
Peoria, Illinois

Gerald N. Wogan
Professor of Toxicology
Head, Department of Nutrition and
  Food Science
Massachusetts Institute of Technology
Cambridge, Massachusetts

TABLE OF CONTENTS

VOLUME I

Chapter 1
Mycotoxins: Occurrence in the Environment 1
Alex Ciegler, H. R. Burmeister, R. F. Vesonder, and C. W. Hesseltine

Chapter 2
Animal Toxicity of Major Environmental Mycotoxins 51
Paul M. Newberne and Adrianne E. Rogers

Chapter 3
Environmental Toxicoses in Humans 107
Ronald C. Shank

Chapter 4
Mycotoxins: Assessment of Risks 141
Ronald C. Shank

Chapter 5
Occurrence of N-Nitroso Compounds in the Environment 155
Ronald C. Shank

Chapter 6
Toxicity and Carcinogenicity of N-Nitroso Compounds 185
Ronald C. Shank and Peter N. Magee

Chapter 7
Human Responses to N-Nitroso Compounds 219
Ronald C. Shank

Chapter 8
N-Nitroso Compounds: Assessment of Carcinogenic Risk 227
Ronald C. Shank

Index 245

VOLUME II

by
William F. Busby, Jr. and Gerald N. Wogan

Chapter 1
Introduction 1

Chapter 2
Aflatoxins 3

Chapter 3
Trichothecenes 29

Chapter 4
Cytochalasins 47

Chapter 5
Luteoskyrin 95

Chapter 6
Cyclochlorotine 101

Chapter 7
Psoralens 105

Chapter 8
Patulin and Penicillic Acid 121

Chapter 9
Ochratoxins 129

Chapter 10
Sterigmatocystins 137

Chapter 11
Zearalenone and Its Derivatives 145

Chapter 12
Summary and Conclusions 155

References 159

Index 195

Chapter 1

MYCOTOXINS: OCCURRENCE IN THE ENVIRONMENT

Alex Ciegler, H. R. Burmeister, R. F. Vesonder, and C. W. Hesseltine

TABLE OF CONTENTS

I.	Factors Affecting Occurrence	2
II.	Mycotoxins of Direct Importance to Humans	4
	A. Alimentary Toxic Aleukia (ATA)	6
	B. Stachybotryotoxicosis	6
	C. Ochratoxin	7
	D. Psoralens	7
	E. Aflatoxins	7
	1. Possible Causes of Fatal Acute Aflatoxicoses in Humans	7
	2. Cirrhosis	10
	3. Aflatoxin Carcinogenesis	10
III.	Mycotoxins of Direct Importance to Food Animals	11
	A. *Fusarium* Toxins	13
	1. Trichothecenes	13
	2. Zearalenone	16
	B. Aflatoxin	18
	1. Swine	18
	2. Cattle	19
	3. Poultry	20
	C. Ochratoxin	22
IV.	Mycotoxin Detection	24
	A. Aflatoxin	24
	1. Analytical	24
	2. Sampling Procedure	26
	3. Official Methods	26
	4. Analyses of M_1 M_2 in Milk and Dairy Products	27
	5. Confirmatory Test for Aflatoxins	28
	6. Biological Tests	28
	B. Zearalenone	29
	C. Trichothecenes	31
	1. Chemical Detection and Physical Methods of Analysis	31
	2. Biological Testing	32
	D. Ochratoxins	33
	1. Chemistry	33
	2. Detection and Analyses	33
	3. Extraction and Purification Procedures	35
	4. Confirmatory Tests	35
	a. Chemical	35
	b. Biological Tests	35
	E. Multitoxin Screening Procedures	36
	References	37

The prevalence of mycotoxins in the environment is difficult to assess, and posing the question qualitatively or quantitatively raises a bewildering array of problems. Among these are what toxins are we to be concerned about, and what are the conditions that lead to their occurrence? There is, of necessity, some subjectivity in any consideration of these points, based on individual assessment of what is often meager or confusing data as well as on a large number of poorly understood variables.

I. FACTORS AFFECTING OCCURRENCE

Factors that influence mold growth on a commodity probably also influence toxin production, e.g., moisture content (MC), relative humidity (RH), temperature, substrate composition, presence of competing microorganisms, fungal strain, etc. Of the physical parameters involved in fungal growth, MC and RH are probably the most important. However, the critical MC varies with the commodity; at a RH of about 65%, the critical MC for sorghum is 14.5%; wheat and corn, 12.5 to 13.5%; peanuts, 8%.[78,79,207] On the basis of moisture requirements and time of appearance in the production cycle of the grain, fungi can be divided into two groupings, field fungi and storage fungi, with some organisms overlapping into both groups. The field fungi which have high moisture requirements invade seeds prior to harvest and are best represented by *Fusaria, Alternaria,* and some *Penicillia* (*Penicillium oxalicum, P. funiculosum,* and *P. citrinum*).[220] However, *Aspergillus flavus* has been found in cottonseeds[22,359] and in corn in the field[193] and must be put into the overlapping category along with *Fusarium moniliforme.* Field fungi require moisture contents of 22 to 23% (wet weight basis) or 30 to 33% on a dry weight basis and RH of 90 to 100%.[175] Field fungi do not readily reinfect grain once it has been dried and rewet or when it has become contaminated by other fungi.[206]

The storage fungi of interest with respect to mycotoxins are primarily the *Aspergilli* and some *Penicillia.* For growth, these organisms require grains whose moisture contents are generally above 15% and in equilibrium with RH of 70 to 90%.[80] Table 1 summarizes MC requirements for some of the more important fungi possibly involved in mycotoxicoses.

It should be recognized that MC in storage bins may not be homogenous for a variety of reasons. Hence, localized regions of a bin may have moisture levels high enough to permit fungal growth and toxin development, often referred to as "hot spots". Additionally, grain entering commercial channels is often mixed to achieve desired safe moisture levels; however, mold development and toxin production can continue during the process of moisture equilibration and can contaminate the entire grain lot. This was shown in a study by Lillehoj et al.[191] who used mixtures of high moisture and dry white and yellow corn, one or the other of which had been contaminated with *A. flavus.*

On dry storage, field flora tend to gradually die away, with little development of storage species. A more complex pattern develops on storage, when the moisture content increases from any of a number of factors; a succession of storage fungi and other microorganisms may ensue. Difinitive research on the source of fungi remains to be carried out, although Tuite[338] has shown that a low level of infestation takes place prior to harvest. An important role as vectors for fungal transmission may be played by insects, but there is a paucity of data. A tentative relationship has been established between insect and *A. flavus* infection of corn and cotton bolls.[123,192,193,371] Additional information has been reported for other ecosystems.[174,202,253,297,315]

Temperature is less restrictive than moisture with respect to fungal growth and toxin production, for most organisms can tolerate and metabolize within a relatively broad temperature range. Further, as those familiar with fermentation processes realize, temperature and other parameters that are optimal for fungal growth may not be optimal

Table 1
MOISTURE CONTENT (MC) LEVELS REQUIRED FOR MOLD GROWTH ON CEREALS

Percent MC	Fungi
10—17	*Aspergillus glaucus* group
14.5—16.5	*A. ochraceus*
15.6—21.0	*Pencillium palitans, P. oxalium, P. viridicatum*
18.0—19.5	*A. flavus*
18.4	*Fusarium moniliforme*
22.2	*F. roseum* var. *graminearum*
21.2—33.0	*Fusarium* spp., *Alternaria* spp.

for secondary metabolite synthesis. Mislivic and Tuite[220] found a possible relationship between temperature and moisture requirements for species of *Penicillium* and their occurrence on corn kernels in the field or in storage. Field fungi tend to have high RH (86 to 90%) and MC requirements, with temperature range for germination and sporulation between about 8 to 36°C, whereas storage fungi in general have lower RH (81 to 83%) needs and a lower temperature range for germination and sporulation (−2 to below 30°C).

Temperature also influences toxin production or accumulation. Burmeister[60] reported excellent T-2 production at 15°C with *F. tricinctum*, but negligible production at higher temperatures. This accords with the observations that *Fusarium* contamination of cereals causes the greatest problem to consuming humans and animals when the weather is cool and wet during the harvest season. Penicillic acid also tends to accumulate to a greater concentration in corn stored at 1 to 10°C when contaminated with "blue-eye" fungi than at higher temperatures, even though the lower temperatures are not optimal for growth.[89]

The importance of substrate as a governing factor in secondary metabolite synthesis is also well known to those versed in fermentation. The cereal grains in general are good substrates for toxin production, whereas those seeds high in protein, e.g., soybeans, peanuts, and cottonseed, appear to support production of certain toxins, but not others. However, one must be careful of overgeneralizing because soybeans, normally a poor substrate for aflatoxin production, can become an adequate medium if the zinc that is bound by phytic acid is released or supplemented with additional zinc.[133]

Substrate may also play a role in selecting for or against toxin producing strains of a given species. This aspect has not been investigated to any degree, although it has been shown in one study that there is a higher proportion of toxin-producing strains of *A. flavus* isolated from peanuts than from cottonseeds, rice, or sorghum.[280] Additionally, we noted that strains of ochratoxin- and citrinin-producing *P. viridicatum* isolated from meat were more unstable than those isolated from grain, and that they rapidly lost toxin-producing ability.[88] We were also able to find a correlation between morphological, physiological, and toxin-producing capability of *P. viridicatum* with source of isolation, indicating, at least in this case, the possibility of a biochemical taxonomy.

Although fungi require air for growth, their oxygen requirements for mycotoxin production are less well known. Landers et al.[187] noted that aflatoxin production by *A. flavus* is progressively reduced upon increasing CO_2 levels from 0.03 to 100%; reduction of O_2 from 5 to 1% similarly reduced aflatoxin production. High aeration requirements for aflatoxin production in submerged culture have been recorded by several investigators.[90,143] However, these and other investigators have not separated aeration requirements for growth from those needed for toxin production. In a study involving ochratoxin production on grains, Lindenfelser and Ciegler[194] noted that ochratoxin synthesis occurred after air in the fermentor had probably been almost exausted. It would seem reasonable to question the need for high aeration rates for secondary metabolite synthesis as examined apart from growth requirements.

Another factor affecting the invasion of seeds by fungi is the degree of kernel damage. Early harvest of high-moisture grains by mechanical picker-shellers tends to create microfissures in the seed coats that normally form a barrier against mold invasion. Insect damage and cracking during heat drying can also increase vulnerability to fungi.

A very important aspect that has been almost completely neglected is the influence of mixed microbial populations, as they might occur naturally, on mycotoxin production. Growth of *A. flavus* with various fungi on peanuts of liquid media resulted in some diminution of aflatoxin synthesis, but the experiment was inconclusive.[23] Recently Mislivic et al.[219] grew a number of mycotoxin-producers in mixed culture: Admixture of *A. flavus* with *A. ochraceus* or *A. versicolor* did not affect aflatoxin production, but eliminated ochratoxin and sterigmatocystin synthesis; mixing *A. flavus* with either *P. citrinum, P. cyclopium,* or *P. urticae* inhibited all toxin synthesis. These experiments bear repetition on cereal grains, although recreation of a natural system is difficult.

Those mycotoxins that have been shown to occur naturally in the environment and are suspected of being involved in mycotoxicoses are listed in Table 2. Mycotoxins that are suspect, but for which insufficient evidence is available, are noted in Table 3. A third category are those toxins shown to be carcinogenic either by use of laboratory animals or via epidemiological evidence from human populations (see Table 4).

II. MYCOTOXINS OF DIRECT IMPORTANCE TO HUMANS*

A causal relationship between human consumption of moldy foods and various illnesses has been long suspected and was actually established with respect to the ergot alkaloids in the 18th century. Subsequent intoxications in the 19th and early 20th centuries implicating moldy food have recently been described.[322] Most notable in this period was "yellow rice disease" in Japan that caused many deaths and was associated with the consumption of imported moldy rice from Southeast Asia.[275] A number of toxin-producing fungi were isolated, including *P. islandicum, P. citreoviride, P. rugulosum,* and *P. citrinum.* These organisms produce a variety of toxic substances, but direct evidence for their implication in the yellow rice syndrome is lacking. This syndrome involved acute cardiac beriberi characterized by vomiting, convulsions, ascending paralysis, and respiratory arrest. Although this syndrome was initially thought to result from a vitamin B_1 deficiency, the Japanese were later able to duplicate symptoms in laboratory animals dosed with citreoviridin produced by *P. citreoviride.*[340] Fatal beriberi disappeared from Japan with the advent of a strict rice inspection system under the "Rice Act" of 1921.

During World War II, a toxicosis causing edema of the legs occurred among Japanese soldiers consuming rice contaminated with *P. islandicum.*[337] This culture produces two hepatotoxic and hepatocarcinogenic metabolites, luteoskyrin and cyclochlorotine (islanditoxin).[116] There have been no known acute intoxications of humans associated with these compounds.

P. citrinum, capable of producing the nephrotoxin citrinin, was isolated from yellow rice imported from Thailand to Japan in 1951, as well as from other areas of the world, but no intoxications have been attributed to it.[275]

---

* Greater detail on this subject is presented in Chapter 4, Environmental Mycotoxicoses in Humans.

TABLE 2
COMPOUNDS PROBABLY INVOLVED IN MYCOTOXICOSES

Toxin	Producing fungi	Susceptible host	Biological effects
Aflatoxin	A. flavus, A. parasiticus	Mammals, fish	Hepatotoxin, cancer
Penitrem A	P. palitans, Penicillum crustosum	Cattle, horse, sheep	Tremorgenic, convulsant
T-2	P. tricinctum	Cattle, man?	Dermal necrosis, hemorrhage
F-2	G. zeae	Swine	Vulvovaginitis, abortion
Slaframine	Rhizoctonia leguminicola	Cattle	Excess salivation
Sporidesmins	Pithomyces chartarum	Cattle, sheep	Hepatotoxin, facial eczema
Ochratoxin A	P. viridicatum, A. ochraceous	Swine, man?	Nephrotoxin
Psoralens	Sclerotinia sclerotiorum	Man	Dermotoxin
Citrinin	P. viridicatum, P. citrinum	Swine	Nephrotoxin
Vomitoxin	F. graminearum	Swine, man?	Vomiting
Maltoryzine	Aspergillus oryzae	Cattle	Death
Unidentified	Phomopsis leptostromiformis	Sheep	Hepatotoxin
Diplodiatoxin	Diplodia maydis	Cattle, sheep	Nephritis, mucoenteritis
Unidentified	Phoma sorghina	Man	Hemorrhagic, bullae mouth, rapid death
Secalonic acids D,F	P. oxalicum	Man	Death
Satratoxins	Stachybotrys atra	Horse, man	Hemorrhage

Table 3
COMPOUNDS SUSPECTED OF BEING MYCOTOXINS

Toxin	Producing fungi	Possible host	Biological effects
Sterigmatocystin	A. flavus	Mammals	Carcinogen
Yellow rice toxins			
Luteoskyrin	P. islandicum	Man	Hepatotoxin
Cyclochlorotine	P. islandicum	Man	Hepatotoxin
Citreoviridin	P. citreoviride	Man	Neurotoxin
Rugulosin	P. rugulosum	Man	Carcinogen
Rubratoxin	P. rubrum	Cattle	Hepatotoxin
Fusaranon-X	F. nivale	Man, swine	Vomiting
Nivalenol	F. nivale	Man, swine	Vomiting
Cytochalasin E	A. clavatus	Man	Death
PR toxin	Penicillium roqueforti	Cattle	Abortion
Patulin	P. urticae	Cattle	Death
Penicillic acid	Penicillium spp.	Farm animals	No data

Table 4
CARCINOGENIC MYCOTOXINS

Mycotoxin	Detected naturally	Target tissue	Type of lesion
Aflatoxin B₁	+	Liver, kidney, trachea, s.c. tissue	Hepatoma, s.c. sarcoma
Aflatoxin G₁	+	Liver, kidney, glandular, stomach, s.c. tissue	Hepatoma
Aflatoxin M₁	+	Liver	Hepatoma
Sterigmatocystin	+	Liver, s.c. tissue	Hepatoma
Luteoskyrin		Liver	Hepatoma
Cyclochlorotine		Liver	Hepatoma
Patulin	+	S.c. tissue	S.c. sarcoma
Penicillic acid	+	S.c. tissue	S.c. sarcoma
Rugulosin		Liver	Hepatoma
Griseofulvin		Liver	Hepatoma

A. Alimentary Toxic Aleukia (ATA)

This is probably the most notorious of the known mycotoxicoses of man having been involved in thousands of deaths in Russia during World War II. Numerous reviews have been published on the subject,[166,167] although the true identity of the toxins involved has only been established recently. The symptoms of the disease resulting from consumption of overwintered cereal grains molded by *F. poae* and *F. sporotrichioides (F. tricinctum)* have been well described and include fever, hemorrhagic rash, bleeding from the nose, throat, and gums, necrotic angina, extreme leukopenia, agranulocytosis, sepsis, and exhaustion of the bone marrow. The death rate has been termed high, but statistical data have not been published.

The toxins involved were initially thought to be steroids (poaefusarin, sporofusarin), but subsequent research has strongly implicated the 12,13-epoxy-Δ^9-trichothecenes. Bamburg et al.[41,43] found that the toxins produced by *F. tricinctum (F. sporotrichioides)* were diacetoxyscirpenol and T-2 toxin and suggested that ATA was caused by trichothecenes and not steroidal toxins. Subsequently, Mirocha and Pathre[215] reported finding only T-2, neosolaniol, T-2 tetraol, and zearalenone as toxic substances in a sample of "poaefusarin" supplied by the Russians. More recently Yagen and Joffe[383] surveyed 131 *Fusarium* isolates from overwintered cereals associated with ATA in the Soviet Union and found that 71 out of 106 *F. sporotrichioides* and 20 of 25 *F. poae* produced T-2 toxin; all the symptoms of ATA could be reproduced in laboratory animals. None of the steroids isolated were toxic. There have been no recent reports of ATA in the Soviet Union or elsewhere.

Reports of emesis in humans consuming moldy cereal grains have been recorded periodically since 1916, particularly in the Soviet Union, where it was known as drunken bread intoxication; an outbreak in the U.S. in 1963 caused nausea, vomiting, abdominal pain, and diarrhea.[97] The causative agent in a similar phenomenon in pigs consuming *F. graminearum* molded corn was shown to result from a new trichothecene, vomitoxin.[364,366] Whether or not the factor causing emesis in humans is identical to that responsible in swine is unknown, particularly since several trichothecenes have been shown to cause vomiting.[113,330,347]

B. Stachybotryotoxicosis

Although this is primarily a disease of horses consuming hay molded with *Stachybotrys alternans*, humans may also be affected, particularly in regions where equine stachybotryotoxicosis is common; inhalation of aerosols from contaminated hay, direct contact, or contact with smoke from burning molded hay may be involved. There may be a severe dermatitis, buccal inflamation, and pain, with burning sensations in the mouth.[103] Peasants in Russia and in the Balkans sleeping on infected hay have experienced a severe dermatitis. The disease probably results from a series of trichothecenes, designated satratoxins C, D, F, G, and H. Satratoxin D was subsequently identified as roridin E.[120,270]

C. Ochratoxin

Only circumstantial evidence implicates ochratoxin A in human mycotoxicoses. There are approximately 20,000 cases of a slowly progressing renal failure in persons living in the Danube River watershed in Romania, Bulgaria, and Yugoslavia. Persons primarily middleage and over are involved in the pathologic changes resembling the endemic nephropathy seen in pigs in Denmark (see Table 5). Ochratoxicosis is suspect, although research on rats has not indicated that long term sublethal dosing is culmulative in its effects.[262] However, *P. verrucosum* var. *cyclopium* isolated from foodstuffs collected in this region, when grown on liquid media and force-fed to rats, produced histological changes in the proximal convoluted tubules similar to those observed in intoxicated humans.[45] Obviously, additional research is required.

D. Psoralens

Two photoreactive furocoumarins, 8-methoxypsoralen and 4,5′,8-trimethylpsorlen, have been isolated from celery infected by *Sclerotinia sclerotiorum*, the causative agent of pink rot.[265,279,376] Celery pickers and others coming in contact with the diseased plant upon subsequent exposure to sunlight may develop a severe dermatitis.[279] There is no evidence that this fungus when growing on other agricultural commodities is capable of producing psoralens or any other toxin.[376]

E. Aflatoxins

The aflatoxins do not appear to represent a primary threat as an acute toxin to man, although some suspect cases have been reported. These are summarized in Table 6.

1. Possible Cases of Fatal Acute Aflatoxicoses in Humans

The outbreak in India and the cases of encephalopathy and fatty degeneration of the viscera (EFDV) in northeast Thailand noted in Table 6[52,178,179,235,295] are worthy of further comment. In late 1974 and early 1975, over 200 villages in western India experienced an outbreak of a disease affecting man and dogs and characterized by jaundice, rapidly developing ascites, portal hypertension, and a high mortality rate, with death usually resulting from massive GI tract bleeding. The disease was confined to the very poor, who were forced to consume badly molded corn containing aflatoxin between 6.25 and 15.6 ppm, an average daily intake per victim of 2 to 6 mg aflatoxin. Analyses of liver, sera, and urine were inconclusive, probably because of a time lag between corn consumption and sample collection.

In northeast Thailand where there is a high aflatoxin incidence in the food, EFDV is a common cause of death among children, in whom it resembles Reye's syndrome. Almost all cases occur in children from rural areas, with incidence increasing during the latter part of the rainy season. There is usually fever, vomiting, hypoglycemia, a rapid unexpected onset of convulsions, coma, and death; death usually occurs within 24 hr of the onset of symptoms.[52,235] Aflatoxin has been found in the livers on biopsy by several investigators.[47,110,295] Aflatoxin has also been found in the liver of an adult living in Missouri, who had carcinoma of the rectum and liver.[240]

Table 5
PATHOLOGICAL CHANGES IN ENDEMIC NEPHROPATHY

	Human	Pig
Tubules	Atrophy	Atrophy
Interstitium	Fibrosis	Fibrosis
Glomeruli	Hyalinization	Sclerosis, atrophy
Blood urea	High	High
Clearance		
Inulin		Depressed
Creatin	Depressed	
Proteinurea	Present	Present

From data presented by Krogh, P., Conf. Mycotoxins in Human and Animal Health, University of Maryland, October 4 to 8, 1976.

Table 6
POSSIBLE CASES OF FATAL ACUTE AFLATOXICOSIS IN HUMANS

Case	Country	Food source involved	Ref.
Boy, 15 years	Uganda	Cassava	290
Child	Germany	Peanuts	51
Girl, 22 months	Czechoslovakia	Rice product?	110
Girl, 12 months		Not known	
Girl, 8 months		Not known	
397 patients, 106 deaths (2:1 male to female)	India	Corn	178, 179
67 children, 1—6 years	Northeast Thailand	Not stated	52, 235
23 children	North Thailand	Not stated	295
3 children	U.S.	Peanuts	142a
Boy, 22 months, girl, 8 months	New Zealand	Rice	47

The cause of Reye's syndrome remains unknown and may represent a response to a variety of insults, including viral.[147] However, Becroft and Webster[47] and others have suggested an aflatoxin etiology; the pertinent facts are reviewed by Harwig et al.[141]

Although aflatoxin can function as an acute toxin, its potential carcinogenicity in humans is generally of greater concern. Aflatoxins have been shown to be carcinogenic for a variety of laboratory animals, including primates[3,264,334] in whom the symptoms and pathology of aflatoxicosis closely resemble some types of human hepatic disease thought to be caused by aflatoxin. Obviously, experimentation on humans is impossible; hence evidence for chronic toxicity in humans is based on epidemiological studies. Some of these studies are listed in Table 7.

The dosage required to elicit toxicosis in primates[130] has been extrapolated to humans by Campbell and Stoloff,[68] who calculated that children eating food contaminated with 1.7 mg aflatoxin per kilogram could within a short time period develop serious liver damage; a single dose of 75 mg/kg could result in death, whereas 51.3 mg/kg would have no apparent effect; daily dosing of 0.34 mg/kg would also have no apparent effect. These figures should be accepted cautiously since they are extrapolations derived from primate data.

Table 7
COMPARISON OF PRIMARY HEPATOMA INCIDENCE (CRUDE RATE)
AND EXPOSURE TO AFLATOXINS IN VARIOUS POPULATIONS[1]

Region	Aflatoxin exposure		Hepatoma incidence (cases/10^5 per year)			Ref.
	Contaminated peanuts		Male	Female	Total	
	Number examined	Percent positive				
Swaziland						
Highlands	37	8			2.2	170
Middlelands	67	25			4.0	
Lowlands	26	54			9.7	
Total population			8.6	1.6		

	Foodstuffs					
	ng B_1/kg body weight/day	Number examined	Percent positive			
Swaziland						
Highlands	8.34	288	3.8	7.02	1.42	238
Middlelands	14.43	288	5.2	14.79	2.21	
Lowlands	53.34	288	11.8	26.65	5.62	
Lebombo	19.89	192	5.7	18.65	0.0	
Total population	18.37	1056	6.7	16.24	2.70	

	Family meals					
	Average B_1 (μg/kg diet)	Number examined	Percent positive			
Kenya (Murang'a)						
Highlands	2.9	808	4.2	3.1	0.	239
Middlelands	3.2	808	6.5	10.8	3.3	
Lowlands	3.7	816	9.6	12.1	5.4	
Total population				10.5	3.8	

	Foodstuffs					
	Average B_1 (μg/kg food)	Number examined	Percent positive			
Uganda						
Karamojang	189	105	36			3.5	9,10
Buganda	39	149	11			2.0	
West Nile, Acholi, Soga, and Ankole	44	128	8			2.3	

	Average daily intake from family meal of aflatoxin B_1, ng/kg, body weight					
	Peak season	Yearly mean				
Thailand						
Singburi area	118—126	51—55			14.0	294, 296
Songhla area	<1—1	0—<1			2.0	

2. Cirrhosis

Aflatoxin has been implicated in cirrhosis in rhesus monkeys[200] and perhaps in Indian childhood cirrhosis, where the disease appears to correlate with religious dietary habits of a middle-income group examined in the Mysore area.[12] The children involved consumed parboiled rice and crude peanut oil which could have been high in aflatoxin content, since alkali refining is not used. In these studies, Amla et al.[11,12] reported that 20 children being treated for kwashiorkor were given 30 to 60 g/day of peanut meal for 5 days to 1 month; the meal was later found to contain 0.3 mg aflatoxin B_1 per kilogram meal. Three children on this supplement for 17 days showed liver cirrhosis. After withdrawal of the meal, there was a gradual transition from fatty to cirrhotic liver over a 1-year period, but characteristic clinical and histopathological signs were not noted before 6 months; histopathology appeared to correlate with duration of toxic meal consumption. This study was subsequently expanded to include 255 cases of childhood cirrhosis[13] and analyses of their urine for aflatoxins B_1 and M; B_1 but not M was found in 7% of the urine samples. Curiously, breast milk samples from 25% of the mothers of affected children showed the presence of an aflatoxin with an R_f of B_1; this is in contrast to all other reports involving milk where aflatoxin M but rarely B_1 was found. Yadagiri and Tulpule,[384] in a similar study involving 16 cirrhotic children, did not find aflatoxin B_1 in the urine, but, unfortunately, HCl had been added to the samples, so that any B_1 present would have been converted to B_{2a} which was not reported. In an earlier study on relating cirrhosis and aflatoxin consumption in India, Robinson[269] also found a variety of violet fluorescing spots in urine and mother's milk, but he lacked the technical facilities for positive identification.

In contrast to the above data, a study in the Philippines, where childhood cirrhosis is not reported as endemic, although aflatoxin was consumed in peanut butter, revealed only aflatoxin M_1 in urine and breast milk.[67] In a personal communication cited by Campbell and Stoloff,[68] Sreenivasamurthy in 1973 noted neither cirrhosis in liver biopsy material nor clinical symptoms when peanut meals with an aflatoxin content averaging 17 μg/kg were used for kwashiorkor therapy. Additionally, van Rensbury[354] believes that the aflatoxin intakes reported by the Indian workers were too low to be responsible for the observed effects, on the basis that aflatoxin is not known to be a potent cirrhogen. However, this does not rule out a possible contributory role of aflatoxin in the desease.

It is difficult to resolve the differences shown between the studies in India and the Philippines, so the problem of aflatoxin acting as a cirrhogen must await further study.

3. Aflatoxin Carcinogenesis

The first evidence that aflatoxin is carcinogenic was established in 1961, when Lancaster et al.[186] reported that 9 of 11 weanling rats fed for 7 months on a purified diet containing 20% toxic peanut meal developed multiple liver tumors, with lung metastases in two animals. Since that report, aflatoxin has been shown to be carcinogenic to a variety of laboratory animals, including nonhuman primates; the subject has been reviewed extensively.[115,144,195,213,233,236,375]

In man, where direct experimentation is out of the question, evidence linking aflatoxin with carcinogenesis depends of necessity on epidemiological investigations. Epidemiological research requires accumulation of data that usually lacks the rigorous controls demanded by most scientific investigations, so that "evidence" must be evaluated cautiously. However, as van Rensburg[354] pragmatically stated, "However laudable caution may be, there comes a point in the accumulation of evidence when it is more prudent to assume that the association is causal until proven otherwise. The *unanswerable**

---

* Italics are ours.

question of whether the hypothesis is 'proven' is irrelevant". We believe this to be the current status implicating aflatoxin in human carcinogensis.

An excellent summary of the epidemiological data up to 1974 has been published by Campbell and Stoloff.[68] A table in their manuscript that compares the primary hepatoma incidence (crude rate) and exposure to aflatoxins in various populations has been expanded to include more recent data (see Table 7).

In addition to the studies summarized in Table 7, epidemiological investigations relating high incidence levels of hepatomas to aflatoxin ingestion have been reported for Mozambique,[355] the Philippines,[59] and Costa Rica.[308] However, a study on varying incidence of esophageal cancer in the Caspian littoral of Iran showed only a low level of aflatoxin in the area.[152]

It becomes apparent from the studies carried out to date that several tentative conclusions can be drawn:

1. Chronic ingestion levels more closely parallel cancer incidence than occasional high exposures.[238,294,354,355]
2. Males appear to be more susceptible than females.[170,238,239,373]
3. Climates that favor growth of *A. flavus* concomitant with poor food storage practices tend to favor increased hepatoma rates, indicating a cause-effect relationship.[238,239,296]
4. Over a wide range, there is a linear relationship between cancer incidence and the logarithm of the level of aflatoxin intake, particularly in males (see Figures 1 and 2).[238,354]

It is obvious that a cause-effect relationship can not be "proven" experimentally or without a shadow of doubt, but the weight of epidemiological evidence would appear to definitely implicate aflatoxin as a carcinogen in humans. In the absence of contrary evidence, most progress can probably be made in solving the disease aspect by at least a tentative acceptance of the aflatoxin hypothesis.

There have been some proposals to link hepatitis virus with primary liver cancer (PLC),[199,256] particularly since the original hypothesis by Oettle[234] narrowed the potential causative agent to this virus and aflatoxin. A causal relationship between chronic hepatitis-B virus infection and PLC is suggested by the frequency with which hepatitis-B surface antigen (HB$_s$A$_g$) and antibody against hepatitis-B core antigen (anti-HB$_c$) are found in the serum of patients with this tumor.[256] However, in countries in which PLC is rare, such as the U.S., there is no obvious relationship between chronic HB$_s$ antigenemia and the occurrence of this tumor. The sigificance of the persistant HB$_s$ antigen and PLC in some parts of the world (Far East, parts of Africa) has not been established. Probably PLC has a multifactorial aetiology, but it is doubtful if the virus plays the primary oncogenic role, based on experimental, clinical, and epidemiological evidence suggesting that androgenic environments favor PLC development.[168] However, it is possible that the virus and aflatoxin may act in concert, this hypothesis is worthy of further investigation.

III. MYCOTOXINS OF DIRECT IMPORTANCE TO FOOD ANIMALS*

Toxins produced by fungi growing on grains used in animal rations may result in a number of deleterious effects that reduce the economic value of the consuming animal. Commonly, there will be a reduction in feed efficiency and a reduced feed intake, ac-

---

* Greater detail on animal mycotoxicoses appears in Chapter 2, Animal Toxicity of Major Environmental Mycotoxins.

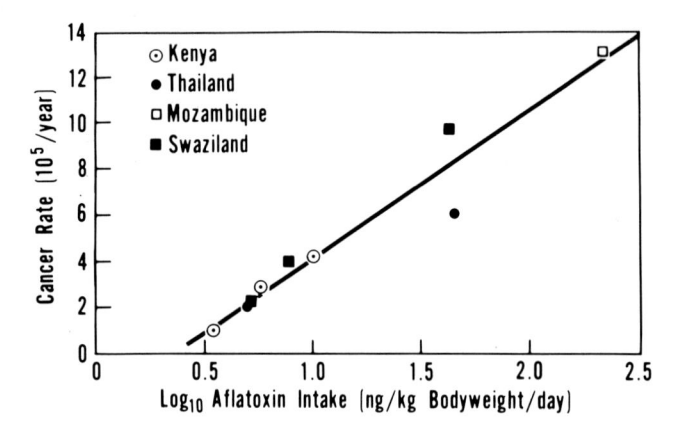

FIGURE 1. Relationship between the level of aflatoxin intake of populations and the primary liver cancer rate.[354]

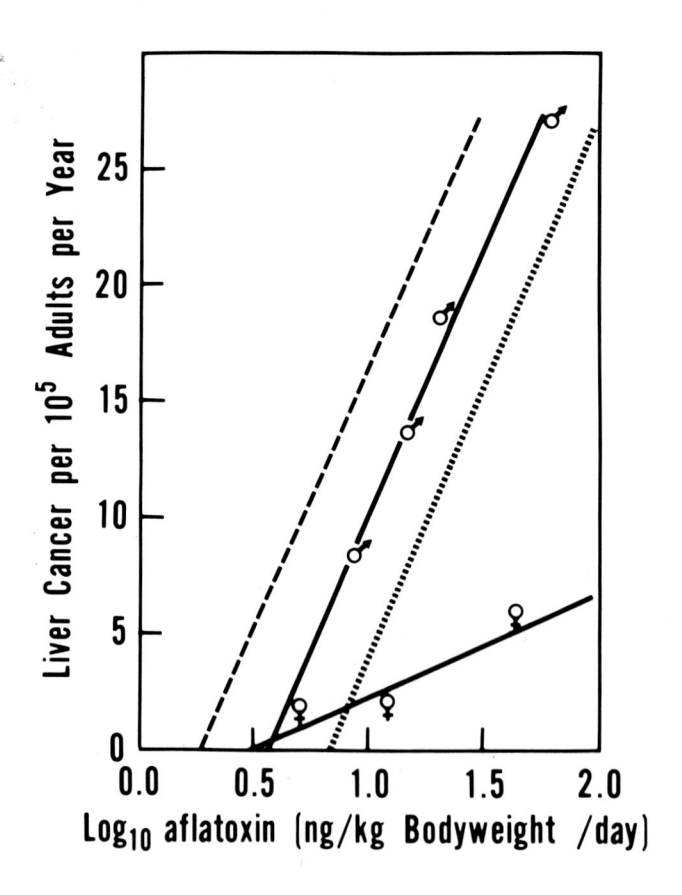

FIGURE 2. Correlation of liver cancer with aflatoxin ingestion. (From Peers, F. G. and Linsell, C. A., *Br. J. Cancer*, 27, 473, 1973. With permission.)

companied by decreased weight gains, reduction in carcass grade, and a lower production of milk or eggs. There also may be fewer healthy offspring from affected animals. Furthermore, the stressed animals become subject to secondary infections, and an impaired immune system may, in some cases, result in poor protection from immunization. Finally, consumption of large doses of potent toxins will cause animal deaths.

A. *Fusarium* Toxins
1. *Trichothecenes*

Documented cases of scabby grain being toxic to humans or animals date back more than 50 years and emanate from Russia, Japan, and the U.S.[102,201,226,276] Examination of the scabby grain led to the association of *Fusarium* molds with the diseased grain and subsequently with the toxicosis. *Gibberella zeae* and its asexual stage *Fusarium graminearum* (syn. *Fusarium roseum*) were always cultured from the diseased grain. According to Mains et al.,[201] the disease on small grains is most serious as a head blight, first becoming evident by the loss of green color of the spikelets. If wet, humid weather prevails, the fungus spreads into the head, followed by the development of a reddish mold. In the 1928 outbreak, Indiana farmers attempted to feed infected barley to hogs, only to find the animals reluctant to eat the feed. If forced by starvation to eat the scabby grain, some of the pigs became sick and vomited. The scabby cereals were eaten by cattle and poultry with no apparent ill effects, but were rejected by horses and mules.

Swine-feeding experiments using various fractions and mixtures of the diseased barley with sound barley or sound barley fermented with *Gibberella saubinetti* clearly demonstrated that pigs had no appetite for this grain.[201] Hoyman,[153] in an unsuccessful attempt to identify the "emetic factor" from scabby barley infected with *G. saubinette*, found that water extracts caused vomiting when adminstered by stomach tube. Effects of good grain spiked with the water extracts and fed to pigs were not reported.

Because of a wet autumn, the harvest of much corn in northern Indiana was delayed into November and December of 1965 and again in 1972. *G. zeae* damage was prevalent, and when more than 5% of the corn kernels were visibly infected, the grain was refused by swine.[339] During these seasons, numerous reports of corn refusal were reported, but only one farmer reported vomiting in a few pigs.

Curtin and Tuite[97] fed experimental animals molded corn to clarify the refusal emesis syndrome. Swine fasted for 24 hr ate sparingly, but not in quantities sufficient to promote growth or emesis. Feed spiked with water extracts of *G. zeae*-molded corn was refused by swine; however, emesis could be induced by oral, i.v., or i.p. administration. Although swine refused good corn spiked with the water extracts, they consumed corn spiked with the methanol-soluble portion of these water extracts. However, pigs eating the methanol soluble fraction of the water extracts promptly vomited, but indicated no loss of appetite. This study suggested that *G. zeae*-molded grain contained two separable products, a water-soluble fraction responsible for feed refusal and a methanol-soluble fraction inducing emesis.

Prentice and Dickson,[254,255] using pigeons as an assay tool, separated by thin layer chromatography (TLC) an emetic factor from *F. graminearum*-infected corn and two emetic factors from culture media fermented with a strain of *F. moniliforme*. The material corresponding to the emetic found in grains was only slightly soluble in water, but was readily soluble in ethanol. When injected into the wing vein of pigeons, the emetic material from infected grain appeared to be free from harmful side effects, whereas the second emetic material from *F. moniliforme* killed the birds within 12 hr after injection. Chemical analysis of the emetics indicated that they were not alkaloids, steriods, terpenes, or carbohydrates. In addition to their chemical studies, Prentice et al.[255] demonstrated that the emetics were produced in laboratory media by strains of several species of *Fusarium* isolated from scabby cereal. Emetic-producing strains were found in the species *F. moniliforme*, *F. roseum*, *F. poae*, *F. culmorum*, and *F. nivale*. A reexamination of these strains by Ellison and Kotsonis[113] revealed that only extracts of the *F. poae* strain produced emesis in pigeons. Two emetic factors, T-2 toxin and acetyl T-2 toxin, were isolated from the *F. poae* culture;[176] both were shown to cause emesis when injected into the wing vein of pigeons. Acetyl T-2 toxin was considerably less toxic to the pigeon than was T-2 toxin. Since only one of the five cultures produced emetic factors, Ellison and Kotsonis[113] surmised that the strains studied may not be identical with the original ones. Each of the "Prentice Strains" was also reexamined by Vesonder et al.[365] These cultures grown on autoclaved corn at 28°C were refused by swine, as were good corn rations after being spiked with a water-methanol extract of each of the five corn-grown cultures.[386] So far, the Prentice-Dickson strains have yielded no substance identifiable as a trichothecene other than the *F. poae* strain shown to produce T-2 toxin and acetyl T-2 toxin and the *F. culmorum* strain that produces vomitoxin.[365]

A portion of the 1972 corn refused by swine was examined by Vesonder et al.[364] The corn was contaminated with *F. graminearum*, *F. moniliforme*, *F. nivale*, *Cepholosporium* sp., and various *Penicillium* sp. Chemical analyses of this corn indicated the presence of no known mycotoxin, but a previously unknown trichothecene extracted from the corn was refused by swine. The compound was subsequently purified, identified, and given the trivial name vomitoxin.[364] *F. graminearum* isolates from this corn also produced vomitoxin on autoclaved grains. Good grain spiked with 40 mg of vomitoxin was rejected by swine, as were extracts from the 1972 corn.[366] Vomiting was also induced by intubation of 7.2 mg of vomitoxin into a 60-lb pig or by forced feeding.[364] Since emesis and rejection were produced by vomitoxin, it was assumed to be the single cause of feed refusal and emesis. The presence of vomitoxin in the 1972 lot of corn was confirmed by Ishii et al.[162] who were also able to purify vomitoxin from corn refused by swine. At about the same time that Vesonder et al[364] were identifying the refusal factor from corn, Morooka et al.[221] were characterizing trichothecenes from cereal grains growing in Japan that were infected with *Fusarium*. They found two trichothecenes, nivalenol and a previously unknown compound, Rd toxin. The Rd toxin was later named deoxynivalenol, since its structure was like nivalenol, but lacked a C-4 hydroxy group.[385] An *F. roseum* strain isolated from toxic cereals produced deoxynivalenol (vomitoxin) and induced vomiting in ducklings and dogs. Deoxynivalenol and vomitoxin were found to have the same chemical structure.[366]

Whether there is a single chemical entity involved in the refusal and emesis of *Fusarium*-infected grain by swine or whether a different factor is responsible for each response requires further investigation. Curtin and Tuite[97] separated two fractions from the 1965 corn crop; on caused feed refusal, and the other caused vomiting. Prentice et al.[255] identified two emetic products from infected grains and *Fusarium* cultures, one highly toxic and the other of low toxicity. Vesonder et al.[364] however, following an extraction procedure different from that of Curtin and Tuite[97] on both naturally contaminated corn and laboratory grown cultures, were able to demonstrate but a single factor, vomitoxin, that caused swine to refuse spike corn and caused emesis when intubated into the stomach. Since field reports of vomiting in swine are not as common as feed rejection, probably because of a limited intake of *Fusarium*-infected grain, and because vomiting is only occasionally induced by feeding *Fusarium*-infected material, the possible presence of other unknown factors required to cause the vomiting in the refusal-emesis syndrome needs to be studied.

In addition to deoxynivalenol and nivalenol, several other trichothecenes have been identified in toxic feedstuffs or have been produced by *Fusarium* sp. cultured from toxic feed products. Most of these trichothecenes produce emesis in experimental animals (see Table 8). Several of these *Fusarium* metabolites, unlike vomitoxin, are potent toxins with LD_{50} values less than 10 mg/kg body weight for laboratory animals. In addition to their extreme toxicity, these substances are highly irritable to dermal tissue.[85] Sensitivity of dermal tissue has led to the use of an animal skin test as a presumptive indicator of some trichothecenes. Eppley et al.[121] also examined North Central State corn from the 1972 crop for the presence of the mycotoxins, aflatoxin, T-2 toxin, and zearalenone. T-2 toxin was not detectable in any of the 173 samples analyzed, but extracts from 93 of the samples caused a dermatic response on rabbit skin. Estimations of skin irritant, using T-2 toxin as an irritant standard,[85] indicated that all positive samples contained a level of toxin not detectable by the chemical procedures used. Abount one fourth of the samples containing a skin irritant also contained some zearalenone.

The trichothecenes consist of about 37 derivatives, but only a few documented cases of these toxins in feeds have been reported. Hsu et al.[156] found T-2 toxin (2 ppm) in moldy silage associated with a lethal toxicosis in dairy cattle, and Mirocha et al.[216] used gas chromatography-mass spectrometry (GC-MS) analysis to identify trichothecenes and zearalenone from nine feeds involved in pathogenicity of swine, cattle, and dogs; diacetoxyscirpinol was identified in two samples that caused hemorrhaging in swine; one sample containing T-2 toxin caused bloody stools of cattle, and deoxynivalenol was present in the six other feeds. Zearalenone was found along with deoxynivalenol in each sample, but of more interest, only the feeds containing deoxynivalenol caused feed refusal or emesis.

The prevalence of *F. tricinctum* on moldy corn that caused a hemorrhagic syndrome in cattle implicated this fungus as a producer of potent mycotoxins.[128] From these strains of *F. tricinctum*, diacetoxyscirpinol and T-2 toxin were purified and identified.[41] Since the mid-1960s Japanese workers have also been intensively studying problems of foods and feeds where Fusaria were prevalent.[331] They have isolated a number of trichochecenes not yet found in *Fusarium* cultures isolated in the U.S. Several of these *Fusarium* metabolites are highly toxic and rather strong emetics. Fusarenon-X, neosolaniol, and nivalenol are among the most interesting of these recently identified mycotoxins.

As mentioned earlier, the trichothecenes of Fusaria are highly toxic and potent skin irritants. Because these fungi commonly grow on grains, they are a potentially serious danger to farm animals; therefore, laboratory studies were undertaken to gather information useful to laboratory investigators and diagnosticians to aid in recognition of unknown nutritional disorders that may be caused by these toxins. T-2 toxin because of its relative ease of production[41,60] has been studied most intensively.

Albino rats fed diets containing 5, 10, and 15 ppm of T-2 toxin were severely stunted in growth, had a reduction in weight gain associated with reduced feed consumption, developed inflammation of the skin around the nose and mouth, and suffered small focal fatty changes and cytoplasmic degeneration in the liver.[205] A long-term feeding study (8 months) where rats consumed 20 times the single LD_{50} levels of toxin administered in low concentrations indicated no cumulative effect of this toxin. T-2 toxin caused extensive inflammation and tissue necrosis when applied to the skin of rats, but failed to produce any papillomas. The long-term feeding study and skin tests indicated that T-2 toxin is probably not a carcinogin.[205] An indication that T-2 toxin can be a teratogen was demonstrated by Stanford et al.,[314] who injected pregnant mice with T-2 toxin and observed gross malformation in a high percentage of the fetuses. In mice, rats, and cats

Table 8
BIOLOGICAL ACTIVITY OF SOME TRICHOTHECENES

Trichothecene	LD$_{50}$ (mg/kg)	Emetic dose (mg/kg)	Animal (route of administration)	Ref.
T-2 toxin	4	—	Rat (oral)	205
	2.75	0.72	Pigeon (oral)	113
		0.15	Pigeon (i.v.)	113
Acetyl T-2 toxin	—	18.2	Pigeon (oral)	113
Diacetoxyscirpino[a]	7.3	—	Rat (oral)	53
	0.75	—	Rat (i.p.)	53
Neosolaniol[a]	14.5	—	Mouse (i.p.)	342
Deoxynivalenol (vomitoxin)	46.0	—	Mouse (oral)	385
	27.0	10.0	Duckling (s.c.)	385
	—	0.1	Dog (s.c.)	385
Nivalenol[a]	4.0	—	Mouse (i.p.)	330
Fusarenon-X	4.4	—	Rat (oral)	347
	—	0.4	Duckling (s.c.)	347

[a] Skin irritant.

administered T-2 toxin and related trichothecenes, severe cellular destruction and karyorrhexis are induced in the actively dividing cells of the bone marrow and mucosal epithelium of the small intestine; these effects are accompanied by a depression in protein synthesis.[343,344]

Consumption of turkey rations containing 1 or 2% of an autoclaved corn fermented with a toxic *F. tricinctum* strain caused death in a large percentage of the birds.[81] Reduced feed efficiency, poor weight gains, and bilateral necrotic lesions at the angles of the mouth were noted in surviving poults. A disease syndrome characterized by raised yellowish-white lesions in the oral cavity was also observed in several commercial broiler flocks and in pigeons in a 1966 outbreak.[381] Chickens fed graded rations containing from 1 to 16 µg/g T-2 toxin developed raised yellowish-white lesions in the mouth parts.[73,382] In addition to the characteristic oral lesions, birds had a reduced growth rate.[73,379] abnormal feathering,[380] hysteroid seizures and an impaired righting reflex,[377] and were more susceptible to *Salmonella* infections.[50] The relative weights of the spleen and bursa of Fabricus were reduced, whereas the relative weight of the pancreas and crop increased in T-2 toxin-consuming birds.[379] No signs of digestive disorder, hemorrhages, or pathological changes in bone marrow, liver, kidney, or heart were observed in chicks fed a ration containing up to 4 µg/g T-2 toxin.[73] Chickens given lethal dosages of T-2 toxin developed asthenia, inappetence, diarrhea, and panting, and the abdominal cavities contained a chalk-like material that covered most of the viscera.[75] In contrast to broiler chickens, the consumption of up to 20 µg/g of T-2 toxin had no apparent effect on any of the vital organs of laying hens.[74,378] The most significant effects on the older chickens was the development of oral lesions, a reduction in feed consumption, lower egg production, and thinner eggshells.[378] Hatchability of fertile eggs of hens fed 2 and 8 ppm was significantly lower than of hens fed a control diet.

2. Zearalenone

Zearalenone or F-2, a resorcylic acid lactone, is a natural metabolite produced by strains of *F. roseum, F. tricinctum, F. oxysporum,* and *F. moniliforme.* When consumed by swine, it can result in an estrogenic syndrome.[210,214] This anabolic compound is generally not lethal to the consuming animal, but it induces a hyperestrogenism. The visible symptoms of this mycotoxicosis include swelling and reddening of the vulva, increase in the size of the uterus, and growth and lactation in mammary glands. Estrogenic stimulation occurs even in males and gilts, as indicated by enlargement of the prepuce and mammary glands.

Reports of the estrogenic syndrome are sporatic but not infrequent and usually follow a season when wet grain is stored. Swine are perhaps the most sensitive of animals to the activity of zearalenone, and its uterotrophic action is most readily seen in this species. A myriad of variables complicate the interpretation of field cases of moldy grain toxicosis; therefore adequate laboratory studies were necessary to elucidate the role of individual fungal metabolites.

Christensen et al.[82] mixed *F. roseum*-invaded corn with a swine ration to provide 500 to 600 ppm of zearalenone and fed it to male and female pigs. All of the gilts on the ration containing *F. roseum*-invaded corn developed swollen vulva and large nipples within 3 to 4 weeks, and three of them developed prolapsed rectums within 7 weeks. Upon return to a control ration, the symptoms of estrogenism regressed. This level of zearalenone also had a pronounced adverse effect on weight gains. The relative weight of the uterine horn of the gilts receiving the rations with *F. roseum* was nearly double that of the controls, and weight of testes of the males on zearalenone was 30% less than those on the control diet.

In southeast Scotland, sows consuming *Fusarium*-molded grain that contained zearalenone developed estrogenic symptoms during the last 2 to 3 months of pregnancy. In practically every farrowing, there were one or more dead piglets that were of normal full-term size. Some litters comprised mainly weak pigs, with splayleg incoordination or partial paralysis of the hind limb. To determine if these symptoms could be reproduced under controlled conditions, two sows were injected at the rate of 5 mg/day with pure zearalenone during the last month of pregnancy. The two pigs farrowed a total of 20 piglets, of which 3 were full-sized but stillborn and another 13 had a splayleg condition.[209] Although this is an interesting finding, additional experimentation is required to confirm the function of zearalenone in the splayleg and stillbirth syndrome. Additional evidence that offspring of animals may be affected by zearalenone was provided by Ruddick et al.,[273] who found that pregnant rats receiving zearalenone developed a higher incidence of fetal skeleton anomalies. Mirocha and Christensen[210] have never been able to induce abortion in swine with pure zearalenone, but they succeeded in doing so when *Fusarium*-invaded corn was added to an otherwise nutritionally adequate diet. However, zearalenone can definitely contribute to infertility in swine by its effect on the ovaries. Pure zearalenone has anabolic properties resulting in weight gains of experimental animals, whereas *Fusarium*-invaded corn, when added to an otherwise balanced diet, will cause the estrogenic syndrome and result in a weight loss.

There is no conclusive evidence that zearalenone causes an estrogenic syndrome in dairy cattle, although vulvo vaginitis in dairy herds is often reported by veterinarians.[210] Zearalenone has been implicated in fertility disturbances of cows where hay invaded by *Fusarium* contained 25 ppm of the estrogen.[271] Animals recovered after returning to a good diet.

Poultry fed rations containing grain colonized by zearalenone-producing Fusaria generally undergo a reduction in the rate of weight gain and egg production. In extreme cases, estrogenic responses can also occur. For a more detailed report see Mirocha et al.[210,214]

Strains of *Fusarium* species produce trichothecenes and zearalenone during storage if conditions are favorable, but very little zearalenone is produced prior to harvest. Caldwell and Tuite[65] inoculated ears of field corn with zearalenone-producing Fusaria to determine if toxin was accumulated prior to harvest. They found no production or an insignificant accumulation of zearalenone 100 days after inoculation at 90% of full silk with eight strains of *F. tricinctum* and 12 strains of *F. roseum*. Production in the field never exceeded 5 ppm and was almost always less. Examination of naturally infected corn ears supported the finding that little zearalenone accumulates in freshly harvested corn.[66] In a comparable study, barley sprayed 6 weeks before harvest with spores of a zearalenone-producing strain of *F. culmorum* had no detectable zearalenone at harvest; however, the moistened grain stored for 4 months at a low temperature developed significant amounts of the estrogen.[131] Since only small amounts of zearalenone are produced in the field, proper drying and storage should minimize this problem. Not all grain, however, can be optimally handled. When moist grain is stored where air is available, zearalenone accumulation may occur. Fusaria grow on stored grains at moisture contents as low as 15% and produce maximum yields of toxin at 12 to 18°C.[122] In general, *F. roseum* strains require low temperatures for maximum zearalenone production, but strains selected from grain sorghum reportedly produce very larqe amounts of toxin at incubation temperatures of 25°C.[281]

B. Aflatoxin

Aflatoxicosis of farm animals, no doubt, occurred long before the recognition that molded feedstuffs affected the thrift of animals. Only the association of obviously molded feeds and sudden illness and death of the consuming animals led to specific molds and their metabolites as the suspected toxic agents. Mostly, even these unfortunate incidences were overlooked because the overall economic impact was not obvious and because epidemiological data linking sporadic outbreaks were not adequately reported or compiled. Occasionally, relatively large, sudden, unexplained losses of farm animals occurred after ingestion of moldy feeds. Such an outbreak amoung cattle and swine in the summers of 1952 to 1953 encouraged Burnside et al.[63] to investigate the molds involved in these poisonings. Most of the mold isolates were *A. flavus* strains. and only cultures of *A. flavus* and *P. rubrum* isolates grown on autoclaved corn caused animal deaths. The symptoms induced by *A. flavus*-molded corn were like the abnormalities seen in farm animals that had consumed the naturally molded corn. British investigators[7,24,49,69,305,311,368] in the early 1960s observed similar symptoms in animals consuming toxic groundnuts and very quickly associated molds and their metabolites as the causative poisons. Several brightly fluorescent compounds were purified from toxic peanut meal and from peanut meal inoculated with cultures of the toxin-producing strains, identified as *A. flavus*.[227] The toxins, separated by TLC and named "aflatoxin", were differentiated into B and G according to their color of fluorescence. Toxic extract killed experimental animals and produced the characteristic liver lesions seen in natural cases of poisoning. Slightly later, toxicological investigations showed that the livers of rats consuming a diet containing toxic peanut meal became grossly abnormal. The liver became brownish-yellow and had numerous lesions, nodules, and cysts. Of 11 rats eating toxic meal, nine developed macroscopic turmors within 30 weeks.[186] Because of its acute toxicity and proven carcinogenic nature, aflatoxin became a mycotoxin of utmost concern.

1. Swine

A number of reports and reviews[5,99,132,172] describing the dietary effects of graded levels of aflatoxin on swine indicate no acute toxic effects below 700 ppb. However, toxic concentrations exceeding 280 ppb caused some depression in growth rates and feed conversion efficiencies of growing pigs,[5,18] but even consumption of rations containing 690 ppb did not seem to affect the performance of pigs during the finishing period when they gained from 120 to 200 lb.[105,106] Substantial improvement in the performance of growing pigs on the 690 ppb ration was also seen when they were placed on a toxin-free ration. Some of the pigs consuming aflatoxin containing feeds developed liver lesions, and those animals consuming the most toxin developed the more severe lesions. Other studies[124,148,171] indicate that even higher levels, near 450 ppb, are required to depress the growth rate and feeding efficiencies of growing swine.

The recognition that aflatoxins were extremely toxic and were carcinogens required information about their effects on the consuming animal and the possibility that they may be transmitted into meat, eggs, or milk. Alterations in liver function, as represented by changes in blood and liver enzymes, occurred in pigs after eating feed containing 450 ppb aflatoxin or more. Pigs consuming a diet with 450 ppb or more toxin developed significantly heavier livers than control animals, but the livers otherwise appeared normal.[171,172] Some pigs fed higher levels of toxin, 615 or 810 ppb, developed pale, yellow livers with a few pale nodules.[124]

Although moderate to high levels of aflatoxin fed to hogs for relatively long periods can induce significant liver damage, tumor induction is a rarity either in experimental animals or in farm animals. An examination of meat-inspection records from 1955 to 1962[293] showed only 187 tumors recorded from more than one half billion swine. Only seven of the 187 were hepatic tumors, and these were benign. The fact that about 90% of the hogs are slaughtered before they become 6 months old is a probable factor in the low tumor incidence. Experimental evidence concerning the transmission of aflatoxin from the diet of pigs to the meat is ambiguous. Animals fed a ration containing 810 ppb for 117 consecutive days were all free of aflatoxin residues when analyzed by currently accepted methods.[171,172] Liver lesions and increases in the liver and kidney weights were indications of toxin stress. In experimental animals, given high aflatoxin dosages, small amounts of toxin accumulate in many tissues, including the liver, kidney, spleen, heart, and muscle.[224] Krogh et al.[184] found a maximum of 54 ppb aflatoxin in the organs and trace quantities in the muscle of bacon pigs fed 120 to 231 days on a diet including 300 or 500 ppb aflatoxin.

Aflatoxin does not appear to affect the reproductive capacity of either boars or gilts. When fed aflatoxin at levels of 450 ppb continuously for more than 8 months, boars produced normal sperm, and gilts farrowed normal litters and lactated in a normal manner, even though minimal liver lesions were diagnosed in each sow.[149] Evidence supporting the observations of Hintz et al.[149] is provided by Armbrecht et al.[19] They fed 1100 ppb aflatoxin to sows for 165 days and found that the sows produced normal litters through four farrowings. Milk of the sows contained aflatoxins B_1, B_2, and M_1. Piglets nursing sows on the aflatoxin diet gained more slowly than piglets in the control group, and traces of toxins were evident in the organs of some piglets. After weaning, however, these animals were reared to market age on a control ration and no organ lesions were found at slaughter.

From the evidence available, it appears that either very large doses, in excess of 1500 ppb, which *A. flavus* is very capable of producing, or a combination of aflatoxin with other fungal and/or plant metabolites is required to produce the acute toxicoses occasionally seen on the farm. Feeding trials also indicate that swine consuming small amounts of aflatoxin, less than 450 ppb, for long periods would show no specific symptoms indicative of aflatoxicosis.

2. Cattle

A few documented cases of acute aflatoxicosis in cattle have been reported. Earliest reports came from England in the early 1960s and later from other countries where extremely toxic peanut meal was used as a protein supplement in animal feeds.[4,91,129,196] Symptoms of aflatoxicosis in cattle are a loss of appetite, poor condition, scours, illness, and death.[91] Postmortem examinations revealed dark, congested carcasses with varying degrees of bruising. The most marked feature in each animal was the appearance of the liver. Livers were hard and pale reddish brown and showed some fibrosis, bile duct proliferation, and chronic endophlebitis of centrolobular and hepatic veins.

Since aflatoxin can be poisonous for cattle, efforts to learn more about its effects on cattle and the degree to which they could resist the toxins were undertaken. Aflatoxin concentrates in feeds up to 300 ppb did not significantly affect 400- to 500-lb steers; however, when rations containing 700 or 1000 ppb were fed for 133 to 196 days, growth-rate inhibition, decreased feed efficiency, and increased liver and kidney weights were observed. Gross liver damage was also evident.[127] Microscopic findings of liver indicated liver cell enlargement, enlarged nuclei, and bile duct proliferation. Although the liver is a primary site for aflatoxin damage, only 44 liver and bilary tract tumors were found in 1.3 million cattle inspected in British abattoirs in 1965 to 1966.[14]

According to Keyl and Booth,[171] there was no chemical evidence of aflatoxin in the meat of steers fed up to 1000 ppb aflatoxin; however, traces of B_1 and M_1 were detected from blood samples. If animals were changed to a control ration 72 hr prior to slaughter, aflatoxin could not be detected, which indicates a rapid clearance from the blood. Transmission of dangerous amounts of aflatoxin into human food via beef products does not appear likely, but aflatoxin M_1 was detectable in the milk of dairy cows that consumed a ration containing 500 ppb aflatoxin B_1.[8,246,350] Aflatoxin M_1 appeared in the milk within 24 hr of feeding a diet containing aflatoxin and usually reached a maximum after 4 days. Detectable toxin disappeared 2 to 4 days after termination of the toxic diet. Traces of aflatoxin M_1 have been found in fluid market milk in South Africa and in Europe.[173] According to Kiermeier,[173] the content limit for aflatoxin B_1 in feedstuffs of the European Common Market Commission can result in the presence of aflatoxin M_1 in milk. For estimating the excretion of aflatoxin M_1 in milk, conversion of roughly 1.5% of the aflatoxin feed intake can be expected.

3. Poultry

Incrimination of toxic peanut meal as the probable cause of a lethal toxicosis, dubbed Turkey "X" disease, and the first comprehensive description of this toxicosis in turkey poults was furnished by Blount.[49] In South England, an unknown endemic disease of poultry had broken out in the summer of 1960. Visibly affected turkeys died within a week, and observation of an acutely affected flock caused Blount to report, "they succumbed literally like flies before your eyes, and their general appearance certainly suggested to me that some form of poisoning had taken place". Inappetence, a reduced growth rate, and a weakness of wings were symptoms of the disease. Postmortem examination of birds revealed enteritis and congested and engorged kidneys and livers, while the body condition appeared to be good.[49,223,305,368]

Consumption of aflatoxin during vaccination impairs the immune system of turkey poults.[241,242] Birds inoculated with fowl cholera on the 9th day after hatching, 2 days after aflatoxin was added to the diet, were given a challenge inoculation on the 28th day. All of the control poults survived the challenge, whereas, 26% of the aflatoxin-consuming birds died after exposure to *Pasturella multocida*. A reduction in relative weight and histologic involution of the thymus in aflatoxin-consuming birds implied that the toxin may exert a direct effect on the immunogenic system.

Ducklings are the most susceptible species to the action of aflatoxin, with an acute LD_{50} for B_1 of 0.335 and for G_1, 0.784 mg/kg.[64] The first signs of acute toxicosis are inappetence and poor growth rate. Conspicuous features occasionally seen in white-skinned birds are a purple discoloration of the legs and feet and lameness. Young birds develop ataxia before death and usually die with a peculiar tetanic spasm.[24] Within 3 or 4 days after consuming toxic meal, hepatic lesions develop; proliferation of bile duct epithelial cells is marked within a week. In chronic cases of poisoning, the liver may exhibit nodular hyperplasia and be extremely cirrhotic.[6,232] Livers of ducklings fed 14 months on a ration containing (about 35 ppb B_1) toxic groundnut meal developed macroscopic lesions. Diseased livers became atrophied and contained numerous yellow focal lesions and nodules.[69] As in other animal species, tolerance of ducks varies with the age of the bird. Bierbower and Armbrecht[48] demonstrated that Pekin-White ducks tolerated a diet containing 33 ppb aflatoxin B_1 for 90 days. Tolerance of ducks to single acute dosages increased at about 2.7 mg/kg body weight each week for the first 3 weeks of age. Birds poisoned by aflatoxin recover slowly when returned to a toxin-free diet, but the healing process was prevented by later repeated poisoning.

In Britain, the toxic peanut meal acquired in 1960 also caused poisoning of chickens. The main effect of aflatoxicosis in chickens was a retardation of growth, but chicks survived on toxic feeds known to kill ducklings. Chicks fed toxic levels of aflatoxin developed marked hepatic lesions. During the first 3 weeks, the livers enlarged and became a pale, yellow-ochre color.[71] Thereafter, there was an apparent reduction in size of the organ, with increased firmness of texture; still later well-defined nodular areas and pectechial hemorrhages were observed.[72]

The first authenticated case of aflatoxicosis in the U.S. appears to be the report of Smith and Hamilton.[309] The case involved 1000 laying hens in North Carolina and occurred after moldy corn was added to their diet. High mortality and symptoms of aflatoxicosis were evident within 48 hr of introduction of the moldy corn to the chicken ration. Analysis of the molded grain indicated an aflatoxin content of about 90 ppm, and a nearly pure growth of *A. flavus* was cultured from the molded corn.[137] Each of the affected birds examined revealed a serious infiltration of the pericardium, a paleness of comb, shank, and bone marrow, an enlarged and pale liver, an enlarged spleen, an enlarged pancreas, a regressed bursa of Fabricus, and a below normal amount of depot fat. In subsequent studies, the addition of aflatoxin to a normal diet resulted in the above symptoms when fed to young birds. Broiler chickens fed a ration containing aflatoxin in excess of 1.25 ppm developed symptoms of aflatoxicosis. The age of broilers did not markedly affect the LD_{50} tolerance level, although earlier reports[24,55] indicated that resistance to aflatoxin increased with age and varied with the breed during the first 8 weeks of growth.[309] The LD_{50} dosage was near 7 to 8 mg/kg at each week of the test for Arbor-Acre 60 X Peterson broilers. Pronounced differences in tolerances of various Leghorn strains to aflatoxin were indicated by LD_{50} values of 6-week-old chickens. About a 2.5-fold tolerance difference was noted between the most resistant and most sensitive strains. Kratzer et al.[177] fed Arbor-Acre chickens 1600 ppb aflatoxin over a period of 60 days without a significant reduction in growth rate. Lesions were present in most livers, but other organs appeared normal. Feed with 400 or 800 ppb resulted in no adverse affects on the growth rate of the livers. Thus, 400 was regarded as a "no effect" level for chicks in their study.

Since chickens are more resistant to aflatoxin than other poultry, they would seem more likely candidates to accumulate aflatoxin residues in their tissues. A number of investigators[177,243,307] found no chemical evidence of aflatoxin in the meat, liver, blood, or eggs of chickens consuming rations containing 1600 to 3000 ppb aflatoxin. In contrast to the reports of no aflatoxin residues in tissues or eggs, Van Zytveld et al.[358] detected fluorescent spots by TLC of muscle and liver extracts of aflatoxin-dosed birds. These fluorescent spots were interpreted as aflatoxin or aflatoxin metabolites, but definitive proof of their identity was not confirmed.

Labeling studies show that ^{14}C aflatoxin was largely excreted by dosed birds, but significant amounts of ^{14}C were detected in the blood, meat and organs.[197,198,278] Further studies of the ^{14}C-labeled metabolites in tissue indicated that about 50% of aflatoxin B_1 was metabolized to the much less toxic aflatoxin B_{2a}.[76] Sawhney et al.[278] detected ^{14}C-labeled aflatoxin or its metabolite in eggs, and direct evidence for toxin transmission into eggs was furnished by Jacobson and Wiseman.[164] Aflatoxin was detected in eggs from hens consuming a diet containing 100 ppb B_1, and nearly one half the eggs of birds on a 200 or 400 ppb ration contained 48 ppb or more. Hens consuming large doses of aflatoxin produced fewer eggs, but the eggs were mostly fertile.[154] Hatchability of the eggs was significantly reduced; however, chicks from these eggs performed about as well as chicks from hens on the normal diet.

Data indicating the absence or presence of aflatoxin residues in meat and eggs of poultry consuming aflatoxin are fairly meager, but a small amount of evidence indicates that residues are present in both. Therefore, precautions are necessary to keep aflatoxin in feeds to a minimum.

C. Ochratoxin

The ochratoxins are a group of isocoumarin derivatives produced by a number of *Aspergillus* species and *P. viridicatum*.[357] In 1965, van der Merwe et al.[353] crystallized and identified the most potent derivative, ochratoxin A, from a toxic *A. ochraceous* strain. Identification, purification, and production of this mycotoxin permitted biological studies in both small animals and farm animals. These studies led to comparisons of symptoms in test animals with farm animals suspected of suffering from natural toxicoses.

Acute biological effects of ochratoxin A in day-old ducklings and rats were first investigated by Theron et al.[260,332] The toxin caused a mild fatty infiltration of the duckling's liver when administered orally in doses of 100 µg. Weanling rats developed widespread hyaline degeneration of the liver cells with focal necrosis. Autopsy of rats that died during an LD_{50} estimation showed enteritis, pale enlarged livers, and signs of degeneration and necrosis in kidneys. Histopathological examinations revealed that the principal damage was a tubular necrosis of the kidney which in the more severe cases appeared to affect the majority of the cells in all the tubules.[258]

Ochratoxin A administered orally to swine, 1 or 2 mg daily, caused emesis, depression, reduced feed intake, weight loss, diarrhea, polyuria, polydipsia, dehydration, and death in 5 to 6 days.[329] Autopsies showed dehydration, enteritis, and pale tan discoloration of livers. An earlier destructive action on the renal tubular epitheluim was evinced by elevation of blood urea, nitrogen, glucosuria, proteinuria, and changes in specific gravity and urinary sediment and mainly by the concentration of urinary enzymes. Renal changes were the most prominant histopathological finding in pigs dosed with ochratoxin A. A morphologically characteristic nephropathy characterized by tubular atrophy and interstitial fibrosis oberved in pigs inspected in slaughterhouses over the past five decades[111,180] was induced by feeding ochratoxin A.[181] Reports of nephropathy of swine are mostly from Denmark, but this renal syndome has been observed in Ireland[58] and in Sweden, Norway, and Germany, according to Elling and Moller.[111] The incidence of nephropathy in Danish swine was somewhat less than one per 14,000, but incidences of 0.1 to 1.5% were observed in some districts in the years 1971 to 1972.

Molded barley containing ochratoxin A (citrinin, viridicatumtoxin, and aflatoxin were not detectable by chemical analyses) was incorporated into a swine diet at levels found in nature, 0, 0.2, 1.0, and 4.0 µg/g, over a 4 month period. The only observable lesion developed was kidney damage, identical to the naturally occurring porcine nephropathy.[181] Weight gains and feed conversions of ochratoxin-exposed pigs were significantly inferior to the control group, and pigs on the 4 µg/g ration drank about twice as much water as the other groups after about the 6th week. Macroscopic renal changes were evident in nearly all the pigs receiving more than 1 µg/g of ochratoxin in their feed, and pigs in all three treatments showed microscopic lesions. Macroscopic lesions in the kidney included a markedly increased relative weight and greyish-yellow color. Microscopically observable changes in renal structure consisted of degeneration of the proximal tubules, interstitial formation of connective tissue, and at later stages development of atrophied, sclerotized glomerula tufts.

Carcasses of the ochratoxin-fed pigs were of good quality and would have passed a meat inspection. Ochratoxin A, however, was found in increasing concentrations through the tissues, muscular, adipose, liver, and kidney. The kidney of pigs fed the largest quantity contained about 50 µg/kg ochratoxin A. Residues in muscle were considerably less than in the kidney, but residues in all tissues increased proportionally with toxin consumption. These residues are of some concern, although their effect on man is unknown. Naturally, effects of mycotoxins cannot be studied in humans, but man is known to suffer nephropathy with symptoms somewhat like those observed in the initial stages of porcine nephropathy.[111,180,181]

Experiments also proved chickens to be highly sensitive to ochratoxin A. Oral median lethal doses vary somewhat with the investigator and experimental procedures. Huff et al.[157] determined a LD$_{50}$ value for day-old Pilch Dekalb chicks of about 2.14 mg/kg, and 3.60 mg/kg for 3-week-old chickens; Peckham et al.[237] earlier reported a LD$_{50}$ value in Babcock-B 300 cockerels of slightly less than 4 mg/kg; and Galtier et al.[125] had a slightly higher LD$_{50}$ estimate of 5.4 mg/kg for day-old chicks, and about 10.67 mg/kg for 10-day-old birds. Chicks dying from acute ochratoxicosis became listless, huddled, and suffered from diarrhea, tremors, and neural disturbances.[125,157,237]

Chicks fed a ration containing graded levels of ochratoxin had a retarded growth rate, and retardation became more severe after long periods of administration. Relative weights of the liver, proventriculus, crop, gizzard, and kidneys increased, whereas the relative weight of the bursa of Fabricus decreased.[158] The most sensitive visible indicator of ochratoxicosis in the chicken is kidney enlargement resulting from edema.[157]

Hamilton et al.[138] made a field survey of mycotoxicoses among U.S. poultry flocks and discovered three outbreaks of ochratoxicosis in turkey flocks totaling about 360,000 birds. Mortality in one flock was nearly 55%. The feed corn contained 16 ppm ochratoxin A and proved negative for other mycotoxins, heavy metals, and polychlorinated biphenyls. Poisoned birds had typical nephropathy symptoms: pale, swollen kidneys, necrosis of the proximal tubules of the kidney, and no change in other organs. Ochratoxicosis was also suspected in four outbreaks in chicken flocks involving about two million birds, where poor growth and poor feed conversion were associated with detectable ochratoxin in their feed. Another case of ochratoxicosis was indicated when 14 birds were rejected because of renal changes observed by Danish inspectors in a poultry slaughterhouse.[112] In four of these birds, a toxic nephropathy was found, characterized by atrophy and degeneration of proximal and distal tubules and interstitial fibrosis. A total of 5 of the rejected birds had residues ranging from 4.3 to 29.2 mg/kg ochratoxin A in muscular tissue.

Weights of ochratoxin-treated chicks remain significantly lower than control birds 20 days after completion of a 10-day treatment, demonstrating the persistence of the toxic effect, and possibly of the toxin itself within the bird. Apparently, ochratoxin is only metabolized slowly and, therefore, accumulates in the body.[125] Krogh et al.[182] fed chickens an ochratoxin-containing diet for 341 days after hatching. The only oberservable lesion was a kidney damage comparable with the naturally occurring avian nephropathy; at slaughter ochratoxin A was found in increasing concentration throughout the tissues, muscular, liver, and kidney. Residues of nearly 50 µg/kg were present in the muscle, even though the birds were in a condition good enough to pass the meat inspector. Residues were not found in eggs from ochratoxin-consuming layers.

Rainbow trout have also been shown to be acutely sensitive to ochratoxins A and B.[101] The occurrences of natural outbreaks in this host, however, have not been recorded.

IV. MYCOTOXIN DETECTION

A. Aflatoxin
1. Analytical

Prior to 1969, analytical methodology for aflatoxins was in a state of flux,[250] however, currently the state of the art is well advanced.[99,169,250,282,319] Throughout the world, aflatoxins are the only contaminates of food being monitored routinely at a level of 10 μg/kg per kilogram[282] Aflatoxins B_1, B_2, G_1 and G_2 (see Figure 3) are most commonly found in food and feed commodities contaminated with *A. flavus;* M_1 and M_2, metabolic byproducts produced by cows on ingestion of feeds containing B_1 (see Figure 3), are found in milk and dairy products. The fluorescence and UV absorption properties of the aflatoxins are summarized in Table 9 along with some of their other physicochemical properties. Spectrophotometric properties, especially the fluorescence exhibited by aflatoxins, are the basis for precise, sensitive transmission, reflectance densiometric, and visual method for detection of as little as 10^{-10} aflatoxins.[250]

Varied methodology has evolved for extraction of aflatoxins depending on the commodity being investigated. Methods generally followed are the established collaborative tested Official Methods of Analysis of the Association of Official Analytical Chemists (AOAC)[36] which include Method 1 [Contaminant Branch (CB)] and Method 2 [Best Food (BF)] for peanuts and peanut products and the cottonseed method. Methods 1 and 2 have been accepted as Official First Action for analysis of corn and other cereal grains.

The AOAC procedures encompass a representative sampling technique (the amount varies from commodity to commodity), solvent extraction (one-phase and two-phase systems), use of precipitating reagents — removal of protein and other interfering substances (pigments), clean-up column chromatography (silica gel, cellulose), and TLC separation of the aflatoxins with measurement by flurodensitometry or visual methods.

The AOAC procedures[25] for aflatoxin assay have been applied with variations in precipitating reagents, extraction solvents, and chromatographic procedures to many commodities either spiked with aflatoxins or produced on various agricultural products experimentally, as well as to agricultural products both for human and animal consumption that were naturally contaminated either in the field or in storage: peanut, peanut meal, cotton seeds, cotton seed meal, copra,[39,114,135,204,208,251,268,306,359,360,370] corn,[298] roasted corn,[247] mixed feeds, nuts,[39,272] spices and herbs,[284] cocoa beans,[161] groundnuts and groundnut products,[94] coffee beans,[189,283] and cereal grains.[145]

Early aflatoxin assay methods were based on the "dilution of extinction" technique.[16,93] These methods involved visual comparison of intensity of fluoresence of sample to internal standard as spots on uniform absorbent layers on TLC plates.[108,114,252] The visual method's accuracy depended on the operator's ability to compare spot sizes of the same concentration of sample to standard.[252] The precision of visual discrimination in judging intensity difference in spots varies from ±20 to 28%.[46]

The need for analytical methods free from the limitations of these two methods led to a spectrophotometric method. Utilizing aflatoxin optical density in methanol at 363 μm, Nabney and Nesbitt[225] were able to analyze contaminated groundnut meals (2.5 to 12.5 ppb B_1). Accurate, precise, and especially more sensitive methods emerged from utilizing the fluorodensitometer, as demonstrated from the work of Engebrecht et al.,[114] Pons et al.,[252] and Jemmali.[165] Ayers and Sinnhuber[40] reported that as low as 8×10^{-5}

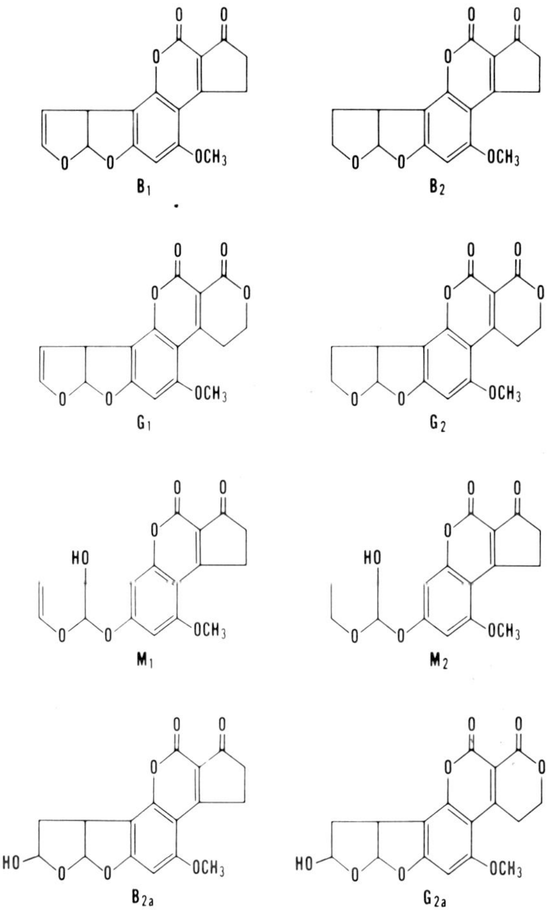

FIGURE 3. Structures of aflatoxins.

Table 9
PHYSIOCHEMICAL PROPERTIES OF AFLATOXINS

Aflatoxin	mp (°C)	MW	$(\alpha)_D$	UV (CH₃OH), 363 nm	Fluorescence emission at 365 nm excitation	IR[a] C = O	Ref.
B₁	268—269 (d)	312	−558	21,800	425	1,760	20, 33
B₂	286—289 (d)	314	−492	24,000	425	1,705	33, 139
G₁	244—246 (d)	328	−556	17,700	450	1,760	20, 33
G₂	237—240 (d)	330	−473	19,300	450	1,690	33, 139
M₁	299 (d)	328	−280	19,000[b]	435	1,760 1,690	151
M₂	293 (d)	330	—	21,000	435	1,760	151
B₂ₐ	240 (d)	330	—	20,400		1,690	107, 109
G₂ₐ	190 (d)	346	—	18,000			107, 109

[a] CHCl₃ γmax

[b] 357 λ EtOH

μg of pure aflatoxin B_1 can be quantitated on TLC plates, using fluorodensitometry with a linear range of 0.25 to 1.5×10^{-5} μg for aflatoxin B_1. The precision of fluorodensitometry ideally varies ±5% for scans of purified aflatoxin.[40] The precision established by colaborative studies for both visual and densitometry measurements of aflatoxins is about 30%;[282] hence, either method of measurement of aflatoxin is acceptable. Some errors may result from the removal of aliquots of toxins from solutions or from the possibility of fluorescent interfering substances with the same migration rate as aflatoxin extracted from the commodity under investigation. Also the aflatoxins have been reported to be photooxidized with UV light.[15]

The use of high-pressure liquid chromatography (HPLC) as a possible alternative to the TLC fluorodensitometry and visual analytical procedures is attractive because of its high resolution potential, its capability to make faster seperation, and perhaps even its improved quantitative accuracy and precision. With this impetus, several investigators reported the use of HPLC in aflatoxins analyses; Seiber and Hsieh[289] showed HPLC to be less sensitive and less selective than TLC fluorodensitometry with porous silica layer (30 to 40 μm) using chloroform or isooctane (2:1). A linear response to B_1 and G_1 in the 400- to 300-ng range resulted with a coefficient of variation of 4.2% for B_1 and 32.3% for G_1. Rao and Anders[263] resolved B_1 and G_1 extracted from molded grains by HPLC on silica gel with the solvents isopropyl ether-tetrahydrofuran or ether-cyclohexane. In 1975, Garner[126] reported HPLC separation of aflatoxins on silica gel with 0.3% v/v methanol in water-saturated dichloromethane. The resolution was stated to be comparable to, if not better than, the TLC fluorodensitometry and visual methods. Recent reports[155,248,291] have shown that HPLC advantages over TLC methods at this time appear to be better resolution, usefulness in preparative work, and minimizing aflatoxin degradation by exposure to light and air. HPLC is not efficient for analysis of a large number of samples, since multiple injection on HPLC is time consuming compared to multiple sample separation on TLC. Interfering substances contained in extracts obtained from naturally contaminated foods and feeds impede the measurement of aflatoxins by HPLC. More work to improve detectors and Preclean up procedures is needed before HPLC is on a par with TLC methods.

2. Sampling Procedure

The accurate measurement of aflatoxins on TLC plates by either visual or fluorodensitometric methods depends on adequate sampling of the agricultural product submitted for analysis. The nonuniform distribution of aflatoxin, as well as the small amount present (usually 1 to 1000 ppb), necessitates that all of the commodity be evaluated to obtain the true level of contamination in a unit lot. Since this is not practical, large representative samples from lots of commodities are obtained by physical selection (probes, stream splitters). Sample sizes from commodity lots to be analyzed are 48 lb for peanuts,[349] 50 to 60 lb for Brazil and pistachio nuts, and 10 to 50 lb for corn. The sample of the product is ground and blended to give homogenous distribution of the toxin, with the particle size adjusted (a disk mill for grains) to pass through a U.S. Standard Number 20 sieve. Water slurries have also been suggested to achieve uniform distribution of particle size and for products high in oil content.[361] Usually 50 g of these homogenous ground samples are taken for toxin analyses.

3. Official Methods

Samples after homogenization are then analyzed for aflatoxins according to the CB or BF methods established for peanuts.[119] The CB method involves addition of 25 g Celite to a 50-g homogenized sample, extraction with $CHCl_3$-H_2O (250 mℓ:25 mℓ in a 500-mℓ glass-stoppered flask on a wrist shaker for 30 min. The solution is filtered, and the first 50 mℓ is collected and chromatographed on a 10-g silica gel column. The column is selectively eluted with 150 mℓ hexane to remove fats, 150 mℓ of ether to remove pigments and interfering substances, and 150 mℓ of $CHCl_3$-CH_3OH (97:3) for elution of the aflatoxins. The $CHCl_3$-CH_3OH eluate is evaporated to near dryness (steam bath or roto evaporator), transferred to a 4-dram vial, and taken to dryness under a gentle stream of nitrogen in a hot-water bath or a heating block. Solvent (benzene-CH_3CN, 98:2) is added to a given volume and then spotted on TLC plates for quantative comparison of fluorescence with standards and confirmational derivatives. The most common solvent system for TLC of aflatoxins is acetone · $CHCl_3$,[38] but if inadequate separation of aflatoxins occurs, many alternative developing solvents are available.[28]

The BF method blends the commodity with a two-phase solvent system, CH_3OH-H_2O/hexane (55:45/100),[35] in a blender for 1 min at high speed. After the phases separate, 25 mℓ of the aqueous CH_3OH phase is pipetted into a separatory funnel and extracted with 25 mℓ $CHCl_3$ by shaking for 30 to 60 sec. The chloroform layer is drained into a 600-mℓ stainless steel beaker (if available), evaporated in a stream of nitrogen, and analyzed by the TLC-densitometric method.

The procedure for cottonseed[32] uses 5% aqueous acetone as extracting solvent and the precipitating agent $Pb(OAc)_4$, whereas the official method for corn uses the same extracting solvent, but $(NH_4)_2SO_4$ as the precipitating reagent. Both methods then follow column chromatographic procedures as prescribed in the official CB method.

4. Analyses of M_1 and M_2 in Milk and Dairy Products

The official method for analytical analyses for M_1 toxin in powdered milk follows the procedure of Purchase and Steyn[261] and for liquid milk follows that of Jacobson et al.[163] The suitability of these two methods for powdered and liquid milk for M_1 analysis was established by Stubblefield et al.[328] in a comparison evaluation with four other published methods, at levels of 0.1 μg/kg for liquid milk and 1 μg/kg in powdered milk. In 1972, an international collaborative study conducted by Purchase et al.[259] showed the precision of the Purchase and Steyn[261] and Jacobson et al.[163] methods to be comparable to official AOAC methods for determination of aflatoxins in agricultural products.

The modified Jacobson method for M_1 quantitation in liquid milk[26] blends 75 mℓ of fluid milk with CH_3OH (300 mℓ) for 3 min; then 25 g of diatomaceous earth is added, followed by additional blending for 30 sec. The homogenate is filtered by suction through 1 cm of diatomaceous earth and washed with 75 mℓ CH_3OH. The filtrate is defatted with hexane, and the aflatoxin is extracted with chloroform.

The official first action method for powdered milk[27] follows the modified procedure of Purchase and Steyn:[261] 25 g of powdered milk is extracted with 200 mℓ of acetone-water (7:3) and filtered, and the residue is rinsed with 50 mℓ acetone-water; the filtrate is evaporated to 25 mℓ, and 10 mℓ of $Pb(OAc)_4$ plus 10 mℓ Na_2SO_4 solutions are added to precipitate insoluble materials; insolubles are centrifuged, and the supernatant is added to a separatory funnel. The precipitate is washed twice with 20 mℓ acetone-water, then the supernatants are combined and extracted with hexane and three 50-mℓ portions $CHCl_3$, dried over Na_2SO_4, evaporated to near dryness, and transferred to a 1-dram vial for TLC analyses.

The official method for dairy products (cheeses and butter) is a modified Pons' procedure.[249,326,328] Aflatoxin is extracted with acetone-water, lead acetate is added to remove soluble protein and phosopholipids, and the solution is extracted with hexane to remove glycerides of fatty acids and other fats; $CHCl_3$ extraction removes M_1. This latter extract is washed with NaCl solution, water is removed by passing over anhydrous Na_2SO_4, and the extract is concentrated to dryness and further purified by cellulose partition column chromatography in 30% aqueous methanol. The elution sequence is hexane-benzene (3:1;

150 mℓ), hexane-ether (2:1, 150 mℓ), and hexane-chloroform (1:1; 200 mℓ). The latter elution solvent is concentrated in a stream of nitrogen on a steam bath in a 1-dram vial for TLC and densitometric measurement of aflatoxins. The TLC solvent for M_1 and M_2 is isopropanol-acetone-$CHCl_3$ (1:1:8).[30]

5. Confirmatory Tests for Aflatoxins

Although the official methods (CB, BF, cottonseed) involving fluorodensitometric and visual estimates of aflatoxin on TLC are precise for detection and quantitation, confirmatory tests (biological and chemical) are necessary to differentiate unequivocally between substances that fluoresce and have chromatographic properties similar to the toxin being analyzed.

In 1964, Andrellos and Reid[16] devised three confirmatory tests to identify B_1: the formation of dimeric acetate with formic thionyl chloride and with acetic thionyl chloride and the formation of B_{2a} and G_{2a} (hemiacetals of B_1 and G_1 (see Figure 3) with trifluoroacetic acid in the presence of water. The hemiacetals are also formed by acid-catalyzed hydration.[57,109] The Andrellos-Reid test was studied collaboratively by 19 laboratories,[320] and no false identifications resulted. In 1970, Pohland et al.[245] revised these confirmatory tests by treating the aflatoxin-containing extract with concentrated HCl and water to yield the water adduct and with concentrated HCl and acetic anhydride to form the epimeric acetate. A reduction in side reactions is the main advantage of this modification. The modification of the Andrellos-Reid test by Pohland was studied collaboratively[321] by nine laboratories and found to be more reliable and simpler than the previous official method. Hence it was adopted by the AOAC.

In 1975, the official confirmatory method was again improved by Przybylski.[257] Aflatoxins B_1 and G_1 are converted to B_{2a} and G_{2a} directly on the TLC plate with the addition of trifluoroacetic acid prior to its development; B_{2a} appears as a blue fluorescent spot at an R_f lower than B_1 and G_1. In addition, the method is highly specific for detecting a false positive analysis for aflatoxins, since aflatoxin blue fluorescence changes to yellow on spraying with 25% H_2SO_4. The modification advanced by Przybylski was studied collaboratively by eight laboratories, and aflatoxin B_1 was confirmed in 17 of 17 samples containing 15 μg of aflatoxin B_1 per kilogram peanut butter and in none of seven aflatoxin-free samples.[312] The method was adopted by the Associate Referee on Confirmational Methods.[34]

The formation of the confirmatory acetate derivative of aflatoxin M_1 in acetic anhydride-pyridine and the hemiacetal of the acetate in HCl-H_2O was described by Stack et al.[313] Both the acetates and the hemiacetal of the acetate derivatives of M_1 can be formed and detected with as little as 1 ng. In an international collaborative study,[327] five false negatives were reported from 29 observers with milk spiked with 30 ng of aflatoxin M_1 for the acetate and hemiacetal of the acetate. Cheese spiked with 50 ng of M_1 gave no false negatives on derivative formation.

Other chemical confirmatory tests for aflatoxins B_1 and B_2 are based on the carbonyl group in the cyclopentanone ring (not adopted by the AOAC) by the formation of oximes and the 2,4-dinitrophenylhydrazone.[95,96] Reduction of B_1 and B_2 with sodium borohydride quantitatively yields trihydroxy compounds.[21] The reduction products have higher R_f values than do B_1 and B_2 in $CHCl_3$-ethylacetate. Aflatoxins B_1 and B_2 form a violet blue, whereas G_1 and G_2 form brown-colored compounds with di-O-ansidine-tetraazolium chloride.[222] Detection limits of 0.03 μg for the former and 0.09 μg for the latter are reported.

6. Biological Tests

In addition to the chemical confirmatory tests, suspected aflatoxin-contaminated products are further confirmed by a number of biological systems.[374] These include the use of chick embryos, ducklings, rats, hamsters, guinea pigs, dogs, trout, catfish cells, brown bullhead, and many others. Biological systems most frequently used are the duckling, chick embryo, and trout.

The chick embryo is the only biological assay adopted by the Association of Official Analytical Chemists.[29] It was shown by Platt et al.[244] that injection of aflatoxin into the 5-day-old chick embryo yolk caused death. Verrett et al.[362] and Choudhary and Manjrekar[77] injected toxin into the air cell or yolk. An LD_{50} at 21-day incubation of 0.048 µg was found for the yolk route and 0.025 µg was found for the air-cell route. Examination of nonsurviving embryos indicated same growth retardation, edema, hemorrhage, underdevelopment of the mesencephalon, mottled and granular liver surface, short legs, and slight clubbing of the down. Injection of 9-day-old chick embryos produced congestion, fatty degeneration, necrosis, and bile duct proliferation. For 10-day-old chick embryos, an LD_{40} of 2 to 50 µg resulted when aflatoxin was injected into the chorioalantoic sac and incubated for 48 hr.

In 1973, Verrett et al.[363] conducted a collaborative study on the chick-embryo bioassay with nine laboratories. An LD_{50} of 0.025 µg per egg was established for pure aflatoxin B_1, and that isolated from peanut butter was 0.045 µg per egg. Further, this bioassay relies on criteria other than the LD_{50}, such as teratogenic effects which lend confirmation to aflatoxin presence, perhaps at a nonlethal dosage level. Higher dosages led to early death, severe growth retardation, and severe edema.

The use of day-old white Peking ducklings approximately 50 g in weight as a bioassay is based on the degree of hyperplasia of the bile duct epithelium resulting from aflatoxin B_1 dosing levels up to 16 µg.[17,70] The toxins or toxic extracts are usually administered by capsule or intubation in propylene glycol or water over a 5-day period. On the seventh day, the bird is sacrificed and liver tissue is examined for degree of bilary proliferation. The lethal effects of aflatoxins to 1-day-old duckling are seen within the first 72 hr. The LD_{50} for day-old ducklings are B_1 (0.36 mg/kg), B_2 (1.69 mg/kg), G_1 (0.78 mg/kg), G_2 (2.45 mg/kg), M_1 (0.33 mg/kg), and M_2 (1.24 mg/kg).[71,151]

Other biological test systems which seem well suited for detection of aflatoxins at a semiquantitative level are the trout and microbial disk assays. Trout are very sensitive to aflatoxin B_1 with an acute LD_{50} of 0.5 mg/kg body weight single dose or 0.3 mg/kg body weight if administered over a 5-day period. Hepatomas occur in trout fed a diet containing aflatoxin B_1 when 0.2 mg of the toxin per kilogram body weight was administered.[136]

In a survey of 329 microorganisms for aflatoxin B_1 sensitivity, Burmeister and Hesseltine[62] found that one strain of *Bacillus brevis* and two of *B. megaterium* were inhibited at levels of 10 and 15 µg/mℓ, respectively. Similar results were reported by Clements.[92]

Brine shrimp, *Artemia samlina*, was described by Brown et al.[56] as a bioassay for aflatoxin B_1. A level of 0.5 µg/mℓ in artificial sea water at 37.5°C resulted in a mortality rate of 60%; 90% mortality resulted at a level of 1 µg/mℓ. Harwig and Scott[142] used brine shrimp larvae in a disc screening method for mycotoxins and reported 21% mortality at 0.2 µg per disc for aflatoxin B_1 and 50% mortality for G_1 at 1.3 µg per disc incubated for 16 hr at 30°C with the larvae.

Zebra fish larvae (*Brachydanio rerio*) were suggested as bioassay test organism by Abedi and McKinley.[1] A level of less than 0.58 µg/mℓ of aflatoxin B_1 or 0.83 µg/mℓ G_1 is acutely toxic to the larvae.

B. Zearalenone

The structure of the potent estrogen, zearalenone, is shown in Figure 4, and in Table 10 its most prominent physicochemical features are listed. It was extracted from 31 commodities[118] by the water-chloroform method of Lee.[188] Eppley[118] used this solvent

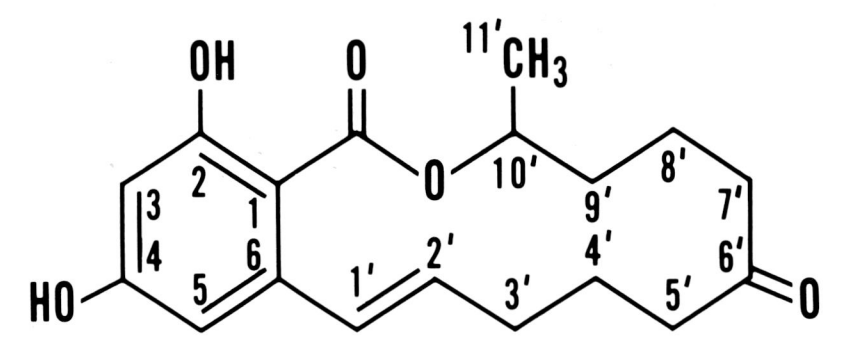

FIGURE 4. Structure of zearalenone.

Table 10
PHYSICAL PROPERTIES OF ZEARALENONE[348]

Property	
mp	164—165°C
Molecular formula	$C_{18}H_{22}O_5$
Mol wt	318
$(\alpha)_{546}^{25}$	−170.5 (CH_3OH)
λ max (nm)	236 (29,700), 274 (13,909), 316 (6,020)
IR (cm⁻)	3,200; 1,688; 1,645
Chemical shifts (δ)	C_3 (6.50 d; J_{AB} 2.5 H_3)
	C_5 (6.39 d; J_{AB} 2.5 H_3)
	$C_{1'}$ (7.01 d; J = 16 H_3)
	$C_{2'}$ (5.63 m); $C_{10'}$ (5.00 m)
	$C_{11'}$ (1.40 d; J = 6 H_3)

system to develop a multiple screening method for zearalenone, aflatoxin, and ochratoxin, with sequential elution of the mycotoxins from a silica gel column. In a collaborative study of the Eppley method for zearalenone conducted by Shotwell et al.[299] 4 of 16 collaborators detected the toxin in yellow corn spiked at levels of 100 μg/kg, and 12 detected it at 300 μg/kg. The between-laboratory coefficient of variations ranged from 53% at 300 μg/kg, to 38.2% at 1000 μg/kg, to 27% to 2000 μg/kg. These wide variations inicate that visual estimation on TLC plates is not consistent and that perhaps other techniques should be used for quantitation. In the same study, two collaborators quantitated zearalenone on TLC plates densitometrically. Zearalenone previously was measured by fluorodensitometry by Jammali.[165]

In 1974, Mirocha et al.[218] described a method for quantitation of zearalenone in a number of grains and feedstuffs which employed either TLC, gas-liquid chromatography (GLC), UV chromatography, GLC-MS, or a combination of all these. The sample (100 g, 30% H_2O), finely ground, was extracted with ethyl acetate by either the Soxhlet techique or the more rapid convenient batch process with equal efficiency. A cleanup of the extract involved sodium hydroxide treatment and partition with water or with acetonitrile-petroleum ether. The solvent system, chloroform-ethanol (97:3) was used for analysis by TLC, with confirmation bases on fluorescence properties. The limit of detection is 0.1 μg, while the sensitivity is 50 ppb. GLC is sensitive to less than 50 ppb and gives the best quantitative results. The limit of detection for zearalenone analysis by UV spectrometry is between 0.1 and 0.5 µg/mℓ GLC and MS can also be used with multiple ion detection. Confirmatory derivatives for zearalenone are dimethoxy and methyl oxime-di-TMS-ether. GLC of free zearalenone, its methyl oxime, and trimethylsilyl derivatives was reported earlier by Vandenheuvel.[351]

Because it is estrogenic to farm animals and possibly to humans, many surveys to determine zearalenone contamination of food and feeds have been conducted. Stoloff and Dalrymple[323] examined products from 82 corn dry milling establishments and detected no zearalenone in any product. Holder et al.[150] described HPLC procedure for determining zearalenone and/or zearalenol in animal chow, with sensitivity levels of 10 ppb. Salient features of the method include methanol extraction of the chow, an initial clean-up on Sephadex LH-20, and a liquid-liquid partitioning in pH 13 and pH 8.3, followed by separation on a column of silica gel.

A number of colorimetric reagents and sprays have been used to provide additional confirmatory evidence for zearalenone. Enhanced fluorescence results with $AlCl_3$.[150] A colored azo derivative is formed with 4 methoxy benzene-diazonium-fluoroborate.[277] Sarudi[277] also observed a selective change of color for the azo derivative with acid and base conditions; an intense yellowish-red color resulted with calcium hydroxide, and a lilac color resulted on spraying with ethanolic-sulfuric acid. Malaiyandi et al.[203] found that zearalenone forms a brick-red product on TLC plates when sprayed with bis-diazotized benzidine; 2 ng was detectable. This spray reagent was reported to be a more specific spray than potassium ferricyanide ferric chloride reported by Mirocha et al.[211]

Occurrence of detectable levels of zearalenone (0.1 to 2909 ppm) in feedstuffs associated with estrogenism in animals was found in 45% of the samples submitted to the University of Minnesota over a 3-year period by Mirocha et al.[212] All of the estrogenic-positive samples without detectable zearalenone concentrations, when administered to female weanling rats, induced a several-fold increase in uterine weights.[214]

Ueno et al.[346] reported that single or repeated administration of zearalenone to immature mice and rats increased uterine weight to several times greater that that of control animals. Oral administration of this estrogen to immature mice and rats is more effective than the s.c. or i.p. route. Ovaricetomized mice were highly sensitive to zearalenone, and the dose response was linear when daily doses of 1 to 2 mg/kg were given for 1 week. This marked uterotrophic response in mice and rats may represent an attractive presumptive bioassay method for evaluation of feedstuffs suspected to be involved in estrogenism in farm animals. Brine shrimp (*Artemia solini* L) larvae are moderately sensitive to zearalenone (10 µg/disc µg per dis produced an 18% mortality).[142]

C. Trichothecenes
1. Chemical Detection and Physical Methods of Analysis

There are 37 trichothecenes produced by species of *Fusarium, Trichoderma, Trichothecium,* and *Stachybotrys*. About half of theses are elaborated by the Fusaria, and the physicochemical and biological properties of these are compended in a recent review of "Fusarium Metabolites" by Vesonder and Hesseltine.[367] For some time, this class of compounds eluded precise chemical analytical methods for quantitation due to their nonfluorescent properties and their lack of intensity in any characteristic absorption band in the UV or IR regions of the spectrum. Most trichothecenes are chemically stable and unaffected upon refluxing in organic solvents. Stability of these sesquiteqenoid-epoxide molecules is partly due to the stability of the 12,13-epoxide to extramolecular and nucelophilic attack. Intramolecular rearrangements occur on prolonged boiling in water (6 hr) by hydration of the olefinic bond at the C_9-C_{10} position to give the rearranged diol due to opening of the epoxide ring. This type of rearrangement is reviewed by Bamburg and Strong.[43]

Visualization of the trichothecenes after treatment with reagents that form colored complexes with epoxides were unsuccessful. However, successful visual location on TLC plates has been made after spraying with *p*-anisaldehyde reagent by Scott et al.;[286] Ueno et al.[344] described the color change of trichothecenes containing a carbonyl at the C_8 position as a nonfluorescent brown spot after spraying with sulfuric acid. Trichothecenes lacking a carbonyl group in the C_8 position fluoresce blue under the long UV wave after spraying with sulfuric acid. These spray reagents are useful during isolation of a particular trichothecene and provide information on degree of purity. An estimation of vomitoxin, 3,7,15-trihydroxy-12,13-epoxytrichothec-9-en-8-one, based on the intensity of the yellow spot produced by Scott's *p*-anisaldehyde reagent, was reported by Vesonder et al.[366]

Spray reagents highly specific for particular trichothecene are Ehrlichs reagent (*p*-dimethylaminobenzaldehyde in hydrochloric acid and ethanol) that gave an intense violet color with diacetoxyscirpenol and related compounds,[98] although Bamburg et al.[44] observed no color change with T-2 toxin, phosphomolybdic reagent, and antimony trichloride disolved in chloroform that produced a blue-purple spot with diacetoxyscirpenol.[128]

Presently quantitation of trichothecenes by GLC and GLC-MS in the selected ion mode is most attractive due to increased sensitivity, precision, and reliability relative to TLC. GLC can be used for quantitation of the known trichothecenes with the exception of roridins and verrucarins.[160] Ikediobi et al.[160] separated and quantitated 13 trichothecenes by GLC. A complication of GLC quantitation lies in eliminating components having retention times similar to the naturally occurring trichothecenes. For example, Mirocha and Pathre[217] found the trimethylsilyl derivative of monoglycerides to have the same retention time as T-2 toxin. It has been suggested[160] that some of these analytical problems could be solved by converting trichothecenes to their parent alcohols by alkaline treatment of the partially purified sample extracts.

We have used a quantitative GLC method for vomitoxin with internal standard. The procedure for obtaining the sample extract was essentially that reported previously.[364,366] A 100-g sample of contaminated corn dried in a force-air oven to a moisture level of less than 10% was extracted twice with butanol in a blender (100 g/200 mℓ). This step is most convenient, since it removes the corn oil, other lipids, and pigments along with other butanol-soluble material. It also extracts T-2 toxin, although not quantitatively, and T-2 toxin can easily be detected in the butanol fraction by the method of Scott et al.[286] The corn freed of butanol solubles is extracted in a blender with 40% aqueous CH_3OH. Acetone can be added to the methanol-water solubles to precipitate sugars and other proteinaceous material, or the MeOH-H_2O soluble fraction may be applied directly to a 10-g silica gel column and sequentially eluted with $CHCl_3$ (150 mℓ) and 4% CH_3OH-$CHCl_3$ (300 mℓ). The latter eluate containing the vomitoxin can be chromatographed on a column packed with 3% OV-101 on Gas Chrom 100/200, after addition of trimethyl silating reagent and internal standard.

Micocha et al.[216] found T-2 toxin vomitoxin, diacetoxyscirpenol, and zearalenone in feedstuffs using a computerized GC-MS operated in the selected ion-monitoring mode. Ethyl acetate-extracted T-2 toxin, diacetoxyscipenol, zearalenone, and the more polar solvent, methanol-water, were used to extract trichothecenes like vomitoxin as described by Vesonder et al.[364]

2. Biological Testing

Methods of bioassay for the trichothecenes are restricted to detection and estimation. Biological test systems include the necrotic skin reaction in laboratory animals,[42,43,128] inhibition of sensitive fungi,[61] inhibitory effect on protein synthesis in rabbit reticulocytes,[341,345] histological detection of the "radiomimetric cellular injury" in poisoned mice,[274] and phytotoxic properties.[53] Although these bioassay are sensitive, the skin reaction of small laboratory animals in response to certain toxins is appealing because of its simplicity, reliability, and sensitivity.

A modified topical rat skin test for T-2 toxin described by Wei et al.[369] is sensitive to 0.05 µg. T-2 applied in methanol or ethyl acetate to shaved backs of 55- to 120-g Albino rats produced a skin reaction in 24 hr. Skin reaction features are reddish weals with application of 0.05 to 0.1 µg; 0.5 to 1.0 µg causes more swollen reddish weals, with nearly white centers; skin treated with 2 to 5 µg becomes mostly white with reddish edges and often shows serious exudate. Chung et al.[85] described a similar method using a rabbit skin test based on the dermatitic properties of T-2 and other toxins as a screening method applicable to corn samples. The method showed that 10-g samples of field corn could be rapidly screened for compounds having dermal toxicity. Extracts for analyses obtained from corn are prepared according to the Eppley method.[118] Diethyl ether and methanol-chloroform eluates from the silica gel column are combined and evaporated, and the resultant oil dissolved in 0.1 mℓ ethylacetate is applied to the closely clipped skin of a 2- to 3-kg rabbit.

Estimation of the degree of skin reaction (erythema, edema, and necrosis) are compared to a graded response with T-2 toxin after 24, 48, and 72 hr. The method[85] is reliable to at least 0.01 µg per test. Rabbits are more sensitive to T-2 toxin than are weanling rats and guinea pigs. However, skin-reaction tests, although quite sensitive, are only presumptive evidence for the presence of trichothecenes which exhibit dermal toxicity.

D. Ochratoxins
1. Chemistry

Ochratoxin A is the major toxic metabolite associated with the *A. ochraceus* group,[83,86,146,352] *P. viridicatum*,[287,288,356] and some other *Penicillium* spp.[87] Ochratoxin A was isolated from maize inoculated wtih *A. ochraceus*[353] and was shown to be a dihydroisocoumarin moiety linked to L-β-phenylalanine.[352] Two minor constituents, ochratoxins B and C, the dechloro and the ethyl ester derivatives of ochratoxin A, respectively, were also isolated. The structures of these minor components, along with ochratoxin A, are shown in Figure 5. In addition, the methyl esters of ochratoxin A and B and ethyl ester of ochratoxin B were isolated from *A. ochraceus*.[316] Physicochemical properties of ochratoxins A, B, and C are summarized in Table 11. Ochratoxins A, B, and C under UV light appear as bright blue-green fluorescent spots in paper chromatograms in the solvent system propanol-3 N aqueous ammonium carbonate (3:1), with R_fs of 0.65, 0.68, and 0.8, respectively.[352] Ochratoxin A, B, and C migrate to R_fs of 0.57, 0.55, and 0.73, respectively, on silica gel TLC plates developed in benzene, methanol, and acetic acid (12:2:1) and appear as bright green fluorescent spots.[352] Ochratoxins are relatively stable to autoclaving (70% destroyed in 3 hr)[335] and the canning process,[140] but are degraded under simulated coffee-roasting conditions (200°C for 5 min).[190]

A combination of nephrotoxic properties and the natural occurrence of ochratoxin A in cereal grains in the U.S.,[302,303] Canada,[287,288] Denmark,[185] and Sweden[183] encouraged development of methods for analysis based on their fluorescence properties.

2. Detection and Analyses

A visual estimation of ochratoxin A based on its fluorescence in UV light[352] was the first described analytical procedure by Steyn and van der Merwe.[318] Ochratoxin A at a level of 0.1 ppm could be detected in spiked corn meal. Silica gel plates spotted with 8 ng of ochratoxin were used for comparison, the fluorescence being enhanced with 0.1 N sodium hydroxide. Two othe TLC methods of ochratoxin A evolved; Scott and Hand[285] described a procedure for grain products spiked with ochratoxin A with a lower detection

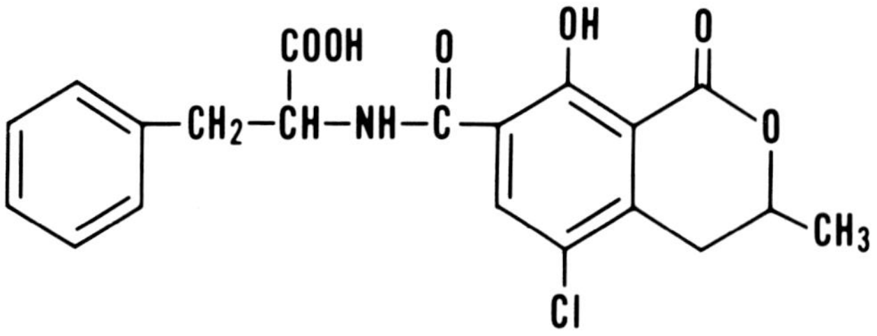

Ochratoxin A

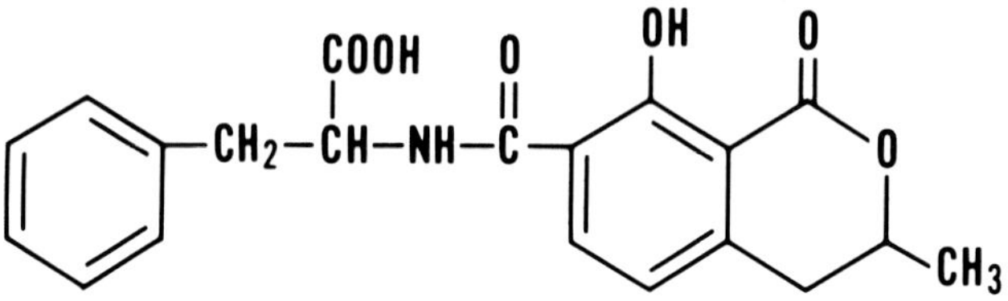

Ochratoxin B

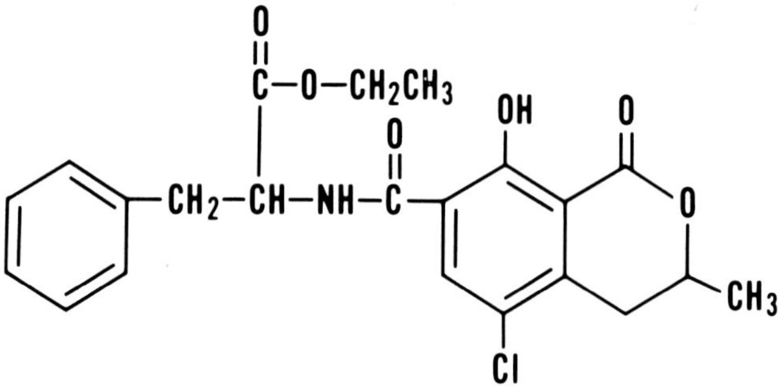

Ochratoxin C

FIGURE 5. Structures of ochratoxins.

Table 11
PHYSICAL PROPERTIES OF OCHRATOXINS

Ochratoxin	Molecular formula	Molecular weight	Melting point	$(\alpha)_D$ $CHCl_3$	UV λ max mm (ε)	Fluorescence λ exc.	Fluorescence λ emm.	Ref.
A	$C_{20}H_{18}Cl\ NO_6$	403	169	−118	213 (36,800) 332 (6,400)	340	475	31,352
B	$C_{20}H_{19}NO_6$	369	221	− 35	218 (37,200) 318 (6,900)	325	475	31,352
C	$C_{22}H_{22}Cl\ NO_6$	431		−100	213 (32,700) 331 (4,100) 378 (2,050)	340	475	31,316

limit of 25 μg/kg and recoveries of 80 to 100%; Nesheim et al.[231] described partition and TLC for extracts obtained from barley amended with 25, 50, and 100 μg/kg of ochratoxin A. A recovery of 81.2% was obtained. In an interlaboratory study, four analysts established a lower detection limit of 12 μg/kg. In a follow-up collaborative study, conducted by Nesheim,[230] analyses for ochratoxins A and B, and their esters, were determined in barley by 13 laboratories. The average recovry of standard ochratoxin was 112% at levels of 45 to 90 μg/kg, with a precision of 27.1%. Similar results were obtained for the ethyl esters of ochratoxin A and B at a level of 120 μg/kg. For ochratoxin B and the esters of ochratoxin A and B, the method was unsatisfactory at the 60 μg/kg level.

Methods prior to the Nesheim method involved multiple assays for zearalenone, aflatoxin, and ochratoxin by Eppley;[118] the TLC method by Steyn[317] for simultaneous separation of several mycotoxins will be discussed under the section, entitled "Multitoxin Screening Procedures."

The precision of visual comparison for fluorescence intensity of spots of ochratoxin A on TLC plates is ± 20%. Methods free of this inherent error are UV spectrophotometry[230] with a sensitivity of around 1 to 5 μg/mℓ[84,100,267] and spectrofluorodensitometry. A densitometric method for ochratoxin A was first described by Nesheim[299] and later advanced by Chu and Butz.[84] A lower limit of 0.5 to 1 ng was found for ochratoxin A on cereals. A recovery of over 85% of the toxin added to cereals was obtained with a precision of 3 to 4% and a sensitivity of 10 to 50 μg/kg.

3. Extraction and Purification Procedures

Ochratoxin A was extracted from moldy corn with $CHCl_3$-CH_3OH (1:1) by the Soxhlet technique. The extract was separated into acidic and neutral fractions with $NaHCO_3$, and the neutral fraction was then acidified and reextracted with $CHCl_3$.[352] the $CHCl_3$ solubles were separated by TLC with benzene-acetic acid (12:2) and examined for ochratoxin under UV light.

The extraction method described by Scott and Hand[285] for cereals uses a two-phase solvent system H_2O-CH_3OH-hexane (45:55:40). Ochratoxin A is eluted from a Celite column with $CHCl_3$-hexane (1:1) and visually estimated on thin layer chromatograms developed in toluene-ethyl acetate-90% formic acid (5:4:1)

The method developed by Nesheim[230] was adopted as the Official First Action[37] for barley. A representative 50-g ground sample placed in a glass-stoppered Erlenmeyer flask is extracted with 25 mℓ 0.1 M H_3PO_4 and 250 mℓ $CHCl_3$ for 30 min on a wrist-action shaker. The extract is chromatographed on a $NaHCO_3$-Celite column to entrap acids; acids are eluted with HCOOH-$CHCl_3$ (1:99), followed by elution of the ochratoxin esters with $CHCl_3$; the samples are taken to dryness and evaluated by the TLC densitometric method.

4. Confirmatory Tests
a. Chemical

In all analyses, the possibility of interfering substances exists. Hence, additional proof is desirable to verify the presence of the mycotoxins under investigaton. Confirmation may be accomplished by developing chromatograms in several solvent systems, utilizing reagents to form derivatives with new migration rates, or producing intensively fluorescent derivatives.

Ochratoxin A fluorescence is intensified under alkaline conditions: treatment of the toxin with ammonia fumes produces a deep-blue fluorescence;[87,117,285,287,336] triethylamine,[285] sodium bicarbonate,[231] ferric chloride,[159] and aluminum chloride[87] also enhance fluorescence. Ochratoxin derivatives useful for confirmatory tests are ethyl esters[28,230,231] and acetates.[357]

b. Biological Tests

Confirmatory — Confirmation of suspected ochratoxin A contamination may be confirmed with biological tests. Emphasis has been placed on biological tests more sensitive than the LD_{50} of 150 μg for ducklings[258] and LD_{50} of 135 to 166 μg for day-old chicks.

A microbiological assay using *B. megaterium* NRRL B-1368 was sensitive to 4 μg/mℓ ochratoxin A.[92] Shotwell et al.[301] confirmed ochratoxin A in the range of 150 ppb for naturally contaminated wheat with *B. megaterium*. Broce et al.[54] described a filter paper method for confirmation of ochratoxin A with the test organism *Bacillus cereus mycoides* LSU. The coefficient of variation was 5.9%, and as little as 1.5 μg per disc could be detected.

Other systems of interest for bioassay of ochratoxin A are a lethal toxicity assay for zebra fish larvae,[2] and brine shrimp larvae assay[142] with an LD_{50} of 10 μg/mℓ, 17% mortality at 0.2 μg per disc.

E. Multitoxin Screening Procedures

General screening procedures applicable to several mycotoxins are desirable, since these toxins can and do occur together in a wide variety of food and feedstuffs throughout the world. Several multitoxin screening methods for simultaneously testing combinations of aflatoxins, zearalenone, ochratoxin, patulin, citrinin, T-2 toxin, penitrem A, penicillic acid, diacetoxyscirpenol, and sterigmatocystin have been described. The Eppley procedure[118] for detecting aflatoxin, zearalenone, and ochratoxin is rapid, sensitive, and has been applied effectively in surveys of grains and dry-milled corn products.[300,304,325] Stoloff et al.[324] described a multiple detection method for 11 mycotoxins spiked onto corn, oats, and barley. The method is based on extracting with CH_3CN-4% KCl (1:1), defatting with isooctane and removing water-soluble components by liquid-liquid transfer of the mycotoxins to $CHCl_3$. Detection limits achieved with multiple toxin assay procedures are not as low as those attained with analytical procedures designed for an individual toxin. The lowest detectable levels of the aflatoxins B_1 and G_1 in corn, wheat, barley, oats, and rye was 20 μg/kg; for ochratoxin A and B, 50 μg/kg for corn, and 100 μg/kg for other grains; for zearalenone, 200 μg/kg for corn, 400 μg/kg for wheat, and 500 μg/kg for other grains, for sterigmatocystin, 60 μg/kg in grains; and patulin varied from 400 μg/kg for wheat to 1000 μg/kg for oats.

Thomas et al.[333] modified the AOAC Method II (BF) for aflatoxins[35] to facilitate the rapid detection of aflatoxins and zearalenone in corn. Centrifugation and pipetting steps were replaced with a filtration step, followed by a hexane wash of the extract. Copper carbonate was added to the extracts to precipitate pigments prior to a half-plate TLC development to reduce the time of analysis. Recoveries of aflatoxins B_1, G_1, and G_2 were low (28 to 60%); B_1 was easily detected at 2 μg/kg, and zearalenone was detected in amounts of 100 μg/kg. The method was only applied to spiked corn samples.

Separation of aflatoxins, ochratoxins, zearalenone, and sterigmatocystin added to commercial oils (olive, peanut, corn, and soybean) was reported by Hagan and Tietjen.[134] They used a TLC plate developed with benzene-hexane (3:1) for initial clean up; a second development in the same direction as the first, using toluene-ethyl acetate-formic acid (5:3:1) or benzene-acetic acid (9:1) separated the toxins. This technique permits the detection of 5 ppb aflatoxin B_1 in corn.

A multitoxin screening method, described by Wilson et al.[372] is applicable for the quantitative recovery of aflatoxins in corn, beans, and peanuts and ochratoxin A in corn and beans; only 50% of ochratoxin A was recovered from peanuts at levels ranging from 59 to 236 μg/kg; zearalenone could not be estimated at levels less than 200 μg/kg; penicillic acid and citrinin could only be determined qualitatively because of streaking on TLC plates and loss of fluorescence of citrinin in the developing solvents tested;

citrinin was not recovered from peanuts at 100- to 4000-μg/kg levels. The method uses 0.5 N phosphoric acid-chloroform (1:10) in the initial extraction; the extract is divided and eluted from two columns to provide a quantitative TLC method for aflatoxin and ochratoxin in corn and dried beans.

Roberts and Patterson[266] described a method for the detection of 12 mycotoxins in mixed animal feedstuff using a dialysis membrane clean up step prior to the multimycotoxin procedures of Stoloff et al.[324] and Scott et al.[287] The method minimizes interference from nonspecific lipid, pigment, and other components of animal feeds. The sensitivity for aflatoxin B_1 is 3 ppb and for ochratoxin A is 80 ppb. The method is less sensitive for sterigmatocystin (350 ppb), patulin (600 ppb), zearalenone (1000 ppb), and for T-2 toxin and diacetoxyscirpenol (1000 to 4000 ppb).

Seitz and Mohr[292] described a simple method for the simultaneous detection of aflatoxin at 5 ppb and zearalenone at 200 ppb in yellow corn which they believed simple enough to be used in marketing laboratories with minimal laboratory facilities. The procedure involves a methanol extraction utilizing a blender and partitioning of fat and pigments into 1,1,2-tri-chlorotrifluoroethane (freon-113) from an aqueous ammonium sulfate layer, followed by extraction of the toxins from the aqueous layer with chlorobenzene. The multiple screening method for detection of mycotoxins appears limited, since toxin recoveries are generally low and, in some cases, recovery is not achieved because more interfering substances are present relative to the specific quantitative procedures.

Analysis of mycotoxins by TLC has been described by Steyn[317] for 11 mycotoxins and by Scott et al.[286] for 18 mycotoxins. The Scott method combines detection under visible light and UV light with a *p*-anisaldehyde spray. Durrchova et al.[104] analyzed 37 mycotoxins and other fungal metabolites using eight solvent systems and various chemical sprays. However, only pure compounds were analyzed.

A field desorption mass spectrometry method for identification of mycotoxins and mycotoxin mixtures and the screening of foodstuffs was described by Sphon et al.[310] How effective this method would be to screen commodities depends on the determination of the minimal detection limits, how extensively the extract must be cleaned, and the factors governing the varying sensitivities observed with actual extracts.

REFERENCES

1. **Abedi, Z. H. and McKinley, W. P.,** Zebra fish eggs and larvae as aflatoxin bioassay test organisms, *J. Assoc. Off. Anal. Chem.,* 51, 902, 1968.
2. **Abedi, Z. H. and Scott, P. M.,** Detection of toxicity of aflatoxins, sterigmatocystin, and other fungal toxins by lethal action on zebra fish larvae, *J. Assoc. Off. Anal. Chem.,* 52, 963, 1969.
3. **Adamson, R. H., Correa, P., Sieber, S. M., McIntire, K. R., and Dalgard, D. W.,** Carcinogenicity of aflatoxin B_1 in Rhesus monkeys: two additional cases of primary liver cancer, *J. Natl. Cancer Inst.,* 57, 67, 1976.
4. **Adamesteanu, I. and Adamesteanu, C.,** Citeva date asupra pericolului micotoxinelor in general si a aflatoxinelor in special in cresterea tineretului animal, *Rev. Zooteh. Med. Vet.,* 9, 51, 1973.
5. **Allcroft, R.,** Aflatoxicosis in farm animals, *Aflatoxin Scientific Background, Control and Implication,* Goldblatt, L. A., Ed., Academic Press, New York, 1969,
6. **Allcroft, R. and Carnaghan, R. B. A.,** Toxic products in groundnuts, biological effects, *Chem. Ind. (London),* 50, 1963.
7. **Allcroft, R. and Lewis, G.,** Groundnut toxicity in cattle: experimental poisoning of calves and a report on clinical effects in older cattle, *Vet. Rec.,* 75, 487, 1963.

8. **Allcroft, R. and Roberts, B. A.**, Toxic groundnut meal: the relationship between aflatoxin B$_1$ intake by cows and excretion of aflatoxin M$_1$ in milk, *Vet. Rec.*, 82, 116, 1968.
9. **Alpert, M. E., Hutt, M. S. R., and Davidson, C. S.**, Hepatoma in Uganda. A study in geographic pathology, *Lancet*, 15, 1265, 1968.
10. **Alpert, M. E., Hutt, M. S. R., Wogan, G. N., and Davidson, C. S.**, The association between aflatoxin content of food and hepatoma frequency in Uganda, *Cancer*, 28, 253, 1971.
11. **Amla, I., Kamala, C. S., Gopalakrishna, G. S., Jayaraj, P., Sreenivasamurthy, V., and Parpia, H. A. B.**, Cirrhosis in children from peanut meal contaminated by aflatoxin, *Am. J. Clin. Nutr.*, 24, 609, 1971.
12. **Amla, I., Kumari, S., Sreenivasamurthy, V., Jayaraj, P., and Parpia, H. A. B.**, Role of aflatoxin in Indian childhood cirrhosis, *Indian Pediatr.*, 7, 262, 1970.
13. **Amla, J., Sreenivasamurthy, V., Jayaraj, P., and Parpia, H. A. B.**, Aflatoxin and Indian childhood cirrhosis — a review, *J. Trop. Pediatr. Environ. Child Health* 19, 28, 1974.
14. **Anderson, L. J., Sandison, A. T., and Jarrett, W. F. H.**, A British abattoir survey of tumours in cattle, sheep and pigs, *Vet. Rec.*, 84, 547, 1969.
15. **Andrellos, P. J., Beckwith, A. C., and Eppley, R. M.**, Photochemical changes of aflatoxin B$_1$, *J. Assoc. Off. Anal. Chem.*, 50, 346, 1967.
16. **Andrellos, P. J. and Reid, G. R.**, Confirmatory tests for aflatoxin B$_1$, *J. Assoc. Off. Anal. Chem.*, 47, 801, 1964.
17. **Armbrecht, B. H. and Fitzhugh, O. G.**, Mycotoxins. II. The biological assay of aflatoxin in Peking White ducklings, *Toxicol. Appl. Pharmacol.*, 6, 421, 1964.
18. **Armbrecht, B. H., Wiseman, H. G., and Shalkop, W. T.**, Swine aflatoxicosis. I. In assessment of growing efficiency and other responses in growing pigs fed aflatoxin, *Environ. Physiol.*, 1, 198, 1971.
19. **Armbrecht, B. H., Wiseman, H. G., and Shalkop, W. T.**, Swine aflatoxicosis. II. The chronic response in brood sows fed sublethal amounts of aflatoxin and the reaction in their piglets, *Environ. Physiol. Biochem.*, 2, 77, 1972.
20. **Asao, T., Büchi, G., Abdel-Kader, M. M., Chang, S. B., Wick, E. L., and Wogan, G. N.**, Aflatoxin B and G, *J. Am. Chem. Soc.*, 85, 1706, 1963.
21. **Ashoor, S. H. and Chu, F. S.**, New confirmatory test for aflatoxins B$_1$ and B$_2$, *J. Assoc. Off. Anal. Chem.*, 58, 617, 1975.
22. **Ashworth, L. J., Jr., McMeans, J. L., and Brown, C. M.**, Infection of cotton by *Aspergillus flavus*: epidemiology of the disease, *J. Stored Prod. Res.*, 5, 193, 1969.
23. **Ashworth, L. J., Jr., Schroeder, H. W., and Langley, B. C.**, Aflatoxins: environmental factors governing occurrence in Spanish peanuts, *Science*, 148, 1228, 1965.
24. **Asplin, F. D. and Carnaghan, R. B. A.**, The toxicity of certain groundnut meals for poultry with special reference to their effect on ducklings and chickens, *Vet. Rec.*, 73, 1215, 1961.
25. **Association of Official Analytical Chemists**, Natural poisons, in *Official Methods of Analysis of AOAC*, 12th ed., Horwitz, W., Ed., 1975, chap. 26 (Methods 26.014 to 26.020).
26. **Association of Official Analytical Chemists**, Natural poisons, in *Official Methods of Analysis of AOAC*, 12th ed., Horwitz, W., Ed., 1975, chap. 26 (Methods 26.079, 26.084 to 26.087).
27. **Association of Official Analytical Chemists**, Natural poisons, in *Official Methods of Analysis of AOAC*, 12th ed., Horwitz, W., Ed., 1975, chap. 26 (Methods 26.088 to 26.089).
28. **Association of Official Analytical Chemists**, Natural poisons, in *Official Methods of Analysis of AOAC*, 12th ed., Horwitz, W., Ed., 1975, chap. 26 (Method 26.010).
29. **Association of Official Analytical Chemists**, Natural poisons, in *Official Methods of Analysis of AOAC*, 12th ed., Horwitz, W., Ed., 1975, chap. 26 (Methods 26.073 to 26.078).
30. **Association of Official Analytical Chemists**, Natural poisons, in *Official Methods of Analysis of AOAC*, 12th ed., Horwitz, W., Ed., 1975, chap. 26 (Method 26.090).
31. **Association of Official Analytical Chemists**, Natural poisons, in *Official Methods of Analysis of AOAC*, 12th ed., Horwitz, W., Ed., 1975, chap. 26 (Method 26.098).
32. **Association of Official Analytical Chemists**, Natural poisons, in *Official Methods of Analysis of AOAC*, 12th ed., Horwitz, W., Ed., 1975, chap. 26 (Methods 26.048 to 26.056).
33. **Association of Official Analytical Chemists**, Natural poisons, in *Official Methods of Analysis of AOAC*, 12th ed., Horwitz, W., Ed., 1975, chap. 26 (Method 26.006).
34. **Association of Official Analytical Chemists**, Natural poisons, in *Official Methods of Analysis of AOAC*, 12th ed., Horwitz, W., Ed., 1975, chap. 26 (Method 26.A17).
35. **Association of Official Analytical Chemists**, Natural poisons, in *Official Methods of Analysis of AOAC*, 12th ed., Horwitz, W., Ed., 1975, chap. 26 (Methods 26.020 to 26.024).
36. **Association of Official Analytical Chemists**, Natural poisons, in *Official Methods of Analysis of AOAC*, 12th ed., Horwitz, W., Ed., 1975, chap. 26.
37. **Association of Official Analytical Chemists**, Natural poisons, in *Official Methods of Analysis of AOAC*, 12th ed., Horwitz, W., Ed., 1975, chap. 26 (Methods 26.091 to 26.098).

38. Associaiton of Official Analytical Chemists, Natural poisons, in *Official Methods of Analysis of AOAC*, 12th ed., Horwitz, W., Ed., 1975, chap. 26 (Method 26.019).
39. Ayers, G. C., Lillard, H. S., and Lillard, D. A., Mycotoxins: detection in foods, *Food Technol.*, 24, 55, 1970.
40. Ayers, J. L. and Sinnhuber, R. O., Fluorodensitometry of aflatoxin on thin-layer plates, *J. Am. Oil Chem. Soc.* 43, 423, 1966.
41. Bamburg, J. R., Riggs, N. V., and Strong, F. M., The structures of toxins from two strains of *Fusarium tricinctum*, *Tetrahedron*, 24, 3329, 1968.
42. Bamburg, J. R. and Strong, F. M., Mycotoxin of trichothecane family produced by *Fusarium tricinctum* and *Trichoderma*, *Phytochemistry*, 8, 2405, 1969.
43. Bamburg, J. R. and Strong, F. M., 12,13-Epoxytrichothecenes, in *Microbial Toxins*, Vol. 7, Kadis, S., Ciegler, A., and Ajl, S., Eds., Academic Press, New York, 1971, 207.
44. Bamburg, J. R., Strong, F. M., and Smalley, E. B., Toxins from moldy cereals, *J. Agric. Food Chem.*, 17, 443, 1969.
45. Barnes, J. M., Carter, R. L., Peristianis, G. C., Austwick, P. K. C., Flynn, F. V., and Aldridge, W. N., Balkan (endemic) nephropathy and a toxin-producing strain of *Penicillium verrucosum* var. *cyclopium*: an experimental model in rats, *Lancet*, 26, 671, 1977.
46. Beckwith, A. C. and Stoloff, L., Fluorodensitometric measurement of aflatoxin thin layer chromatograms, *J. Assoc. Off. Anal. Chem.*, 51, 602, 1968.
47. Becroft, D. M. O. and Webster, D. R., Aflatoxins and Reye's syndrome, *Br. Med. J.*, 4, 117, 1972.
48. Bierbower, G. W. and Armbrecht, B. H., Chronic duck aflatoxicosis induced by feeding mixed and purified aflatoxin, *Environ. Physiol. Biochem.*, 2, 68, 1972.
49. Blount, W. P., Turkey "X" disease, *Turkeys*, 9, 52, 1961.
50. Boonchuvit, B., Hamilton, P. B., and Burmeister, H. R., Interaction of T-2 toxin with *Salmonella* infection of chickens, *Poult. Sci.*, 54, 1693, 1975.
51. Bösenberg, H., Diagnostische moglichkeiten zun nachweis van aflatoxin-vergiftungen, *Zentralbl. Bakteriol. Parasitenkd., Infektionskr. Hyg. Abt. 1, Orig. Reihe A*, 220, 252, 1072.
52. Bourgeois, C., Olson, L., Comer, D., Evans, H., Keschamras, N., Cotton, R., Grossman, R., and Smith, T., Encephalopathy and fatty degeneration of the liver: a clinicopathologic analysis of 40 cases, *Am. J. Clin. Pathol.*, 56, 558, 1971.
53. Brian, P. W., Dawkins, A. W., Grove, J. F., Hemming, H. G., Lowe, D., and Norris, G. L. F., Phytotoxic compounds produced by *Fusarium equiseti*, *J. Exp. Bot.*, 12, 1, 1961.
54. Broce, D., Grodner, L. M., Rosumond, L., Killerbrew, R. L., and Bonner, F. L., Ochratoxins A and B confirmation by microbiological assay using *Bacillus cereus* mycoides, *J. Assoc. Off. Anal. Chem.*, 53, 616, 1970.
55. Brown, J. M. M. and Abrahms, L., Biochemical studies on aflatoxicosis, *Onderstepoort J. Vet. Res.*, 32, 119, 1965.
56. Brown, R. F., Wildman, G. D., and Eppley, R. M., Temperature dose relationships with aflatoxin on the brine shrimp, *Artemia salina*, *J. Assoc. Off. Anal. Chem.*, 51, 905, 1968.
57. Büchi, G., Foulkes, M., Kurono, M., Mitchell, G. F., and Schneider, R. S., The total synthesis of racemic aflatoxin B_1, *J. Am. Chem. Soc.*, 89, 6745, 1967.
58. Buckley, H. G., Fungal nephrotoxicity in swine, *Ir. Vet. J.*, 25, 194, 1971.
59. Bulatao-Jayme, J., Almero, E. M., and Salamat, L., Epidemiology of primary liver cancer in the Philippines with special consideration of a possible aflatoxin factor, *J. Philipp. Med. Assoc.*, 52, 129, 1976.
60. Burmeister, H. R., T-2 toxin production by *Fusarium tricinctum* on solid substrate, *Appl. Microbiol.*, 21, 739, 1971.
61. Burmeister, H. R., Ellis, J. J., and Hesseltine, C. W., Survey for fusaria that elaborate T-2 toxin, *Appl. Microbiol.*, 23, 1165, 1972.
62. Burmeister, H. R. and Hesseltine, C. W., Survey of the sensitivity of microorganisms to aflatoxin, *Appl. Microbiol.* 14, 403, 1966.
63. Burnside, J. E., Sippel, W. L., Forgacs, J., Carll, W. T., Atwood, M. B., and Doll, E. R., A disease of swine and cattle caused by eating moldy corn. II. Experimental production with pure culture of molds, *Am. J. Vet. Res.*, 18, 817, 1957.
64. Butler, W. H., Aflatoxin, *Mycotoxins*, in Purchase, I. F. H., Ed., Elsevier Scientific Publ. Co., Amsterdam, 1974.
65. Caldwell, R. W. and Tuite, J., Zearalenone production in field corn in Indiana, *Phytopathology*, 60, 1696, 1970.
66. Caldwell, R. W. and Tuite, J., Zearalenone in freshly harvested corn, *Phytopathology*, 64. 752, 1974.
67. Campbell, T. C., Caedo, J. P., Jr., Bulatao-Jayme, J., Salamat, L., and Engel, R. W., Aflatoxin M_1 in human urine, *Nature (London)*, 227, 403, 1970.

68. **Campbell, T. C. and Stoloff, L.,** Implication of mycotoxins for human health, *J. Agric. Food Chem.,* 22, 1006, 1974.
69. **Carnaghan, R. B. A.,** Heptic tumours in ducks fed a low level of toxic groundnut meal, *Nature (London),* 208, 308, 1965.
70. **Carnaghan, R. B. A., Hartley, R. D., and Kelley, G.,** Toxicity and fluorescence properties of the aflatoxins, *Nature (London),* 200, 1101, 1963.
71. **Carnaghan, R. B. A., Hebert, C. N., Patterson, D. S. P., and Sweasey, D.,** Comparative biological and biochemical studies in hybrid chicks. II. Susceptibility to aflatoxin and effects on serum, protein constituents, *Br. Poult. Sci.,* 8, 279, 1967.
72. **Carnaghan, R. B. A., Lewis, G., Patterson, D. S. P., and Allcroft, R.,** Biochemical and pathological aspects of groundnut poisoning in chickens, *Pathol. Vet.,* 3, 601, 1966.
73. **Chi, M. S., Mirocha, C. J., Kurtz, H. J., Weaver, G., Bates, F., and Shimoda, W.,** Subacute toxicity of T-2 toxin in broiler chicks, *Poult. Sci.,* 56, 306 1977.
74. **Chi, M. S., Mirocha, C. J., Kurtz, H. J., Weaver, G., Bates, F., and Shimoda, W.,** Effects of T-2 toxin on reproductive performance and health of laying hens, *Poult. Sci.,* 56, 628, 1977.
75. **Chi, M. S., Mirocha, C. J., Kurtz, H. J., Weaver, G., Bates, F., Shimoda, W., and Burmeister, H. R.,** Acute toxicity of T-2 toxin in broiler chicks and laying hens, *Poult. Sci.,* 56, 103, 1977.
76. **Chipley, J. R., Mabee, M. S., Applegate, K. L., and Dreyfuss, M. S.,** Further characterization of [^{14}C] aflatoxin B$_1$ in chickens, *Appl. Microbiol.,* 28, 1027, 1974.
77. **Choudhary, P. G. and Manjrekar, S. L.,** Observation on biological activity of pure aflatoxin in chick embryos, *Indian Vet. J.,* 44, 543, 1967.
78. **Christensen, C. M.,** Moisture content, moisture transfer and invasion of sorghum seeds by fungi, *Phytopathology,* 60, 280, 1970.
79. **Christensen, C. M.,** Loss of viability in storage: microflora, *Seed Sci. Technol.,* 1, 547, 1973.
80. **Christensen, C. M.,** Microflora, in *Storage of Cereal Grains and Their Products,* Christensen, C. M., Ed., American Association of Cereal Chemists, St. Paul, Minn., 1974, 158.
81. **Christensen, C. M., Meronuck, R. A., Nelson, G. H., and Behrens, J. C.,** Effects on turkey poults of rations containing corn invaded by *Fusarium tricinctum* (Cds.) Sny. and Hans, *Appl. Microbiol.,* 23, 177, 1972.
82. **Christensen, C. M., Mirocha, C. J., Nelson, G. H., and Quast, J. F.,** Effects on young swine of consumption of rations containing corn invaded by *Fusarium roseum, Appl. Microbiol.,* 23, 202, 1972.
83. **Chu, F. S.,** Comparative study of the interaction of ochratoxins with swine serum albumin, *Biochem. Pharmacol.,* 23, 1105, 1974.
84. **Chu, F. S. and Butz, M. E.** Mycotoxins: spectrophotofluorodensitometric measurement of ochratoxin A in cereal products, *J. Assoc. Off. Anal. Chem.,* 53, 1253, 1970.
85. **Chung, C. W., Trucksess, M. W., Giles, A. L., Jr., and Friedman, L.,** Rabbit skin test for estimation of T-2 toxin and other skin-irritating toxins in contaminated corn, *J. Assoc. Off. Anal. Chem.,* 57, 1121, 1974.
86. **Ciegler, A.,** Bioproduction of ochratoxin A and penicillic acid by members of the *Aspergillus ochraceus* group, *Can. J. Microbiol.,* 18, 631, 1972.
87. **Ciegler, A., Fennell, D. I., Mintzlaff, H. J., and Leistner, L.,** Ochratoxin synthesis by *Penicillium* species, *Naturwissenschaften,* 59, 1311, 1972.
88. **Ciegler, A., Fennell, D. I., Sansing, G. A., Detroy, R. W., and Bennett, G. A.,** Mycotoxin-producing strains of *Penicillium viridicatum:* classification into subgroups, *Appl. Microbiol.,* 26, 271, 1973.
89. **Ciegler, A. and Kurtzman, C. P.,** Penicillic acid production by blue-eye fungi on various agricultural commodities, *Appl. Microbiol.* 20, 761, 1970.
90. **Ciegler, A., Peterson, R. E., Lagoda, A. A., and Hall, H. H.,** Aflatoxin production and degradation by *Aspergillus flavus* in 20-liter fermentors, *Appl. Microbiol.,* 14, 826, 1966.
91. **Clegg. F. G. and Bryson, H.,** An outbreak of poisoning in store cattle attributed to Brazilian groundnut meal, *Vet. Rec.,* 74, 992, 1962.
92. **Clements, N. L.,** Mycotoxin Inhibition of *Bacillus megaterium,* presented at the 81st Annual Meeting of Official Analytical Chemists, October 9 to 12, 1967, Washington, D.C., Abstract 169, 1967.
93. **Coomes, T. J., Crowther, P. C., Frances, B. G., and Stevens, L,** The detection and estimation of aflatoxin in goundnuts and groundnut material, *Analyst,* 90, 492, 1965.
94. **Coomes, T. J. and Sanders, J. C.,** The detection and estimation of aflatoxin in groundnuts and groundnut material. *Analyst (London),* 88, 209, 1963.
95. **Crisan, E. V. and Grefig, A. T.,** The formation of aflatoxin derivatives, *Contrib. Boyce Thompson Inst.,* 24, 3, 1967.
96. **Crisan, E. V. and Mazzucca, E.,** Separation of aflatoxin on selectively deactivated silicic acid, *Contrib. Boyce Thompson Inst.,* 23, 361, 1967.
97. **Curtin, T. M. and Tuite, J.,** Emesis and refusal of feed in swine associated with *Gibberella zea* infected corn, *Life Sci.,* 5, 1937, 1966.

98. **Dawkins, A. W.** Phytotoxic compounds produced by *Fusarium equisiti*. II. The chemistry of diacetoxyscirpenol, *J. Chem. Soc.*, 116, 1966.
99. **Detroy, R. W., Lillehoj, E. B., and Ciegler, A.,** Aflatoxin and related compounds, in *Microbial Toxins*, Vol. 6, Ciegler, A., Kadis, S., and Ajl, S. J., Eds., Academic Press, New York, 1971, 3.
100. **Doster, R. C., Arscott, G. H., and Sinnhuber, R. O.,** Comparative toxicity of ochratoxin A and crude *Aspergillus ochraceus*. Culture extract in Japanese quail (*Coturnix coturnix japonica*), *Poult. Sci.*, 52, 2351, 1973.
101. **Doster, R. C., Sinnhuber, R. O., and Wales, J. H.,** Acute intraperitoneal toxicity of ochratoxins A and B in rainbow trout (*Salmo gairdneri*), *Food Cosmet. Toxicol.*, 10, 85, 1972.
102. **Dounin, M.,** The fusariosis of cereal crops in European Russia 1923, *Phytopathology*, 16, 305, 1926.
103. **Drobotko, V. G.,** Stachybotryotoxicosis, a New Disease of Horses and Humans, report presented at the Academy of Science, U.S.S.R.; as cited in Rodricks, J. V. and Eppley, R. M., *Mycotoxins*, Purchase, I. F. M., Ed, Elsevier, New York, 1974.
104. **Durrčhová, Z., Betina, V. and Nemec, P.,** Systematic analysis of mycotoxins by thin-layer chromatography, *J. Chromatogr.*, 116, 141, 1976.
105. **Duthie, I. F., Lancaster, M. C., Taylor, J., Lomax, E. B., and Clarkson, H. M.,** Toxic groundnut meal in feeds for pigs. II. The effects of consuming toxic groundnut meals during part of the growing period or during the finishing period, *Vet. Rec.*, 82, 427, 1968.
106. **Duthie, I. F., Lancaster, M. C., Taylor, J., Thomas, D. C., Shacklady, C. A., Ottfield, P. H., and Fuller-Lewis, E.,** Toxic groundnut meal in feeds for pigs. I. A trial made at two laboratories with pigs from about 40 to 200 lb live-weight fed to a restricted scale, *Vet. Rec.*, 79, 621, 1966.
107. **Dutton, M. F. and Heathcote, J. G.,** Two new hydroxyaflatoxins, *Biochem. J.*, 101, 21, 1966.
108. **Dutton, M. F. and Heathcote, J. G.** Methods apparatus: new products research, process development and design, *Chem. Ind. (London)*, 418, 1968.
109. **Dutton, M. F. and Heathcote, J. G.,** The structure, biochemical properties and origin of the aflatoxins B_{2a} and G_{2a}, *Chem. Ind. (London)*, 418, 1968.
110. **Dvoračková, I., Brodsky, F., and Cerman, J.,** Aflatoxin and encephalitic syndrome with fatty degeneration of viscera, *Nutr. Rep. Int.*, 10, 89, 1974.
111. **Elling, F. and Moller, T.,** Mycotoxic nephropathy in pigs, *Bull. W. H. O.*, 49, 411, 1973.
112. **Elling, F., Hald, B., Jacobsen, C., and Krogh, P.,** Spontaneous toxic nepropathy in poultry associated with ochratoxin A, *Acta Pathol. Microbiol. Scand. Sect. A*, 83, 739, 1975.
113. **Ellison, R. A. and Kotsonis, F. N.,** T-2 as an emetic factor in moldy corn, *Appl. Microbiol.*, 26, 540, 1975.
114. **Engelbrecht, R. H., Ayers, J. L., and Sinnhuber, R. O.,** Isolation and determination of aflatoxin B_1 in cottonseed meals, *J. Assoc. Off. Anal. Chem.*, 48, 815, 1965.
115. **Enomoto, M. and Saito, M.,** Carcinogens produced by fungi, *Annu. Rev. Microbiol.*, 26, 279, 1972.
116. **Enomoto, M. and Ueno, I.,** *Penicillium islandicum* (toxic yellowed rice)-leuteoskyrin islanditoxincyclochlorotine, in *Mycotoxins*, Purchase, I. F. H., Ed., Elsevier, New York, 1974, 303.
117. **Eppley, R. M.,** Report of the joint American Oil Analytical Chemists-American Oil Chemical Society aflatoxin committee, *J. Assoc. Off. Anal. Chem.*, 52, 975, 1969.
118. **Eppley, R. M.,** Screening method for zearalenone, aflatoxin, and ochratoxin, *J. Assoc. Off. Anal. Chem.*, 51, 74, 1968.
119. **Eppley, R. M.,** A versatile procedure for assay and the preparatory separation of aflatoxins from peanut products, *J. Assoc. Off. Anal. Chem.*, 49, 1218, 1966.
120. **Eppley, R. M. and Bailey, W. J.,** 12, 13-Epoxy-Δ^9-trichothecenes as the probable mycotoxins responsible for stachybotrytoxicosis, *Science*, 181, 758, 1973.
121. **Eppley, R. M., Stoloff, L., Trucksess, M. W., and Chung, C. W.,** Survey of corn for *Fusarium* toxins, *J. Assoc. Off. Anal. Chem.*, 57, 632, 1974.
122. **Eugenio, C. P., Christensen, C. M., and Mirocha, C. J.,** Factors affecting production of the mycotoxin F 2 by *Fusarium roseum Phytopathology*, 60, 1055, 1970.
123. **Fennell, D. I., Lillehoj, E. B., and Kwolek, W. F.,** *Aspergillus flavus* and other fungi associated with insect-damage field corn, *Cereal Chem.*, 52, 314, 1975.
124. **Gagne, W. E., Dungworth, D. E., and Moulton, J. E.,** Pathologic effects of aflatoxin in pigs, *Pathol. Vet.*, 5, 370, 1968.
125. **Galtier, P., More, J., and Alvinerie. M.,** Acute and short-term toxicity of ochratoxin A in 10-day old chicks, *Food Cosmet. Toxicol.*, 14, 129, 1976.
126. **Garner, R. C.,** Aflatoxin separation by high-pressure liquid chromatography, *J. Chromatogr.* 103, 186, 1975.
127. **Garrett, W. N., Heitman, H., Jr., and Booth, A. N.,** Aflatoxin toxicity in beef cattle, *Proc. Soc. Exp. Biol. Med.*, 127, 188, 1968.

128. **Gilgan, M. W. Samlley, E. B., and Strong, F. M.,** Isolation and partial characterization of a toxin from *Fusarium tricinctum* on moldy corn, *Arch. Biochem. Biophys.*, 114, 1, 1966.
129. **Gopal, T., Zaki, S., Narayanaswamy, M., and Premlata, S.,** Aflatoxicosis in dairy cattle, *Indian Vet. J.*, 45, 707, 1968.
130. **Gopalan, C., Tulpule, P. G., and Krishnamurthi, D.,** Induction of hepatic carcinoma with aflatoxin in the Rhesus monkey, *Food Cosmet. Txoicol.*, 10, 519, 1972.
131. **Gross, V. J. and Robb, J.,** Zearalenone production in barley, *Ann. Appl. Biol.*, 80, 211, 1975.
132. **Gumbmann, M. R., and Williams, S. N.,** Biochemical effects of aflatoxin in pigs, *Toxicol. Appl. Pharmacol.*, 15, 393, 1969.
133. **Gupta, S. K., and Venkitasubramanian, T. A.,** Production of aflatoxin on soybeans, *Appl. Microbiol.*, 29, 834, 1975.
134. **Hagan, S. N. and Tietjen, W. H.,** A convenient thin layer chromatographic cleanup procedure for screening several mycotoxins in oils, *J. Assoc. Off. Anal. Chem.*, 58, 620, 1975.
135. **Hald, B. and Krogh, P.,** Occurrence of aflatoxin in imported cottonseed products, *Nord. Veterinaermed.*, 22, 39, 1970.
136. **Halver, J. E.,** Aflatoxicosis and rainbow trout hepatoma, in *Mycotoxins in Foodstuffs*, Part III, Wogan, G. N., Ed., MIT Press, Cambridge, Mass., 1964, 209.
137. **Hamilton, P. B.,** A natural and extremely severe occurrence of aflatoxicosis in laying hems. *Poult. Sci.*, 50, 1880, 1971.
138. **Hamilton, P. B., Huff, W. E., Harris, J. R., and Wyatt, R. D.,** Outbreaks of ochratoxicosis in poultry, Abstr. Annu. Meet. Am. Soc. Microbiol., New Orleans. 1977.
139. **Hartley, R. D., Nesbitt, B. F., and O'Kelly, J.,** Toxic metabolites of *Aspergillus flavus*, *Nature (London)*, 198, 1056, 1963.
140. **Harwig, J., Chen, Y. K., and Collins, D. L.,** Stability of ochratoxin A in beans during canning, *Can. Inst. Food Sci. Technol. J.*, 7, 288, 1974.
141. **Harwig, J., Przybylski, W., and Moodie, C. A.,** A link between Reye's syndrome and aflatoxin?, *Can. Med. Assoc. J.*, 113, 281, 1975.
142. **Harwig, G. and Scott, P. M.,** Brine shrimp (*Artemia salina* L.) larvae as a screening system for fungal toxins, *Appl. Microbiol.* 21, 1011, 1971.
142a.**Hayes, A. W.,** personal communication.
143. **Hayes, A. W., Davis, N. D., and Diener, U. L.,** Effect of aeration on growth and aflatoxin production by *Aspergillus flavus* in submerged culture, *Appl. Microbiol.*, 14, 1019, 1966.
144. **Heathcote, J. G. and Hibbert, J. R.,** Biological activity and electronic structure of the aflatoxins, *Br. J. Cancer*, 29, 470, 1974.
145. **Hesseltine, C. W.,** Natural occurrence of mycotoxins in cereals, *Mycopathol. Mycol. Appl.*, 53, 141, 1974.
146. **Hesseltine, C. W., Vandergraft, E. E., Fennell, D. I., Smith, M. L., and Shotwell, O. L.,** Aspergilli as ochratoxin producers, *Mycologia*, 64, 539, 1972.
147. **Hilty, M. D.,** Etiology of Reye's syndrome in *Reye's Syndrome*, Pollack, J. D., Ed., Grune and Stratton, New York, 1975, 383.
148. **Hintz. H. F., Booth, A. N., Cuculla, A. F., Gardner, H. K., and Heitman, H., Jr.,** Aflatoxin toxicity in swine, *Proc. Soc. Exp. Biol. Med.*, 124, 266, 1967.
149. **Hintz, H. F., Hietman, H., Jr., Booth, A. N., and Gagne, W. E.,** Effects of aflatoxin on reproduction in swine, *Proc. Soc. Exp. Biol. Med.*, 126, 146, 1967.
150. **Holder, C. L., Naney, C. R. and Bowman, M. C.,** Trace analysis of zearalenone and/or zearalanol in animal chow by high pressure liquid chromatography and gas liquid chromatography, *J. Assoc. Off. Anal. Chem.*, 60, 272, 1977.
151. **Holzapfel, C. W., Steyn, P. S., and Purchase, I. F. H.,** Isolation and structure of aflatoxins M_1 and M_2, *Tetrahedron Lett.*, 2799, 1966.
152. **Hormozdiari, H., Day, N. E., Aramesh, B., and Mahboubi, E.,** Dietary factors and esophageal cancer in the Caspian littoral of Iran, *Cancer Res.*, 35, 3493, 1975.
153. **Hoyman, W. G.,** Concentration and characterization of the emetic principal present in barley infected with *Gibberella saubinette, Phytopathology*, 31, 871, 1941.
154. **Howarth, B., Jr. and Wyatt, R. D.,** Effect of dietary aflatoxin on fertility, hatchability, and progeny performance of broiler breeder hens, *Appl. Microbiol.*, 31, 680, 1976.
155. **Hsieh. D. P. H., Fitzell, D. L., Miller, J. L., and Seiber, J. N.,** High-pressure liquid chromatography of oxidative aflatoxin metabolites, *J. Chromatogr.* 117, 474, 1976.
156. **Hsu, I., Smalley, E. B., Strong, F. M., and Ribelin, W. E.,** Identification of T-2 toxin in moldy corn associated with a lethal toxicosis in dairy cattle, *Appl. Microbiol.*, 24, 684, 1972.
157. **Huff, W. E., Wyatt, R. D., and Hamilton, P. B.,** Nephrotoxicity of dietary ochratoxin A in broiler chickens, *Appl. Microbiol.*, 30, 48, 1975.

158. **Huff, W. E., Wyatt, R. D., Tucker, T. L., and Hamilton, P. B.,** Ochrotoxicosis in the broiler chichen, *Poult. Sci.,* 53, 1585, 1974.
159. **Hutchinson, R. D., Steyn, P. S., Thompson, D. C.,** The isolation and structure of 4-hydroxyochratoxin A and 7-carboxy-3,4-dihydro-8-hydroxy-3-methylisocoumarin from *Penicillium viridicatum, Tetrahedron Lett.* 43, 4033, 1971.
160. **Ikediobi, C. D., Hsu, I. C., Bamburg, J. R., and Strong, F. M.,** Gas-liquid chromatography of mycotoxins of the trichothecene group, *Anal. Biochem.,* 43, 327, 1971.
161. **International Union of Pure and Applied Chemistry,** Recommended method for aflatoxins in cocoa beans, *IUPAC Inf. Bull. Tech. Rep.,* 8, 1, 1973.
162. **Ishii, K., Ando, Y., and Ueno, Y.,** Toxicological approaches to the metabolites of Fusaria. IX. Isolation of vomiting factor from moldy corn infected with *Fusarium* species, *Chem. Pharm. Bull.,* 23, 2162, 1975.
163. **Jacobson, W. C., Harmeyer, W. C., Wiseman, H. G., and Dairy, G.,** Determination of aflatoxins B_1 and M_1 in milk, *J. Dairy Sci.,* 54, 21, 1971.
164. **Jacobson, W. C. and Wiseman, H. G.,** The transmission of aflatoxin B_1 into eggs, *Poult. Sci.,* 53, 1743, 1974.
165. **Jemmali, M.,** Evaluation de differentes mycotoxines (aflatoxins, ochratoxines, zearalenone) sur couches minces per fluorodensitometric per reflectance, *Ann. Biol. Anim. Biochim. Biophys.,* 14, 845, 1974.
166. **Joffe, A. Z.,** Alimentary toxic aleukia, in *Microbial Toxins,* Vol. 7, Kadis, S. Ciegler, A., and Ajl, S., Academic Press, New York, 1971,
167. **Joffe, A. Z.,** Toxicity of *Fusarium poae* and *F. sporotrichioides* and its relation to alimentary toxic aleukia, in *Mycotoxins,* Purchase, I. F. H., Ed., Elsevier, New York, 1974, 229.
168. **Johnson, F. L., Feagler, J. R., Lerner, K. G., Majerus, P. W., Siegel, M., Hartman, J. R., and Thomas, E. D.,** Association of androgenicanabolic steroid therapy with development of hepatocellular carcinoma, *Lancet,* 16, 1273, 1972.
169. **Jones, B. D.,** Methods of Aflatoxin Analysis, Report G 70, Tropical Products Institute, London,
170. **Keen, P. and Martin, P.,** Is aflatoxin carcinogenic in man? The evidence in Swaziland, *Trop. Geogr. Med.,* 23, 44, 1971.
171. **Keyl, A. C. and Booth, A. N.,** Aflatoxin effects in livestock, *J. Am. Oil Chem. Soc.,* 48, 599, 1971.
172. **Keyl, A. C., Booth, A. N., Masri, M. S., Gumbmann, M. R., and Gagne, W. E.,** Chronic effects of aflatoxin in farm animal feeding studies, in Proc. 1st U.S.-Japanese Conf. Toxic Microorganisms, U.S. Department of Interior and U.J.N.R., Panel on Toxic Microorganisms, Washington, D.C., 1968, 72.
173. **Kiermeier, F.,** Aflatoxin residues in fluid milk, *Pure Appl. Chem.,* 35, 271, 1973.
174. **Kiyomoto, R. K. and Ashworth, L. J., Jr.,** Status of cotton boll rot in the San Joaquin Valley of California following simulated pink bollworm injury, *Phytopathology,* 64, 259, 1974.
175. **Koehler, B.,** Fungus growth in shelled corn as affected by moisture, *J. Agric. Res.,* 56, 291, 1968.
176. **Kotsonis, F. N., Ellison, R. A., and Smalley, E. B.,** Isolation of acetyl T-2 toxin from *Fusarium poae, Appl. Microbiol.,* 30, 493, 1975.
177. **Kratzer, F. H., Bandy, D., Wiley, M., and Booth, A. N.,** Aflatoxin effects in poultry, *Proc. Soc. Exp. Biol. Med.,* 131, 1281, 1969.
178. **Krishnamachari, K. A. V. R., Bhat, R. V., Nagarajan, V., and Tilak, T. B. G.,** Investigations into an outbreak of hepatitis in parts of Western India, *Indian J. Med. Res.,* 63, 1036, 1974.
179. **Krishnamachari, K. A. V. R., Bhat, R. V., Nagarajan, V., and Tilak, T. B. G.,** Hepatitis due to aflatoxicosis. An outbreak in Western India, *Lancet,* 10, 1061, 1975.
180. **Krogh, P.,** Mycotoxic nephropathy, in *Mycotoxins,* Purchase, I. F. H., Ed., Elsevier, Amsterdam, 1974,
181. **Krogh, P., Axelsen, N. A., Elling, F., Gyrd-Hansen, N., Hald, B., Hyldgaard-Jensen, J., Larsen, A. E., Madsen, A., Mortensen, H. P., Moller, T., Petersen, O. K., Ravnskov, U., Rostgaard, M., and Aalund, O.,** Experimental porcine nephropathy: changes of renal function and structure induced by ochratoxin A contaminated feed, *Acta Pathol. Microbiol. Scand. Sect. A Suppl.,* 246, 1, 1974.
182. **Krogh, P., Elling, F., Hald, B., Jylling, B., Peterson, V. E., Shadhauge, E., and Svendsen, C. K.,** Experimental avian nephropathy, *Acta Pathol. Microbiol. Scand. Sect. A,* 84, 215, 1976.
183. **Krogh, P., Hald, B., Englund, P., Rutquist, L., and Swahn, O.,** Contamination of Swedish cereals with ochratoxin A, *Acta Pathol. Microbiol. Scand. Sect. B,* 82, 301, 1974.
184. **Krogh, P., Hald, B., Hasselager, E., Madsen, A., Mortensen, H. P., Larsen, A. E., and Campbell, A. D.,** Aflatoxin residues in bacon pigs, *Pure Appl. Chem.,* 35, 275, 1973.
185. **Krogh, P., Hald, B., and Pedersen, E. J.,** Occurrence of ochratoxin A and citrinin in cereals associated with mycotoxin porcine nephropathy, *Acta Pathol. Microbiol. Scand. Sect. B,* 81, 689, 1973.
186. **Lancaster, M. C., Jenkins, F. P., and Philp, J. McL.,** Toxicity associated with certain samples of groundnut, *Nature (London),* 192, 1095, 1961.
187. **Landers, K. E., Davis, N. D., and Diener, U. L.,** Influence of atmospheric gasses on aflatoxin production by *Aspergillus flavus* in peanuts, *Phytopathology,* 57, 1086, 1967.

188. **Lee, W. V.**, Quantitative determination of aflatoxin in groundnut products, *Analyst (London)*, 90, 305, 1965.
189. **Levi, C. P. and Borker, E.**, Survey of green coffee for potential aflatoxin contamination, *J. Assoc. Off. Anal. Chem.*, 51, 600, 1968.
190. **Levi, C. P., Trenk, H. L., and Mohr, H. K.**, Study of the occurrence of ochratoxin A in green coffee beans, *J. Assoc. Off. Anal. Chem.*, 57, 866, 1974.
191. **Lillehoj, E. B., Fennell, D. I., and Hesseltine, C. W.**, *Aspergillus flavus* infection and aflatoxin production in mixtures of high moisture and dry maize, *J. Stored Prod.*, 12, 11, 1976.
192. **Lillehoj, E. B., Kwolek, W. F., Fennell, K. I., and Milburn, M. S.**, Aflatoxin incidence and association with bright greenish-yellow fluorescence and insect damage in a limited survey of freshly harvested high-moisture corn, *Cereal Chem.*, 52, 403, 1975.
193. **Lillehoj, E. B., Kwolek, W. F., Peterson, R. E., Shotwell, O. L., and Hesseltine, C. W.**, Aflatoxin contamination, fluorescence, and insect damage in corn infected with *Aspergillus flavus* before harvest, *Cereal Chem.*, 53, 505, 1976.
194. **Lindenfelser, L. A. and Ciegler, A.**, Solid substrate fermentor for ochratoxin A Production, *Appl. Microbiol.*, 29, 323, 1975.
195. **Lensell, C. A. and Peers, F. G.**, The aflatoxins and human liver cancer, in *Current Problems in the Epidemiology of Cancer and Lymphomas*, Grundmann, E. and Tulinius, H., Eds., Springer-Verlag, New York, 1972,
196. **Loosmore, R. H. and Markson, L. M.**, Poisoning of cattle by Brazilian groundnut meal, *Vet. Rec.*, 73, 813, 1961.
197. **Mabee, M. S. and Chipley, J. R.**, Tissue distribution and metabolism of aflatoxin B_1 ^{14}C in layer hens, *J. Food Sci.*, 38, 566, 1973.
198. **Mabee, M. S. and Chipley, J. R.**, Tissue distribution and metabolism of aflatoxin B_1 ^{14}C in broiler chickens, *Appl. Microbiol.*, 25, 763, 1973.
199. **Macnab. G. M., Urbanowicz, J. M., Geddes, E. W., and Kew, M. C.**, Hepatitis-B surface antigen and antibody in Bantu patients with primary hepatocellular cancer, *Br. J. Cancer*, 33, 544, 1976.
200. **Madhavan, T. V., Tulpule, P. G., and Gopalan, C.**, Aflatoxin-induced hepatic fibrosis in Rhesus monkeys, *Arch. Pathol.*, 79, 466, 1965.
201. **Mains, E. B., Vestal, C. M., and Curtis, P. B.**, Scab of small grains and feeding troubles in Indiana in 1928, *Proc. Indiana Acad. Sci.*, 39, 101, 1930.
202. **Majchrowicz, I. and Yendol, W. G.**, Fungi isolated from the gypsy moth, *J. Econ. Entomol.*, 66, 823, 1973.
203. **Malaiyandi, M., Barrette, J. P., and Waverock, R.**, Bis diazatized benzidine as a spray reagent for detecting zearalenone on thin layer chromatographic plates, *Anal. Chem.*, 59, 959, 1976.
204. **Marsh, P. B., Simpson, M. E., Graig, H. O., Donoso, G., and Ramay, G. H.**, Occurrence of aflatoxins in cotton seeds at harvest in relation to location of growth and field temperature, *J. Environ. Qual.*, 2, 276, 1973.
205. **Marasas, W. F. O., Bamburg, J. R., Smalley, E. B., Strong, F. M., Ragland, W., and Degurse, P. E.**, Low-dose, long term effects on trout and rats of T-2 toxin produced by the fungus *Fusarium tricinctum*, *Toxicol. Appl. Pharmacol.*, 15, 471, 1969.
206. **Martin, P. M. D. and Gilman, G. A.**, A Consideration of the Mycotoxin Hypothesis with Special Reference to the Mycoflora of Maize, Sorghum, Wheat and Groundnuts, Publication G105, Trop. Prod. Inst., England, 1976.
207. **McDonald, D.**, The effect of wetting dried groundnuts on fungal infection of kernels, *Samaru Agric. Newsl.*, 10, 4, 1968.
208. **McKinney, J. D.**, Use of zinc acetate in extract purification for aflatoxin assay of cottonseed products, *J. Am. Oil Chem. Soc.*, 52, 213, 1975.
209. **Miller, J. K., Hacking, A., Harrison, J., and Gross, V. J.**, Stillbirths, neonatal mortality and small litters in pigs associated with the ingestion of *Fusarium* toxin by pregnant sows, *Vet. Rec.*, 93, 555, 1973.
210. **Mirocha, C. J. and Christensen, C. M.**, Oestrogenic mycotoxins synthesized by *Fusarium*, in *Mycotoxins*, Purchase, I. F. H., Ed., North-Holland, Amsterdam, 1974, 129.
211. **Mirocha, C. J., Christensen, C. M., Davis, G., and Nelson, G. H.**, Detection of diethylstilbestrol contamination in swine feedstuff, *J. Agric. Food Chem.*, 21, 135, 1973.
212. **Mirocha., C. J., Christensen, C. M., and Nelson, G. H.**, Physiologic activity of some fungal estrogens produced by *Fusarium*, *Cancer Res.*, 28, 2319, 1968.
213. **Mirocha, C. J., Christensen, C. M., and Nelson, G. H.**, Toxic metabolites produced by fungi implicated in mycotoxicoses, *Biotechnol. Bioeng.*, 10, 468, 1968.
214. **Mirocha, C. J., Christensen, C. M., and Nelson, G.H.**, F-2 (zearalenone). Estrogenic mycotoxin from *Fusarium*, in *Microbial Toxins*, Vol. 6, Ciegler, A., Kadis, S., and Ajl, S., Eds., Academic Press, New York, 1971, 107.

215. **Mirocha, C. J. and Pathre, S. V.**, Identification of the toxic principle in a sample of poaefurarin, *Appl. Microbiol.*, 26, 719, 1973.
216. **Mirocha, C. J., Pathre, S. V., Schauerhamer, B., and Christensen, C. M.**, Natural occurrence of *Fusarium* toxins in feedstuff, *Appl. Environ. Microbiol.*, 32, 553, 1976.
217. **Mirocha, C. J. and Pathre, S. V.**, Substances interfering with the gas-liquid chromatographic determination of T-2 mycotoxin, *J. Assoc. Off. Anal. Chem.*, 59, 222, 1976.
218. **Mirocha, C. J., Schauerhamer, B., and Pathre, S. V.**, Isolation, detection, and quantitation of zearalenone in maize and barley, *J. Assoc. Off. Anal. Chem.*, 57, 1104, 1974.
219. **Mislivec, P. B., Bruce, V. R., and Trucksess, M. W.**, Effect of three toxic mold species on aflatoxin production by *Aspergillus flavus*, Abstr. Annu. Meet. American Society of Microbiology, May 8 to 13, New Orleans, 248.
220. **Mislivic, P. B. and Tuite, J.**, Species of *Penicillium* occurring in freshly harvested and in stored dent corn kernels, *Mycologia*, 62, 67, 1970.
221. **Morooka, N., Uratsuje, N., Yoshizawa, T., and Yamamoto, H.**, Studies on the toxic substances in barley infected with *Fusarium* spp., *J. Food Sanit.*, 13, 368, 1972.
222. **Mücke, W. and Keirmeier, F.**, Spruhreagens zum nachweis von aflatoxiner zeit, *Z. Lebensm. Unters. Forsch.*, 146, 329, 1971.
223. **Muller, R. D., Carlson, C. W., Semenick, G., and Harskfeld, G. S.**, The response of chicks, ducklings, goslings, pheasants, and poults to graded levels of Aflatoxin, *Poult. Sci.*, 44, 1346, 1970.
224. **Murthy, T. R. K., Jammali, M., Henry, Y., and Frayssinet, C.**, Aflatoxin residues in tissues of growing swine: effect of separate and mixed feeding of protein and protein free portions of the diet, *J. Anim. Sci.*, 41, 1339, 1975.
225. **Nabney, G. and Nesbitt, B. F.**, A spectrophotometric method for determining the aflatoxins, *Analyst (London)*, 90, 155, 1965.
226. **Naumov, N. A.**, Intoxicating bread, *Min. Zeml. (Russia), Trudy Ruiri Miwel i. Fitopatol. Uchen, Kom.*, 216, 1916.
227. **Nesbitt, B. F., O'Kelly, J., Sargeant, K., and Sheridan, A.** Toxic metabolites of *Aspergillus flavus*, *Nature (London)*, 195, 1062, 1962.
228. **Nesheim, S. J.**, Isolation and purification of ochratoxins A and B and preparation of the methyl and ethyl esters, *J. Assoc. Off. Anal. Chem.*, 52, 975, 1969.
229. **Nesheim, S. J.**, Screening method for zearalenone, aflatoxin and ochratoxins, *J. Assoc. Off. Anal. Chem.*, 51, 74, 1968.
230. **Nesheim, S. J.**, Analysis of ochratoxins A and B and their esters in barley using partition and thin layer chromatography. II. Collaborative study, *J. Assoc. Off. Anal. Chem.*, 56, 822, 1973.
231. **Nesheim, S. J., Hardin, N. F., Francis, O. J., and Langham, W. S.**, Analysis of ochratoxins A and B and their esters in barley, using partition and thin layer chromatography. I. Development of the method, *J. Assoc. Off. Anal. Chem.*, 56, 817, 1973.
232. **Newberne, P. M., Carlton, W. W., and Wogan, G. N**, Hepatomas in rats and hepatorenal injury in duckling fed peanut meal and *Aspergillus flavus* extract, *Pathol. Vet.*, 1, 105, 1964.
233. **Newberne, P. M. and Rogers, A. E.**, Animal model of human disease. Primary hepatocellular carcinoma, *Am. J. Pathol.*, 72, 137, 1973.
234. **Oettle, A. G.**, Cancer in Africa, especially in regions south of the Sahara, *J. Natl. Cancer Inst.*, 33, 383, 1964.
235. **Olson, L. C., Bourgeouis, C. H., Cotton, R. B., Harikul, S., Grossman, R. A., and Smith, T. J.**, Encephalopathy and fatty degeneration of the viscera in Northeastern Thailand. Clinical syndrome and epidemiology, *Pediatrics*, 47, 707, 1971.
236. **Ong, T.-M.**, Aflatoxin mutagenesis, *Mutat. Res.*, 32, 35, 1975.
237. **Peckham, J. C., Doupnik, B., Jr., and Johns, O. H., Jr.**, Acute toxicity of ochratoxin A and B in chicks, *Appl. Microbiol.*, 21, 492, 1971.
238. **Peers, F. G., Gilman, G. A., and Linsell, C. A.**, Dietary aflatoxins and human liver cancer. A study in Swaziland, *Int. J. Cancer*, 17, 167, 1976.
239. **Peers, F. G. and Linsell, C. A.**, Dietary aflatoxins and liver cancer: a population based study in Kenya, *Br. J. Cancer*, 27, 473, 1973.
240. **Phillips, D. L., Yourtee, D. M., and Searles, S.**, Presence of aflatoxin B_1 in human liver in the United States, *Toxicol. Appl. Pharmacol.*, 36, 403, 1976.
241. **Pier, A. C. and Heddleston, K. L.**, The effect of aflatoxin on immunity in turkeys. I. Impairment of actively acquired resistance and bacterial challenge, *Avian Dis.*, 14, 797, 1970.
242. **Pier, A. C., Heddleston, K. L., Cysewski, S. J., and Patterson, J. M.**, Effect of aflatoxin on immunity in turkeys. II. Reversal of impaired resistance to bacterial infection by passive transfer of plasma, *Avian Dis.*, 16, 381, 1972.

243. **Platonow, N.,** Investigations of the possibility of the presence of aflatoxin in meat and liver of chickens fed toxic groundnut meal, *Vet. Rec.,* 77, 1028, 1965.
244. **Platt, B. S., Stewart, R. J. C., and Gupta, S. R.,** The chick embryo as a test organism for toxic substances in food, *Proc. Nutr. Soc.,* 30, 21, 1962.
245. **Pohland, A. E., Yin, L., and Dantzman, J. G.,** Rapid chemical confirmatory method for aflatoxin B_1. II. Collaborative study. *J. Assoc. Off. Anal. Chem.,* 53, 101, 1970.
246. **Poland, A. E., Hayes, J. R., and Campbell, T. C.,** Consumption and fate of aflatoxin B_1 by lactating cows, *J. Agric. Food Chem.,* 22, 635, 1974.
247. **Pons, W. A., Jr.,** A quantitative method for determination of aflatoxin B_1 in roasted corn, *J. Assoc. Off. Anal. Chem.,* 58, 745, 1975.
248. **Pons, W. A., Jr.,** Resolution of aflatoxins B_1, B_2, G_1, and G_2 by high-pressure liquid chromatography, *J. Assoc. Off. Anal. Chem.,* 59, 101, 1976.
249. **Pons, W. A., Jr., Cucullu, A. F., and Lee, L. S.,** Method for the determination of aflatoxin M_1 in fluid milk and milk products, *J. Assoc. Off. Anal. Chem.,* 56, 1431, 1973.
250. **Pons, W. A., Jr. and Goldblatt, L. A.,** Physiochemical assay of aflatoxins, in *Aflatoxin,* Goldblatt, L. A., Ed., Academic Press, New York, 1969, 77.
251. **Pons, W. A., Jr. and Goldblatt, L. A.,** The determination of aflatoxins in cotton seed products, *J. Am. Oil Chem. Soc.,* 42, 471, 1975.
252. **Pons, W. A., Jr., Robertson, J. A., and Goldblatt, L. A.,** Objective fluorometric measurement of aflatoxin on TLC plates, *J. Am. Oil Chem. Soc.,* 43, 665, 1966.
253. **Porter, D. M. and Smith, J. C.,** Fungal colonization of peanut fruit as related to Southern corn rootworm injury, *Phytopathology,* 64, 249, 1974.
254. **Prentice, N. and Dickson, A. D.,** Emetic material associated with *Fusarium* species in cereal grains and artificial media, *Biotechnol. Bioeng.,* 10, 413, 1968.
255. **Prentice, N., Dickson, A. D., and Dickson, J. G.,** Production of emetic materials by species of *Fusarium, Nature (London),* 184, 1319, 1959.
256. **Prince, A. M., Szmuness, W., Michon, J., Demaille, J., Diebolt, G., Linhard, J., Quenum, C., and Sankale, M.,** A case control study of the association between primary liver cancer and hepatitis-B infection in Senegal, *Int. J. Cancer,* 16, 376, 1975.
257. **Przybylski, W.,** Formation of aflatoxin derivatives on thin layer chromatographic plates, *J. Assoc, Off. Anal. Chem.,* 58, 163, 1975.
258. **Purchase, I. F. H. and Nel, W.,** Recent advances in research on ochratoxin: toxicological aspects, in Mateles, R. I. and Wogan, G. N., Eds., *Biochemistry of Some Foodborne Microbial Toxins,* MIT Press, Cambridge, 1967, 153.
259. **Purchase, I. F. H., Stubblefield, R. D., and Altenkirk, B. A.,** Collaborative study of the determination of aflatoxin M_1 in milk, IUPAC Inf. Bull. Tech. Rep. 11,
260. **Purchase, I. F. H., and Theron, J. J.,** Acute toxicity of ochratoxin A to rats, *Food Cosmet. Toxicol.,* 6, 479, 1968.
261. **Purchase, I. F. H. and Steyn, P. S.,** Estimation of aflatoxin M in milk, *J. Assoc. Off. Anal. Chem.,* 50, 363, 1967.
262. **Purchase, I. F. H. and van der Watt, J. J.,** The long term toxicity of ochratoxin A to rats, *Food Cosmet. Toxicol.,* 9, 681, 1971.
263. **Rao, Gundu, H. R., and Anders, M. W.,** Aflatoxin detection by high-speed liquid chromatography and mass spectrometry, *J. Chromatogr.,* 84, 402, 1973.
264. **Reddy, J. K., Svoboda, D. J., and Rao, M. S.,** Induction of liver tumors by aflatoxin B_1 in the tree shrew (*Tupaia glis*), a nonhuman primate, *Cancer Res.,* 36, 151, 1976.
265. **Richards, D. E.,** The isolation and identification of the toxic coumarins, in *Microbial Toxins,* Vol. 8, Kadis, S., Ciegler, A., and Ajl, S., Eds., Academic Press, New York, 1972, 3.
266. **Roberts, B. A. and Patterson, D. S. P.,** Detection of twelve mycotoxins in mixed animal feedstuff, using a novel membrane cleaning procedure, *J. Assoc. Off. Anal. Chem.,* 58, 1178, 1975.
267. **Roberts, J. C. and Woollven, P.,** Studies in mycological chemistry. XXIV. Synthesis of ochratoxin A, a metabolite of *Aspergillus ochraceus* Wilh., *J. Chem. Soc. C.,* 278, 1970.
268. **Roberton, G. A., Lee, L. S., Cucullu, A. F., and Goldblatt, L. A.,** Determination of aflatoxins in individual peanuts and peanut sections, *J. Am. Oil Chem. Soc.,* 43, 83, 1966.
269. **Robinson, P.,** Infantile cirrhosis of the liver in India with special reference to probable aflatoxin etiology, *Clin. Pediatr. (Philadelphia),* 6, 57, 1967.
270. **Rodricks, J. V. and Eppley, R. M.,** *Stachybotrys* and stachybotryotoxicosis, in *Mycotoxins,* Purchase, I. F. H., Ed., Elsevier, New York, 1974. 181.
271. **Roine, K., Korpinen, E. L., and Kallela, K.,** Mycotoxicosis as probable cause of infertility in dairy cows, *Nord. Veterinaermed.,* 23, 628, 1971.

272. **Romer, F. R.,** Screening method for the detection of aflatoxins in mixed feeds and other agricultural commodities with subsequent confirmation and quantitative measurement of aflatoxins in positive samples, *J. Assoc. Off. Anal. Chem.,* 58, 500, 1975.
273. **Ruddick, J. A., Scott, P. M., and Harwig, J.,** Teratological evaluation of zearalenone administered orally to the rat, *Bull. Environ. Contam. Toxicol.,* 15, 678, 1976.
274. **Saito, M., Enomoto, M., and Tatsumo, T.,** Radiomimetic biological properties of the new scirpene metabolites of *Fusarium nivale, Gann,* 60, 599, 1969.
275. **Saito, M., Enomoto, M., and Tatsumo, T.,** Yellowed rice toxins. Luteoskyrin and related compounds, chlorine containing compounds and citrinin, in *Microbial Toxins,* Vol. 6, Ciegler, A., Kadis, S., and Ajl, S., Eds., Academic Press, New York, 1971, 299.
276. **Saito, M. and Tatsuno, T.,** Toxins of *Fusarium nivale,* in *Microbial Toxins: A Comprehensive Treatise,* Vol. 7, Kadis, S., Ciegler, A., and Ajl, S. J., Eds., Academic Press, New York, 1971, 293.
277. **Sarudi, J. I.,** Methode zum nachweis des zearalenon (F-2) toxins, *Z. Lebensm. Unters. Forsch.* 154, 65, 1974.
278. **Sawhney, D. S., Vadehra, D. V., and Baker, R. C.,** The metabolism of ^{14}C aflatoxin in laying hens, *Poult. Sci.,* 52, 1302, 1973.
279. **Scheel, L. D., Perone, V. B., Larkin, R. L., and Kupal, R. E.,** The isolation and characterization of two phototoxic furanocoumarins (psoralens) from diseased celery, *Biochemistry,* 2, 1127, 1963.
280. **Schroeder, H. W. and Boller, R. A.,** Aflatoxin production of species and strains of the *Aspergillus flavus* group isolated from field crops, *Appl. Microbiol.,* 25, 885, 1973.
281. **Schroeder, H. W. and Hein, H., Jr.,** A note on zearalenene in grain sorghum, *Cereal Chem.,* 52, 751, 1975.
282. **Schuller, P. L., Harvesty, W., and Stoloff, L.,** Review of aflatoxin methodology. A review of sampling plans and collaboratively studied methods of analysis for aflatoxins, *J. Assoc. Off. Anal. Chem.,* 59, 1315, 1976.
283. **Scott, P. M.,** Note on analysis of aflatoxins in green coffee, *J. Assoc. Off. Anal. Chem.,* 51, 609, 1968.
284. **Scott, P. M., Barry, P. C., and Kennedy, B.,** Analysis of spices and herbs for aflatoxins, *Can. Inst. Food Sci. Technol. J.,* 8, 124, 1975.
285. **Scott, P. M. and Hand, T. B.,** Method for the detection and estimation of ochratoxin A in some cereal products, *J. Assoc. Off. Anal. Chem.,* 50, 366, 1967.
286. **Scott, P. M., Lawrence, J. W., and Van Walbeck, W.,** Detection of mycotoxins by thin-layer chromatography. Application to screening of fungal extracts, *Appl. Microbiol.,* 20, 839, 1970.
287. **Scott, P. M., Van Walbeck, W., and Anyeti, D.,** Mycotoxins (ochratoxin A, citrinin, and sterigmatocystin) and toxigenic fungi in grains and other agricultural products, *J. Agric. Food Chem.,* 20, 1103, 1972.
288. **Scott, P. M., Van Walbeck, W., Harwig, J., and Fennell, D. I.,** Occurrence of a mycotoxin, ochratoxin A, in wheat and isolation of ochratoxin A and citrinin-producing strains of *Penicillium viridicatum, Can. J. Plant Sci.,* 50, 583, 1970.
289. **Seiber, T. N. and Hsieh, P. H.,** Application of high-speed liquid chromatography to the analysis of aflatoxin, *J. Assoc. Off. Anal. Chem.,* 56, 857, 1973.
290. **Serck-Hanssen, A.,** Aflatoxin-induced fatal hepatitis? A case report from Uganda, *Arch. Environ. Health,* 20, 729, 1970.
291. **Seitz, L. M.,** Comparison of methods for aflatoxin analysis by high-pressure liquid chromatography, *J. Chromatogr.,* 104, 81, 1975.
292. **Seitz, L. M. and Mohr, H. E.,** Simple method for simultaneous detection of aflatoxin and zearalenone in corn, *J. Assoc. Off. Anal. Chem.,* 59, 106, 1976.
293. **Shalkop, W. T. and Armbrecht, B. H.,** Carcinogenic response of brood sows fed aflatoxin for 28-30 months, *Am. J. Vet. Res.,* 35, 623, 1974.
294. **Shank, R. C., Bhamarapravati, N., Gordon, J. E., and Wogan, G. N.,** Dietary aflatoxins and human liver cancer. IV. Incidence of primary liver cancer in two municipal populations of Thailand, *Food Cosmet. Toxicol.,* 10, 171, 1972.
295. **Shank, R. C., Bourgeous, C. H., Keschamras, N., and Chandavimol, P.,** Aflatoxins in autopsy specimens from Thai children with an acute disease of unknown aetiology, *Food Cosmet. Toxicol.,* 9, 501, 1971.
296. **Shank, R. C., Gordan, J. E., Wogan, G. N., Nondasuta, A., and Subhamani, B.,** Dietary aflatoxins and human liver cancer. III. Field survey of rural Thai families for ingested aflatoxins, *Food Cosmet. Toxicol.,* 10, 71, 1972.
297. **Sherwood, R. F. and Peberdy, J. F.,** Production of the mycotoxin, zearalenone, by *Fusarium graminearum* on stored grain. II. Grain stored at reduced temperatures, *J. Sci. Food Agric.,* 25, 1081, 1974.
298. **Shotwell. O. L.,** Aflatoxin in corn, *J. Am. Oil Chem. Soc.,* 54, 216, 1977.
299. **Shotwell, O. L., Goulden, M. L., and Bennett, G. A.,** Determination of zearalenone in corn: collaborative study, *J. Assoc. Off. Anal. Chem.,* 59, 666, 1976.

300. **Shotwell, O. L., Goulden, M. L., and Hesseltine, C. W.**, Survey of U.S. wheat for ochratoxin and aflatoxin, *J. Assoc. Off. Anal. Chem.*, 59, 122, 1976.
301. **Shotwell, O. L., Hesseltine, C. W., and Goulden, M. L.**, Note on the natural occurrence of ochratoxin A, *J. Assoc. Off. Anal. Chem.*, 52, 81, 1969.
302. **Shotwell, O. L., Hesseltine, C. W., and Goulden, M. L.**, Ochratoxin A: occurrence as natural contaminant of a corn sample, *Appl. Microbiol.*, 17, 765, 1969.
303. **Shotwell, O. L., Hesseltine, C. W., Goulden, M. L., and Vandergraft, E. E.**, Survey of corn for aflatoxin, zearalenone, and ochratoxin, *Cereal Chem.*, 47, 700, 1970.
304. **Shotwell, O. L., Hesseltine, C. W., Vandergraft, E. E., and Goulden, M. L.**, Survey of corn from different regions for aflatoxin, ochratoxin, and zearalenone, *Cereal Sci. Today*, 16, 266, 1971.
305. **Siller, W. G. and Ostler, D. C.**, The histopathology of an entero-hepatic syndrome of turkey poults, *Vet. Rec.*, 73, 134, 1961.
306. **Simpson, M. E. and Marsh, P. B.**, The geographical distribution of *Aspergillus flavus*, boll rot in the U.S. cotton crop of 1970, *Plant Dis. Rep.*, 55, 510, 1971.
307. **Sims, W. M., Jr., Kelley, D. C., and Sandford, P. E.**, A study of aflatoxicosis in laying hens, *Poult. Sci.*, 49, 1082, 1970.
308. **Sittenfeld de Vargas, A. M.**, Relacion Entre las Tasas de Mortalidad por Cancer Hapatico en Costa Rica y Condiciones Optimas para las Produccion de Aflatoxinas por *Aspergillus flavus*, personal communication, prepublication manuscript.
309. **Smith, J. W. and Hamilton, P. B.**, Aflatoxicosis in the broiler chicken, *Poult. Sci.*, 49, 207, 1970.
310. **Sphon, J. A., Dreifuss, P. A., and Schulten, H. R.**, Mycotoxins. Field desorption mass spectrometry of mycotoxins and mycotoxin mixtures, and its application as a screening technique for foodstuffs, *J. Assoc. Off. Anal. Chem.*, 60, 73, 1977.
311. **Spensley, P. C.**, Aflatoxin, the activity principle in turkey "X" disease, *Endeavour*, 22, 75, 1963.
312. **Stack, M. E. and Pohland, A. E.**, Collaborative study of a method for chemical confirmation of the identity of aflatoxin, *J. Assoc. Off. Anal. Chem.*, 58, 110, 1975.
313. **Stack, M. E., Pohland, A. E., Dantzman, J. G., and Nesheim, S. J.**, Derivative method for chemical confirmation of the identity of aflatoxin M., *J. Assoc. Off. Anal. Chem.*, 55, 313, 1972.
314. **Stanford, G. K., Hood, R. D., and Hayes, A. W.**, Effect or prenatal administration of T-2 toxin to mice, *Res. Commun. Chem. Pathol. Pharmacol.*, 10, 743, 1975.
315. **Stephenson, L. W. and Russell, T. E.**, The association of *Aspergillus flavus* with hemipterous and other insects infecting cotton bracts and foliage, *Phytopathology*, 64, 1502, 1974.
316. **Steyn, P. S. and Holzapfel, C. W.**, The isolation of the methyl and ethyl esters of ochratoxins A and B, metabolites of *Aspergillus ochraceous*, *J. S. Afr. Chem. Inst.*, 20, 186, 1967.
317. **Steyn, P. S.**, The separation and detection of several mycotoxins by thin-layer chromatography, *J. Chromatog.*, 45, 473, 1969.
318. **Steyn, P. S. and Van der Merwe, K. J.**, Detection and estimation of ochratoxin A, *Nature (London)*, 211, 418, 1966.
319. **Stoloff, L.**, Analytical methods for mycotoxins. Analytical methods for aflatoxins, *Clin. Toxicol.*, 5, 465, 1972.
320. **Stoloff, L.**, Collaborative study of a method for the identification of aflatoxin B_1 by derivative formation, *J. Assoc. Off. Anal. Chem.*, 50, 354, 1967.
321. **Stoloff, L.**, Rapid chemical confirmatory method for aflatoxin B_1. II Collaborative study, *J. Assoc. Off. Anal. Chem.*, 53, 102, 1970.
322. **Stoloff, L.**, Occurrence of mycotoxins in foods and feeds. Mycotoxins and other fungal related food products, *Adv. Chem. Ser.*, 149, 23, 1970.
323. **Stoloff, L. and Dalrymple, B.**, Aflatoxin and zearalenone occurrence in dry-milled corn products, *J. Assoc. Off. Anal. Chem.*, 60, 579, 1977.
324. **Stoloff, L., Nesheim, S. J., Yin, L., and Rodricks, J. V.**, A multimycotoxin detection method for aflatoxins ochratoxins, zearalenone, sterigmatocystin, and patulin, *J. Assoc. Off. Anal. Chem.*, 54, 91, 1971.
325. **Stoloff, L., Henery, S., and Francis, O. J., Jr.**, Survey for aflatoxins and zearalenone in 1973 crop corn stored on farms and in country elevators, *J. Assoc. Off. Anal. Chem.*, 59, 118, 1976.
326. **Stubblefield, R. D. and Shannon, G. M.**, Aflatoxin M_1: analysis in dairy products and distribution in dairy foods made from artificially contaminated milk, *J. Assoc. Off. Anal. Chem.*, 57, 847, 1974.
327. **Stubblefield, R. D. and Shannon, G. M.**, Collaborative study of methods for the determination and chemical confirmation of aflatoxin M_1 in dairy products, *J. Assoc. Off. Anal. Chem.*, 57, 852, 1974.
328. **Stubblefield, R. D., Shannon, G. M., and Shotwell, O. L.**, Aflatoxin M_1 in milk: evaluation of method, *J. Assoc. Off. Anal. Chem.*, 56, 1106, 1973.
329. **Szczech, G. M., Carlton, W. W., Tuite, J., and Caldwell, R.**, Ochratoxin A toxicosis in swine, *Vet. Pathol.*, 10, 347, 1973.
330. **Tatsuno, T.**, Toxicologic research on substances from *Fusarium nivale*, *Cancer Res.*, 28, 2393, 1968.

331. **Tatsuno, T., Ohtsubo, K., and Saito, M.,** Chemical and biological detection of 12,13-epoxytrichothecenes isolated from *Fusarium* species, *Pure Appl. Chem.*, 35, 309, 1973.
332. **Theron, J. J., Van der Merwe, K. J., Liegenberg, N., Joubert, H. J. B., and Nel, W.,** Acute liver injury in ducklings and rats as a result of ochratoxin poisoning, *J. Pathol. Bateriol.*, 91, 521, 1966.
333. **Thomas, R., Eppley, R. M., and Trucksess, M. W.,** Rapid screening method for aflatoxin and zearalenone in corn, *J. Assoc. Off. Anal. Chem.*, 58, 114, 1975.
334. **Tilak, T. B. G.,** Induction of cholangiocarcinoma following treatment of a Rhesus monkey with aflatoxin, *Food Cosmet. Toxicol.*, 13, 247, 1975.
335. **Trenk, H. L. Butz, M. E., and Chu, F. S.,** Production of ochratoxins in different cereal products by *Aspergillus ochraceus, Appl. Microbiol.*, 21, 1032, 1971.
336. **Trenk, H. L. and Chu, F. S.,** Improved detection of ochratoxin A on thin layer plates, *J. Assoc. Off. Anal. Chem.*, 54, 1307, 1971.
337. **Tsunoda, H.,** Microorganisms which deteriorate stored cereals and grains, in Proc. 1st U.S.-Japanese Conf. Toxic Microorg., Honolulu, 1968, U.S. Department of the Interior and U.J.N.R. Panels on Toxic Microorganisms, Washington, D.C., 1970, 143.
338. **Tuite, J.** Low incidence of storage molds in freshly harvested seed of soft red winter wheat, *Plant Dis. Rep.*, 43, 70, 1959.
339. **Tuite, J., Shanner, G., Rambo, G., Foster, J., and Caldwell, R. W.,** The *Gibberella* ear rot epidemics of corn in Indiana in 1965 and 1972, *Cereal Sci. Today*, 19, 238, 1974.
340. **Ueno, Y.,** Citreoviridin from *Penicillium citreoviride* Biourge, in *Mycotoxins*, Purchase I. F. H., Ed., Elsevier, New York, 1974, 283.
341. **Ueno, Y., Hosoya, M., and Ishikowa, Y.,** Inhibitory effects of mycotoxin on the protein synthesis in rabbit reticulocytes, *J. Biochem. (Tokyo)*, 66, 419, 1969.
342. **Ueno, Y., Ishii, K., Sakai, K., Kanaeda, S., Tsunoda, H., Tanaka, T., and Enomoto, M.,** Toxicological approaches to the metabolites of Fusaria. IV. Microbial survey on "Bean-hull poisoning of horses" with the isolation of toxic trichothecenes, neosolaniol and T-2 toxin of *Fusarium solani* M-1-1, *Jpn. J. Exp. Med.*, 42, 187, 1972.
343. **Ueno, Y., Nakajima, M., Sakai, K., Ishii, K., Sato, N., and Shimoda, N.,** Comparative toxicology of trichothec mycotoxins: inhibition of protein synthesis in animal cells, *J. Biochem. (Tokyo)*, 74, 285, 1973.
344. **Ueno, Y., Sato, N., Ishii, K., Tsunoda, H., and Enomoto, M.,** Biological and chemical detection of trichothecene mycotoxins of *Fusarium* species, *Appl. Microbiol.*, 25, 699, 1973.
345. **Ueno, Y. and Shimada, N.,** Reconfirmation fo the specific nature of reticulocytes bioassay system to tricothec mycotoxins of *Fusarium* spp., *Chem. Pharm. Bull.*, 22, 2744, 1974.
346. **Ueno, Y., Shimada, N., Yagasaki, S., and Enomoto, M.,** Toxicological approaches to the metabolites of Fusaria. VII. Effects of zearalenone on the uteri of mice and rats, *Chem. Pharm. Bull.*, 22, 2835, 1974.
347. **Ueno, Y., Ueno, I., Iitoi, Y., Tsunoda, H., Enomoto, M., and Ohtsubo, K.,** Toxicological approaches to the metabolites of *Fusaria*. III. Acute toxicity of fusaranon-X, *Jpn. J. Exp. Med.*, 41, 521, 1971.
348. **Urry, W. H., Wehrmeister, H. L., Hodge, E. B., and Hidy, P. H.,** The structure of zearalenone, *Tetrahedron Lett.*, (27), 3109, 1966.
349. **U.S. Department of Agriculture,** Peanuts: 1977 crop. Incoming and outgoing quality regulations and identification, *Fed. Resist.*, 43, 32278, 1977.
350. **van der Linde, G. A., Ferns, A. M., and van Esch, G. J.,** Experiments with cows fed groundnut meal containing aflatoxin, in *Mycotoxins in Foodstuffs*, Part III, Wogan, G. N., Ed., MIT Press, Cambridge, Mass., 1965, 247.
351. **Vandenheuvel, W. J. A.,** Gas-liquid chromatographic behavior of the zearalenones, a new family of biologically active material products, *Sep. Sci.*, 3, 151, 1968.
352. **Van der Merwe, K. J., Steyn, P. S., and Fourie, L.,** Mycotoxins. II. The constitution of ochratoxins A, B, and C, metabolites of *Aspergillus ochraceus* Wilh., *J. Chem. Soc.*, 382, 7083, 1965.
353. **Van der Merwe, K. J., Steyn, P. S., Fourie, L., Scott, D. B., and Theron, J. J.,** Ochratoxin A, a toxic metabolite produced by *Aspergillus ochraceous* Wilh., *Nature (London)*, 205, 1112, 1965.
354. **Van Rensburg, S. J.,** Role of Epidemiology in the Elucidation of Mycotoxin Health Risks, Conf. on Mycotoxins in Human and Animal Health, University of Maryland, College Park, October 4 to 8, 1976.
355. **Van Rensburg, S. J., Kirsipuu, A., Coutinho, L. P., and van der Watt, J. J.,** Circumstances associated with the contamination of food by aflatoxin in a high primary liver cancer area, *S. Afr. Med. J.*, 49, 877, 1975.
356. **Van Walbeek, W., Scott, P. M., Harwig, J., and Lawrence, J.,** *Penicillium viridicatum* Westling: a new source of ochratoxin A, *Can. J. Microbiol.*, 15, 1281, 1969.
357. **Van Walbeek, W., Scott, P. M., and Thatcher, F. S.,** Mycotoxins from food-borne fungi, *Can. J. Microbiol.*, 14, 131, 1968.
358. **Van Zytveld, W. A., Kelley, D. C., and Dennis, S. M.,** Aflatoxicosis: the presence of aflatoxin or their metabolites in livers and skeletal muscles of chickens, *Poult. Sci.*, 49, 1350, 1970.

359. **Vedanayagam, H. S., Indulkar, A. S., and Rao, S. R.,** Aflatoxins and *Aspergillus flavus* in Indian cottonseed, *Indian J. Exp. Biol.,* 9, 410, 1971.
360. **Velasco, J. and Whitaker, T. B.,** Sampling cottonseed lots for aflatoxin contamination, *J. Am. Oil Chem. Soc.,* 52, 191, 1975.
361. **Velasco, J. and Whitaker, T. B.,** Use of water slurries in aflatoxin analysis, *J. Agric. Food Chem.,* 24, 86, 1975.
362. **Verrett, M. J., Marlisc, J. P., and McLaughlin, J.,** Use of the chicken embryo in the assay of aflatoxin toxicity, *J. Assoc. Off. Agric. Chem.,* 47, 1103, 1964.
363. **Verrett, M. J., Winbush, J., Reynolds, E. T., and Scott, W. F.,** Collaborative study of the chicken embryo bioassay for aflatoxin B$_1$, *J. Assoc. Off. Anal. Chem.,* 56, 901, 1973.
364. **Vesonder, R. F., Ciegler, A., and Jensen, A. H.,** Isolation of the emetic principle for *Fusarium*-infected corn, *Appl. Microbiol.,* 26, 1008, 1973.
365. **Vesonder, R. F., Ciegler, A., and Jensen, A. H.,** Production of refusal factors by *Fusarium* strains on grains, *Appl. Environ. Microbiol.,* 34, 105, 1977.
366. **Vesonder, R. F., Ciegler, A., Jensen, A. H., Rohwedder, W. K., and Weisleder, D.,** Co-identity of the refusal and emetic principle from *Fusarium*-infected corn, *Appl. Environ. Microbiol.,* 31, 280, 1976.
367. **Vesonder, R. F. and Hesseltine, C. W.,** Metabolites of *Fusarium,* in *Fusarium,* Nelson, P. E., Ed., Pennsylvania State University Press, University Park, in press.
368. **Wannop, C. C.,** The histopathology of turkey "X" disease in Great Britian, *Avian Dis.,* 3, 371, 1961.
369. **Wie, R.-D., Smalley, E. B., and Strong, F. M.,** Improved skin test for detection of T-2 toxin, *Appl. Microbiol.,* 23, 1029, 1972.
370. **Whitten, M. E.,** Occurrence of Aflatoxins in Cottonseed and Cottonseed Products, Mycotoxin Res. Sem. Proc., Washington, D.C., June 8-9 (1967).
371. **Widstrom, N. W., Lillehoj, E. B., Sparks, A. N., and Kwolek, W. F.,** Corn earworm damage and aflatoxin B on corn ears protected with insecticide, *J. Econ. Entomol.,* 69, 677, 1976.
372. **Wilson, D. M., Tabor, W. H., and Trucksess, M. W.,** Screening method for the detection of aflatoxin, ochratoxin, zearalenone, penicillic acid, and citrinin, *J. Assoc. Off. Anal. Chem.,* 59, 125, 1976.
373. **Wogan, G. N.,** Experimental toxicity and carcinogenicity of aflatoxins, in *Mycotoxins in Foodstuffs,* Part III, Wogan, G. N., Ed., MIT Press, Cambridge, 1965, 209.
374. **Wogan, G. N. and Pong, R. S.,** Aflatoxins, *Ann. N.Y. Acad. Sci.,* 174, 623, 1970.
375. **Wogan, G. N.,** Aflatoxin carcinogenesis, in *Methods in Cancer Research,* Busch, H., Ed., Academic Press, New York, 1973, .
376. **Wu, C. M., Koehler, P. E., and Ayres, J. C.,** Isolation and identification of xanthotoxin (8-methoxypsoralen) and bergapten (5-methoxpsoralen) from celery infected with *Sclerotinia sclerotiorum, Appl. Microbiol.,* 23, 852, 1972.
377. **Wyatt, R. D., Colwell, W. M., Hamilton, P. B., and Burmeister, H. R.,** Neural disturbances in chickens caused by dietary T-2 toxin, *Appl. Microbiol.,* 26, 757, 1973.
378. **Wyatt, R. D., Doerr, J. A., Hamilton, P. B., and Burmeister, H. R.,** Egg production, shell thickness, and other physiological parameters of laying hens affected by T-2 toxin, *Appl. Microbiol.,* 29, 641, 1975.
379. **Wyatt, R. D., Hamilton, P. B., and Burmeister, H. R.,** The effect of T-2 toxin in broiler chickens, *Poult. Sci.,* 52, 1853, 1973.
380. **Wyatt, R. D., Hamilton, P. B., and Burmeister, H. R.,** Altered feathering of chick caused by T-2 toxin, *Poult. Sci.,* 54, 1042, 1975.
381. **Wyatt, R. D., Harris, J. R., Hamilton, P. B., and Burmeister, H. R.,** Possible outbreaks of fusariotoxicosis in avians, *Avian Dis.,* 16, 1123, 1972.
382. **Wyatt, R. D., Weeks, B. A., Hamilton, P. B., and Burmeister, H. R.,** Severe oral lesions in chickens caused by ingestion of dietary fusariotoxin T-2, *Appl. Microbiol.,* 24, 251, 1972.
383. **Yagen, B. and Joffe, A. Z.,** Screening of toxic isolates of *Fusarium poae* and *Fusarium sporotrichioides* involved in causing alimentary toxic aleukia, *Appl. Environ. Microbiol.,* 32, 423, 1976.
384. **Yadagiri, B. and Tulpule, P. G.,** Aflatoxin in buffalo milk, *Indian J. Dairy Sci.,* 27, 293, 1974.
385. **Yoshizawa, T. and Morooka, N.,** Studies on the toxic substances in the infected cereals, III. Acute toxicities of new trichothecene mycotoxins: deoxynivalinol and its monoacetate, *J. Food Hyg. Soc. Jpn.,* 15, 261, 1974.

Chapter 2

ANIMAL TOXICITY OF MAJOR ENVIRONMENTAL MYCOTOXINS

Paul M. Newberne and Adrianne E. Rogers

TABLE OF CONTENTS

I. Aflatoxins 52
    A. Acute Toxicity 52
        1. Fowl 52
        2. Rats 53
        3. Dogs 53
        4. Guinea Pigs 53
        5. Monkeys 53
        6. Metabolism 54
        7. Teratogenicity 55
    B. Chronic Toxicity and Carcinogenicity 55
        1. Mice 55
        2. Rats 55
        3. Fish 57
        4. Monkeys 58
        5. Ducks 58

II. Ochratoxins 58
    A. Acute Toxicity 58
        1. Pigs, Dogs 59
        2. Rats, Mice 59
        3. Poultry 59
        4. Fish 60
        5. Teratogenicity 60
    B. Chronic Toxicity 60

III. Trichothecene Toxins 60
    A. Acute Toxicity 60
        1. Guinea Pigs, Rabbits 61
        2. Dogs 61
        3. Cats 61
        4. Rats 61
        5. Swine 61
        6. Cattle 61
        7. Poultry 62
    B. Chronic Studies with Trichothecene Toxins 62

IV. Sterigmatocystin 63
    A. Acute Toxicity 63
    B. Chronic Toxicity 64

V. Zearalenone 65

References 101

I. AFLATOXINS

Evaluation of the carcinogenicity of aflatoxins by the International Association for Research on Cancer (IARC), World Health Organization (WHO), Food and Drug Administration (FDA), and other groups has led to formulation of strict regulations limiting the amount of aflatoxin permitted in foods. The assessments of risk have been based on epidemiological data discussed elsewhere in this volume and on extensive data from studies using animals. Carcinogenicity of one or more aflatoxins has been demonstrated in the rat, mouse, rainbow trout, salmon, guppy, marmoset, rhesus monkey, tree shrew, duck, and ferret. Except for studies in mice, carcinogenic exposure occurred through the gastrointestinal (GI) tract, emphasizing the probable risk from eating contaminated food. Acute aflatoxin toxicity to the liver, kidney, and other organs has been demonstrated in many species.

A. Acute Toxicity

Acute toxicity of aflatoxins was first recognized in domestic animals given contaminated feed. Hepatic necrosis, fatty infiltration, bile duct proliferation, and hepatic failure were observed in turkey poults, ducklings, chickens, and pigs fed peanut meal or other feed contaminated with aflatoxins, usually in amounts that could be measured in milligrams per kilogram. With the identification of aflatoxins in animal feeds, it was realized that they or other mycotoxins probably had been responsible for outbreaks of acute hepatic disease which had caused serious losses of domestic animals in the past. Episodes of acute hepatic necrosis in dogs and of subacute hepatic disease in calves have also been attributed to aflatoxicosis.[1] Early descriptions of aflatoxicosis and its reproduction in experimental animals have been reviewed.[1-3]

The structural and functional liver damage observed in field outbreaks has been experimentally reproduced in most laboratory and in several domestic animals; the following paragraphs summarize much of this work. LD_{50}s are known for many species (see Table 1).* The principal target organ is the liver, but fatty infiltration and focal necrosis occur also in heart and kidney. Necrosis of spleen and pancreas has been reported in some cases. Cerebral edema may be prominent in aflatoxin-poisoned monkeys, and gall bladder edema and hemorrhage are distinctive features of acute aflatoxicosis in dogs. Hemorrhage occurs in many organs, probably secondary to failure of hepatic production of blood clotting factors.[1,2,4,5] Clotting time was prolonged in goats, rats, and ducks given small doses of aflatoxin B_1 (AFB_1) that did not cause histological changes in the liver.[6]

Animals that survive exposure to single or repeated toxic doses of aflatoxin exhibit liver regeneration with marked increases in hepatocellular DNA synthesis and mitosis. (Early proliferation of bile ducts is also prominent in most species, but usually recedes, even with continued aflatoxin treatment.) Fibrosis and cirrhosis have not been reported in laboratory rodents, but have been described in turkeys, ducklings, calves, pigs, and monkeys also after field or experimental exposure to aflatoxins.[15-22] The other naturally occuring aflatoxins are less toxic than AFB_1.

1. Fowl

Day-old ducklings are highly sensitive to toxicity of AFB_1. In fact, ducklings provided the earliest bioassay for aflatoxin contamination of foods; detection of 50 μg/kg was possible. A single LD_{50} of AFB_1 results in liver injury with bile retention (see Figure 1),** periportal hepatic necrosis, failure to remove the hepatic lipid normally present at hatching (see Figure 2), and extensive proliferation of bile duct cells (see Figure 3).

*All tables are found starting on page 66.
**All figures are found starting on page 71.

7. Teratogenicity

Teratogenicity of AFB₁ was demonstrated in hamsters, but has not been found in other species.[44]

B. Chronic Toxicity and Carcinogenicity

The delayed results of a single large or repeated small doses of aflatoxin include hepatocyte regeneration, bile duct proliferation, and, sometimes, fibrosis in certain species; however, the major late effect is development of hepatocarcinoma or, occasionally, renal, colon, or other carcinomas. Aflatoxin B₁ carcinogenicity has been studied intensively, and carcinogenicity of aflatoxins G₁ and B₂ (AFG₁, AFB₂) and of the metabolites AFM₁ and AFQ₁ have been demonstrated in rats or rainbow trout or both.[3,7,45-47]

A linear dose response has been found for hepatocarcinoma incidence in rats and rainbow trout treated with AFB₁.[3,48] Male Fischer rats fed a purified diet containing 1 μg AFB₁ per kilogram of diet had 10% incidence of hepatocarcinoma at 2 years, and 100% of rats fed 100 μg AFB₁ per kilogram of diet had hepatocarcinoma by 88 weeks.[49] A dose-related response is evident also for AFG₁, although it has not been as clearly defined. In a proposal to lower the tolerance for aflatoxin in foods from 20 to 15 μg/kg, the FDA used data from several studies in rats to calculate the lifetime risk of developing hepatocarcinoma from ingesting diets containing smaller amounts of AFB, i.e., 0.1 or 0.3 μg/kg, levels which may be present in human diets in the U.S. The range of risks calculated is large, but approximates the estimated lifetime risk in the U.S. population of developing hepatocarcinoma which is 161 per 100,000 (see Table 5).[50]

Species and strain of animal determine susceptibility to aflatoxin carcinogenesis. Mice are highly resistant; the only definitive induction of hepatic tumors was in neonates given AFB, by i.p. injection.[51] In earlier studies, a few hepatic tumors (see Figure 25) developed in mice fed aflatoxin-contaminated diets, but malignancy of the tumors was not demonstrated, and the aflatoxins were not clearly characterized.[16] At the opposite end of the spectrum of susceptibility, rainbow trout developed 33% tumor incidence after 1 year's exposure to diet containing 0.5 μg AFB₁ per kilogram;[52] trout developing from eggs or embryos exposed to 500 μg/kg in water for 1 hr had tumor incidence up to 58% after 1 year.[53] A summary of results of studies of aflatoxin carcinogenesis in several species follows.

1. Mice

C3H × C57BL mice fed aflatoxins B₁ and G₁ (mixed) or pure B₁ in a purified diet t 100 to 150 mg/kg did not develop hepatocarcinomas. The i.p. administration of AFB₁, 1.25 to 6 μg/g body weight distributed over 1 to 5 doses in the first 7 days of life to C3H × C57BL mice, induced hepatoma incidence of 23 to 100% at 82 weeks.[51] Histologically the tumors were well differentiated; no metastases were observed despite replacement of the liver by tumor nodules. Injection site sarcomas and pulmonary adenomas have been reported after i.p. or s.c. injection.[3]

2. Rats

Biochemical, histological, histochemical, and ultrastructural aspects of the pathogenesis of hepatocarcinomas induced by aflatoxins have been studied intensively in rats; exposure has been primarily through the GI tract, although parenteral exposure also has been used. Although few direct comparisons have been made, Fischer rats appear more susceptible to hepatocarcinogenesis than either Sprague-Dawley or Wistar rats.[3,54,55] Porton and Wistar rats develop renal carcinomas in response to AFB₁ more frequently than do other strains; in one direct comparison, Fischer rats did not develop renal tumors under the same conditions.[56-58]

Wogan's 1973 summary of studies on AFB carcinogenesis[3] includes data on the intensity and duration of exposure required for tumor induction. Dietary content as low as 1 µg per AFB per kilogram feed induced hepatocarcinoma in male Fischer rats (10% incidence at 2 years); rats fed a diet containing 1 mg/kg for only 2 weeks had 7% hepatocarcinomas at 82 weeks. A tumor incidence of 50 to 100% was induced by diets containing 15 µg aflatoxin per kilogram (fed for 68 to 80 weeks) or 1 mg/kg (fed for 2 or more weeks).[49] Intragastric administration of AFB_1 in dimethylsulfoxide (DMSO), which allows more precise control of dosage and reduces the probability of environmental contamination and personnel exposure, has been used in recent studies and is effective in a broad range of dosage. Several multiple-dose regimens have produced tumor incidences of between 10 and 100% at approximately 9 to 18 months. A total dose of 300 to 500 µg per rat, given in divided doses over 3 to 4 weeks beginning at 4 to 8 weeks of age, has been used in many experiments to induce tumors in Fischer rats (see Table 6).

Two long-term studies of rats given a single dose of AFB_1 have been reported. In the first, 7 of 16 rats given 7.65 mg/kg had hepatocarcinomas at 1 to 2 years,[60] but in the second, no tumors were found in rats given 5 mg/kg and observed up to 69 weeks.[61]

Histologic changes in the liver of rats fed AFB_1 can be found at termination of short-term or chronic, long-term treatment; some of these changes progress over the following months to tumor development. In addition to bile duct proliferation which occurs during treatment and may be regressing by its termination, the most prominent early change is focal and diffuse hepatocyte hyperplasia; this is easily recognized in autoradiographs from rats given ^{3}H-thymidine (see Figure 26) and can be detected also in routine H&E-stained sections by the presence of increased numbers of mitotic figures and increased cytoplasmic basophilia in groups of hepatocytes (see Figure 27). Subsequently the basophilic foci enlarge; foci of enlarged eosinophilic hepatocytes appear which have vacuolated cytoplasm, clumping of nuclear chromatin, and focal lymphocytic infiltration. DNA synthesis and mitosis are increased. The initial changes are microscopic, but macroscopic nodules occur after 4 to 8 months. These are followed by uni- or multicentric hepatocellular carcinoma (see Figure 28) which may be trabecular (see Figure 29) or anaplastic (see Figure 30). Both types frequently metastasize (see Figure 31).[62,63] Several histochemically demonstrable enzymes (acid phosphatase, glucose-6-phosphatase, succinic dehydrogenase, ATPase) are decreased (see Figure 10) in the hyperplastic areas, but one, gamma-glutamyl transpeptidase, is increased.[62-64]

The nodules of these abnormal hepatocytes apparently follow one of several paths; they may persist, revert to normal, become vacuolated and degenerate, or progress to malignancy. The course of events cannot be proved in histologic sections, but changes seen in the abnormal foci and nodules or in parts of them suggest that these sequences do occur. A detailed correlation of histologic and ultrastructural changes in the liver of AFB_1-treated rats has been published and supports the progression of both proliferative and degenerative changes.[64,65]

Renal tumors induced by AFB_1 in 25 to 50% of male Wistar rats fed 0.25 to 3 mg AFB per kg of a purified diet were well-differentiated adenomas (see Figure 32) and adenocarcinomas, occurring bilaterally in approximately half the tumor-bearing animals. Fischer rats treated concurrently did not develop renal tumors,[57] although they are susceptible to renal carcinogenicity of AFG_1 as are MRC outbred rats.[58]

Colon carcinomas (see Figures 33 and 34) have been induced by AFB_1 in Sprague-Dawley and Fischer rats, usually in low incidence;[66] a 20% incidence after prolonged exposure to a high dose (2 mg/kg in diet) has been reported,[67] but the possible influence of dietary vitamin A on these results was apparently unrecognized by the authors.

Tumors of other organs have occurred sporadically in aflatoxin-treated animals, but the significance of the association is difficult to assess. The tumors include gastric adenocarcinoma, Harderian gland tumors, and squamous carcinoma of esophagus or tongue. Subcutaneous administration induced local sarcomas.[68]

Certain dietary components affect AFB_1 carcinogenesis. Before aflatoxins themselves were recognized, it was noted that choline, methionine, and other lipotropes protected animals against the induction of hepatocarcinomas by diets containing peanut meal; later studies using purified AFB_1 confirmed the effect.[1,16] However, the dietary interactions are complex, and recent results indicate that components other than lipotropes are involved.[69] A diet only marginally deficient in lipotropes markedly enhanced AFB_1 carcinogenesis, but supplementation with choline and methionine did not decrease the enhancement. Review of the earlier studies suggested that the source of dietary fat and the balance of amino acids may significantly influence expression of the lipotropic effect on AFB_1 carcinogenesis. Experiments in which the effect was demonstrated utilized diets that contained lard or vegetable oils as major lipid sources,[70-72] and experiments in which little or no effect was demonstrated utilized beef fat.[63,69,73]

The protein sources used in the two groups of experiments were similar, but proportions derived from different sources varied, and the amino acid balance in the first group of experiments is closer to the rat requirements.

Vitamin B_{12} enhanced aflatoxin carcinogenesis in male Fischer rats despite its lipotropic action and metabolic interrelationships with choline and methionine.[74] Rats were fed mixed aflatoxin for 33 weeks (95% AFB_1 and AFG_1 in a ratio of 1:1) at 1 mg/kg of diets that contained approximately 5 or 20% casein, with or without vitamin B_{12} (50 μ/kg). Tumor incidences were low in all groups, but were significantly higher in rats fed 20% casein and vitamin B_{12} than in all other groups. Protein deficiency did not affect overall tumor incidence, although deficient animals had a higher incidence of hyperplastic nodules.

Deficiency of vitamin A did not significantly affect hepatic tumor incidence in AFB_1-treated rats, but did increase the incidence of colon tumors.[54,66]

It is thought that nutrition and factors such as hormones influence aflatoxin toxicity and carcinogenesis by affecting DNA synthesis, cell division and differentiation, and aflatoxin metabolism and excretion. Animals severely restricted in caloric intake are less susceptible to the action of many carcinogens. The retardation of growth induced by severe nutritional deficiencies, however, may explain this reduced tumor incidence.

Male rats are more susceptible than females to both toxicity and carcinogenicity of aflatoxins, and hormonal manipulations alter the response of experimental animals. Sprague-Dawley males treated with diethylstilbesterol and fed AFB_1 had a lower hepatocarcinoma incidence that did pair-fed controls.[75] Hypophysectomy inhibited AFB_1 hepatocarcinogenesis, but the effect may have been nonspecific, resulting from inhibition of somatic growth;[76] ACTH treatment inhibited AFB_1 carcinogenesis in Fischer rats, while treatment with growth hormone or insulin had no significant effect. Apparently, differences in body weight were not sufficient to account for the differences in tumor induction.[77]

3. Fish

Rainbow trout are highly sensitive to aflatoxin carcinogenicity. Hepatic tumors in hatchery-reared trout (see Figure 35) were reported and related to aflatoxin contamination of food at about the same time as outbreaks of aflatoxicosis were recognized in poultry.[78] A logarithmic dose response relationship to incidence of liver tumors in trout was established by feeding diets that contained 4 to 20 μg AFB_1 per kilogram. Extrapolations to lower doses predicted 0.1 μg/kg to be the minimum effective level for tumor induction.[79] Morphologic studies of tumor development, reported even before the etiology of the tumors was established, showed changes resembling the ones that had lead to hepatocellular carcinoma in rats (see Figure 35).[80] Because of their sensitivity to aflatoxins, trout were used to assay human urine samples for carcinogenic aflatoxin metabolites.[81]

Cyclopropene-containing fatty acids (CPFA), which occur in cottonseed and other plant oils, enhance AFB_1, AFM_1, and AFQ_1 carcinogenesis in rainbow trout. Addition of 100 to 200 mg CPFA per kilogram of diet increased hepatic tumor incidence from 20 to 100% of trout fed 4 µg AFB_1 per kilogram of diet for 9 months, and 13 to 67% of trout fed 4 µg AFM_1 per kilogram of diet for 12 months.[82,83] CPFA alone may induce tumors in trout, however, and the significance of the combined effect requires further evaluation.[45] Aroclor 1254, given at 100 mg/kg of diet, protected rainbow trout against both toxicity and carcinogenicity of AFB_1 given at 6 µg/kg of diet.[84] SKF 525A protected them against tumor induction by AFB_1 given i.p. in divided doses at a total dose of 2.5 mg/kg over 25 weeks.[85]

AFB_1 has induced hapatic tumors in Coho salmon and guppies. Salmon fed 12 µg AFB_1 per kilogram of diet which also contained 50 mg/kg CPFA had 40 to 50% hepatic tumor incidence at 20 months, but salmon fed 1, 5, or 20 µg/kg AFB_1 without CPFA did not develop tumors.[52,79,83] Guppies fed 6 mg/kg diet had 40 to 50% hepatic tumor incidence at 9 to 11 months.[86]

4. Monkeys

Hepatocarcinomas have been reported in aflatoxin-treated male and female Rhesus monkeys from three different laboratories. A male treated with 100 µg mixed aflatoxins (44% B_1) i.p. 5 days/week for 11 months, followed by oral administration of 200 µg for 4.5 years, had metastatic hepatocarcinoma 8 years after treatment was begun.[87] Cholangiocarcinoma developed in a female given 50 to 100 µg mixed aflatoxins per day i.m. for 5.5 years and then observed for 5 additional years.[88]

From a group of 20 treated 2 years or longer with AFB_1, 3 monkeys developed cancer. Hepatocellular carcinoma was reported in 2 females, 1 given 842 mg AFB_1, total, p.o., over 74 months from the age of 8 days and the other given 99 mg AFB_1, total, i.p. and p.o., for 46 months from the age of 1 day. Hepatic hemangioendothelial sarcoma occurred in a male given 119 mg AFB_1, total, i.p. and p.o., for 50 months.[89]

Hepatocarcinoma has been induced in two lower primates, marmosets (*Saguinas oedipomidas*) and tree shrews (*Tupaia glis*), given diets that contained 2 mg AFB_1 per kilogram.[90,91] In the tree shrews, AFB_1 induced fatty liver and premalignant changes similar to those described in hepatocytes of rats. A possible enhancement of carcinogenesis by viral hepatitis was sought in the study of marmosets, but no obvious influence was found in the small number of survivors.

5. Ducks

Hepatocarcinoma and cholangiocarcinoma were induced in Khaki Campbell ducks fed aflatoxin-contaminated peanut meal (calculated to contain 30 g/kg AFB_1, from 7 days of age for 14 months. Tumor incidence was 73%.[92] Two ducks fed meal that contained 3.5 mg/kg aflatoxin for 16 months developed hepatocarcinomas.[16] Hepatocarcinoma occurs in ducks near Shanghai, China, a region where human incidence of this tumor is elevated; this suggests the presence of an environmental carcinogen, possibly aflatoxin.[93]

II. OCHRATOXINS

A. Acute Toxicity

Ochratoxins, metabolites of several species of fungi in the genera *Pencillium* and *Aspergillus*, are toxic to experimental and domestic animals. The primary target organ is the kidney. Of the seven ochratoxins known, only ochratoxin A is thought to be significant in outbreaks of ochratoxicosis in domestic animals. It has been isolated from cereal grains and legumes and has been found in kidney, liver, and muscle of pigs and in muscle of poultry fed contaminated feed.[94-96]

Renal toxicity of ochratoxin A has been demonstrated in pigs, rats, mice, dogs, ducklings, hens, and rainbow trout. In the field, the toxin was first recognized in pigs, and many experimental studies have used them.[94-97]

1. Pigs, Dogs

Krogh and co-workers[94] fed female pigs naturally contaminated barley for 9 days, 68 days, or 3 to 4 months. The experimental diets provided approximately 8, 40, or 160 μg ochratoxin A per kilogram body weight per day. Weight gain and feed efficiency were decreased in all three groups, and renal tubular function was decreased after 3 to 4 months. The decrease correlated with ochratoxin A dosage. Urinary excretion of leucine amino peptidase, an enzyme of the proximal tubular cell brush border, was increased.[95] In the two higher-dose groups, biochemical enzyme abnormalities were detectable in the kidney after 9 days. Gross and histologic abnormalities were found at 68 days in pigs fed the highest level of ochratoxin A; after 3 to 4 months, pigs in the two higher-dose groups had gross and microscopic abnormalities of the kidneys. Pigs given the lowest exposure had only histologic abnormalities.

The kidneys of ochratoxin-treated pigs were enlarged, grey yellow, and markedly increased in firmness. (see Figure 36) Microscopically the lesions ranged from dilatation and focal degeneration of proximal tubules to severe tubular atrophy, with increased thickness of the basement membrane, extensive interstitial fibrosis, and glomerular atrophy and sclerosis (see Figures 37 and 38) The renal medulla and blood vessels appeared normal.

Ultrastructural studies revealed a reduced brush border, reduced numbers of mitochondria, and increased numbers of lysosomes. Histochemical studies were consistent with the ultrastructural changes.[98]

Gut, liver, and lymphoreticular tissues were damaged in pigs and beagle dogs exposed to ochratoxin at doses 10 times those used in the above studies. Beagles given 200 to 400 μg daily ochratoxin for 1 to 3 weeks had abnormal renal function, urinary excretion of renal enzymes, and tubular necrosis.

Dogs given daily doses of 0.2 mg ochratoxin per kilogram weight died after 10 to 14 days with hemorrhagic enteritis, necrosis of lymph nodes, and degeneration of proximal renal tubules.[99] Ultrastructural abnormalities included cytoplasmic disarray, mitochondrial swelling, and presence of myelin figures in the proximal tubules.[100]

2. Rats, Mice

Rats and mice are sensitive to the renal toxicity of ochratoxin A. The i.p. LD_{50} for mice is 22 mg/kg body weight, and the oral LD_{50} for Wistar rats is 20 to 30 (males) and 20 to 21 (females) mg/kg.[96] Both species developed proximal renal tubular necrosis and hepatic periportal necrosis. Rats exposed to ochratoxin A in food at 0.2 to 5 mg/kg of diet for 3 months had histologic and functional evidence of tubular damage.[101] Male Wistar rats, 180 to 300 g, given 5 or 15 mg ochratoxin A per kilogram body weight orally for 3 days, had decreased renal clearance of PAH and inulin; the basement membrane of the proximal convoluted tubules was thickened 24 hr after the third dose.[97]

3. Poultry

Oral LD_{50} values for chicks may increase with age. In 1-day-old chicks, the 7 day median lethal dose (MLD_{50}) for ochratoxin A was 3.3 to 3.9 mg/kg; for ochratoxin B, it was 54 mg/kg.[102] In 10-day-old chicks, the LD_{50} was 10.67 mg/kg.[103] Avian species vary in sensitiviity to ochratoxin A. Comparisons in the same laboratory gave LD$_{50}$s for 3-day-old birds of 3.4 mg/kg in chicks, 5.9 mg/kg in turkeys, and 16.5 mg/kg in Japanese quail.[104]

As in other species, renal toxicity is the major finding and includes tubular degeneration with fat deposition.[103] Long-term exposure of chickens to ochratoxin A at 0.3 or 1.0 mg/kg in feed induced both glomerular and tubular dysfunction and tubular degeneration.[94] Changes in renal function and structure resemble those seen in pigs.[96] Focal hepatic necrosis and damage also occur.[103]

4. Fish

Rainbow trout given an approximate LD$_{50}$ of 4.7 mg/kg, i.p., had both glomerular and proximal renal tubular necrosis.[105]

5. Teratogenicity

Ochratoxin A was teratogenic in mice given 1 dose of 5 mg/kg, i.p., in rats given repeated doses of 0.75 or 1 mg, orally, and hamsters given one dose of 5 to 20 mg/kg.[106-108]

B. Chronic Toxicity

Long-term studies to determine whether ochratoxin A is carcinogenic in laboratory animals have been inconclusive; more comprehensive studies are needed.[96]

III. TRICHOTHECENE TOXINS

The trichothecenes are a group of chemically related fungal metabolites produced by various molds of the species *Fusarium, Cephalosporium, Myrothecium, Stachybotrys, Trichoderma,* and *Verticimonosporium*. The toxins are associated with several mycotoxicoses in both animals and humans. The chemistry of these toxic metabolites is addressed in Volume II of this series. The fungi grow in several staple agricultural products; toxic metabolites are elaborated and bound to plant tissues. Ingested, these contaminated foods cause serious intoxication. It has now been established that red-mold toxicosis in Japan, moldy corn toxicosis in the U.S., stachybotryotoxicosis in northern Europe, and alimentary toxic aleukia in the Soviet Union are all caused by the trichothecenes produced by one or more of the various fungi noted above.[109,110]

More than 40 kinds of trichothecenes have been classified into four groups according to their chemical structure and fungal origin. The toxicological effects of the trichothecenes are inhibition of protein synthesis, inhibition of DNA synthesis, vomiting, inflammation of the skin, and leukopenia.[111-118]

The toxicological effects closely correlate with the chemical characteristics of the various toxins although some of the metabolites (e.g., T-2 toxin) appear to be transformed in the body into the proximate toxin that produces the biologic response. In general, species affected by these toxins exhibit clinical symptoms and lesions referred to above, selected examples of which will be described below.

A. Acute Toxicity

The acute toxicity of the naturally occurring trichothecenes is shown in tables found in Volume II.

Signs and symptoms of toxicologic response to the trichothecenes in field cases have been reproduced under laboratory conditions in mice, rats, guinea pigs, avians, horses, dogs, and cats. The major features of the toxicologic response, however, vary considerably depending upon the species of animal. Table 7 lists some of the effects of exposure to trichothecene mycotoxins.

Young animals are more susceptible to the toxins than adults. In observations so far, there has been little indication of sex difference. Hemorrhage into the intestine, karyorrhexis, and cellular necrosis of actively dividing cells of the thymus, ovary, spleen, and testis are characteristic.[115] Shortly after dosing, dogs, cats, and ducklings exhibit emesis and vomiting, regardless of the route of administration. There is also a marked leukocytosis. Cats, which appear to be particularly sensitive to the T-2 toxin, exhibit destruction of the bone marrow, as well as the hemorrhage into lung, brain, and intestine that are seen in guinea pigs and rabbits. A leukopenic response is characteristic of most animals following intoxication with the trichothecenes.

Meningeal hemorrhage with extension into the brain and bleeding into the lungs and intestine are toxicologic features found in both experimental animals and in human cases reported from the Soviet Union, where people became ill after eating overwintered grain. These cases are described in detail by Joffe.[119]

1. Guinea Pigs, Rabbits

Guinea pigs and rabbits exposed to crude culture filtrates of the various fungi that produce the trichothecenes exhibit edema, inflammation, dilated blood vessels, and hemorrhages subcutaneously and in the intestinal wall (see Figures 39 to 41). Necrosis of the intestinal crypts and acutely dying cells are also prominent.

2. Dogs

Dogs exhibit a cessation of stomach motor activity, increased body temperature and pulse rate, depressed leukocyte count, stomatitis and GI hemorrhage, degeneration of the liver and kidney, and changes in the epicardium and CNS.

3. Cats

Cats are excellent subjects for experimental work with the trichothecenes. Exposure to the toxins causes decreases in hemoglobin, red blood cells, leukocytes, and neutrophils and an increase in lymphocytes. Vomiting, hyperemia, and hemorrhage of the tissues of the GI tract, kidney, and adrenal glands are characteristic in the cat along with damage to the bone marrow and consequent progressive leukopenia. Acute necrosis of cells of intestinal crypts is marked (see Figure 42).

4. Rats

Rats exposed to trichothecenes exhibit a prolonged prothrombin time, inflammation of the GI tract, necrosis and sloughing of tissues of the oral cavity, stomach, and intestines, and a decrease in circulating white blood cells concomitant to damaged bone marrow. Necrosis of the skin where the material is applied is characteristic of all the trichothecene compounds.[110]

5. Swine

Swine are susceptible to the effects of the trichothecenes and to crude cultures of molds or moldy feeds that contain them. The response of swine includes incoordination, loss of appetite, frequent urination, and muscle tremor. Respiration rate is increased, and in some cases, the hind quarters become paralyzed prior to death. Erythrocytes, white cells, and hemoglobin levels increase, and thrombocytes decrease; a catarrhal hemorrhagic gastroenteritis occurs. More chronic signs of intoxication include necrotic changes of the tissues around the snout, lips, and mucous membranes of the mouth. The animals cease to eat; weakness and depression are progressive, and there is a pronounced leukopenia. Extensive hemorrhage in the epiglottis, heart muscle, mucosal surface of the stomach, skeletal muscle, and intestines have been described by Smalley and Strong.[120]

6. Cattle

A field case attributed to *Fusarium* toxicosis in cattle was described by Hsu et al.[121] A total of 7 of 35 cows died over a period of 4 months after continuous consumption of a diet that consisted of 60% moldy corn, soybean meal, and feed supplements. At necropsy there were extensive hemorrhages of the mucosal surfaces of all viscera. Frequent abortions had been observed in the herd. The concentration of T-2 toxin found in the feed was about 2 mg/kg or 2 ppm.

In one experiment,[122] a 295-kg cow was given T-2 toxin according to the following schedule: 72 mg, intramuscularly, daily for 3 days; 35 mg, intramuscularly, daily for 4 days; 18 mg/day, intramuscularly, for 33 days; 35 mg orally for 1 day; and 30 mg daily, intramuscularly, for 22 days. Following this exposure (a total of 645 mg over 62 days), the cow died. During the period of exposure, she lost 50 kg of body weight and gradually refused feed. Difficult respiration with epistaxis and bloody diarrhea were observed just prior to death. The prothrombin time increased from 55 to 200 sec. At necropsy there were hemorrhages in the epicardium and in the rumen, abomasum, and intestinal walls. There were also hemorrhages from the turbinates and into the cervical and mesenteric lymph nodes. The most massive hemorrhage, however, was in the lumen of the large intestine. Histologically, there was severe fatty degeneration of the liver.

7. Poultry

Pure T-2 toxin adminstered to poultry resulted in decreased growth rate. Yellow-white lesions appeared on the hard palate and along the margins of the tongue;[123] the oral lesions became more severe with further administration of the compound, resulting in difficulty in eating and decreased food intake. The tissues of the oral cavity became intensely inflamed, and focal necrosis (see Figures 43 and 44) was observed histologically. The size of the spleen decreased, with depletion of lymphocytes.

B. Chronic Studies with Trichothecene Toxins

Chronic evaluation of the effects of trichothecene toxins in animals has received very little attention compared to the research into acute toxicity.[120,124] Though few, long-term studies indicate that at least the T-2 toxin is not carcinogenic.

Saito and Ohtsubo[125] fed rats and mice with rice made moldy with Fusaria and observed hyperplasia of the bone marrow, atypical hyperplastic epithelia in the gastric and intestinal mucosa, and intrahepatic bile duct hyperplasia. In addition, atrophic or hypoplastic changes in the thymus, spleen, bone marrow, and testicles were observed in more than half the animals. Chronic bronchitis and bronchopneumonia were the most frequent causes of death in these studies. The toxicity of the rice inoculated with *Fusarium nivale* and *Fusarium graminearum* varied considerably from one lot to another, however, making it very difficult to determine a dose-response relationship.

Ohtsubo and Saito gave rats and mice weekly doses of Fusarenon-X, orally for 50 weeks or by s.c. injections for 10 to 22 weeks. Their histological findings resembled those described in the earlier experiment with moldy rice.[124] Table 8[126] indicates that unusual tumors were observed at various sites, although their numbers were quite small.

In view of this finding, investigators repeated the study using pure Fusarenon-X in rats and T-2 toxin in mice. Table 9 illustrates the observations of the study with Fusarenon-X. Tumors occurred in all groups, including the control animals. A scirrhous adenocarcinoma of the stomach, the first tumor to be observed, was in a rat that died after 319 days of being fed 50 μg of Fusarenon-X daily. Most of the tumors, however, were found in the last half period or at the end of the study by microscopic examination. One case of transitional cell carcinoma of the urinary bladder was observed at day 719. Because of low incidence and lack of dose-response information, it is difficult to conclude that the carcinomas, though rare in the rat, were induced by Fusarenon-X. Fusarenon-X inhibits growth in rats and reduces resistance to infection, even at dietary levels as low as 2.5 ppm; however, its carcinogenicity is doubtful.

In the studies using T-2 toxin in mice, an effort was made to determine whether or not the irritating effect of T-2 toxin, characteristic when applied to the skin, might cause GI necrosis and lead to more serious, irreversible changes. (Earlier investigators had observed that crude extracts from cultures of *Fusarium poae* and *Fusarium sporotrichioides* produced forestomach lesions in rats). Results were negative; a few hepatomas, uterine sarcomas, and thymomas were observed in the treated groups (see Table 10). One case each of leukemia and reticulum cell sarcoma was observed in the control group. The authors state that these tumors are not unusual in the mice they were using. Thus, T-2 toxin does not appear to be carcinogenic under the conditions of this study.

These results are in agreement with those of Marasas and others,[127] who gave diets containing 15 and 5 ppm T-2 toxin for 19 days, followed by 10 ppm T-2 toxin over a period of 8 months. These investigators reported no effects on the stomach. From studies conducted thus far, it appears quite unlikely that chronic trichothecene toxicosis results in induction of malignant tumors, and further it is unlikely that chronic toxicosis from the trichothecene family occurs in man or in livestock. Further work along these lines is indicated.

IV. STERIGMATOCYSTIN

Sterigmatocystin, a pale yellow compound first isolated as a yellow-crystalline material from the mycelial mats of *Aspergillus versicolor*, is now known to be elaborated by several strains of *A. versicolor*,[128-130] as well as by *Aspergillus rugulosus, Aspergillus nidulans*, and a species of *Bipolaris*. The molecule contains a xanthone nucleus linked in an angular fashion to a dihydrofuro-benzofuran ring system, the first molecule in mycotoxin chemistry found to contain this unique functionality. The hepatocarcinogenicty of sterigmatocystin in rats was established by Purchase and van der Watt.[131,132] The toxin also has the capacity to produce a series of skin lesions, including squamous cell carcinomas, after it is applied to the shaved skin of animals

Of the numerous mycotoxins that have been studies, sterigmatocystin has received suprisingly little attention, perhaps because of its limited distribution. In addition to the principle toxin, several derivatives have been identified as products of cultures producing sterigmatocystin. These are described in detail in the chapter on food-borne mycotoxins and alimentary mycotoxicoses. Sterigmatocystin has several biological properties similar to those of aflatoxin B_1; both toxins induce nucleolar lesions characteristic of DNA-dependent, RNA synthesis inhibitors and both are carcinogenic to the liver or rats. Sterigmatocystin has been found only on a small number of samples of molded wheat and green coffee beans.

A. Acute Toxicity

Mice, which are resistant to AFB_1 toxicosis, also resist the acute toxic effects of sterigmatocystin. The LD_{50} for mice is in excess of 800 mg/kg.[133,134] In Wistar rats, the 10-day LD_{50} was 166 mg/kg body weight in males and 120 mg/kg in females. Dickens and co-workers[68] have demonstrated that sterigmatocystin is carcinogenic at the site of s.c. administration. Studies using male vervet monkeys indicated that injury to the liver and kidneys became more severe as dosage was increased.[5] The major metabolite in monkeys has been isolated and identified; more than 50% of the recovered material was a glucuronic acid conjugate found in the urine.[135]

Some of the sterigmatocystin derivatives have been shown to have varying, but generally low, toxicity to mice and ducklings. On the other hand, one of them, aspertoxin, is severely toxic to developing chick embryos.[136] A fine structural nucleolar change is induced in rat liver and in primary tissue culture cells from 10-day-old chicken embryonal liver by o-acetyl-sterigmatocystin and related compounds. In 3 days after the injection of sterigmatoxystin, widespread liver necrosis was observed in rats (see Figure 45).[137] However, if rats were treated with 8 mg o-acetyl-sterigmatocystin per kilogram body weight and 3 mg AFB_1 per kilogram body weight, they exhibited massive liver necrosis. The solubility of sterigmatocystin appears to play an important role in determining toxicity.[138]

van der Watt and Purchase[5] examined the acute effects of sterigmatocystin in male vervet monkeys given graded doses of sterigmatocystin by intraperitoneal injection (see Table 11). The LD_{50} was 32 mg/kg body weight, with 95% confidence limits of 15.0 to 70.0 mg/kg. At necropsy, glomerular degeneration and edema of Bowmans space in the kidney were observed. There were also hyaline droplet degeneration (see Figure 46), fatty changes, and necrosis of the renal tubular epithelium. Cells of the glomerulosa of the adrenal cortex were severely damaged, with loss of cytoplasm and pycnosis of nuclei.

The monkeys' livers exhibited ballooning degeneration, fatty change, and focal necrosis, most evident in the centrilobular zone. Hemorrhagic necrosis was present at higher dose levels (see Figure 47). Bile stasis was prominent, particularly in animals that survived longer to develop icterus.

B. Chronic Toxicity

Long-term studies in experimental animals have indicated that sterigmatocystin is carcinogenic.[139] Two groups of 3-week-old IRC mice of both sexes were fed a diet containing 5 mg pure sterigmatocystin per kilogram or 5 mg *A. versicolor* culture per kilogram. Diets containing the materials were fed for 2 weeks, and then uncontaminated control diets were fed for the next 2 weeks; this regimen was continued for 58 weeks. Table 12 gives details of results of the study. Pulmonary adenomas (see Figure 48) were found in 21 of 25 mice in the group fed the pure sterigmatocystin and in 33 of 35 mice given the *A. versicolor* culture. The biologic significance of the adenomas is not clear since they did not appear to affect lifespan. These adenomas were observed in animals of this study surviving more than 30 weeks. Only 4 of 37 controls had pulmonary adenomas. Adenocarcinomas of the lung occurred in nine out of the 25 mice fed sterigmatocystin, with none observed in controls. Tumors at other sites were not increased in any of the mice.

Tumors of the lung and liver were induced in newborn mice injected with sterigmatocystin at 5, 1, or 0.5 $\mu g/g$ of body weight.[140]

Sterigmatocystin, administered by stomach tube or in the diet to Wistar rats for 52 weeks, with dosage ranging from 0.15 to 2.25 mg per animal per day, resulted in hepatocellular carcinoma in 39 of 50 survivors within 123 weeks of the study. No liver tumors were observed in 19 controls given sunflower oil by stomach tube or in any of those fed a basal diet alone.[132]

Purchase and van der Watt[141] used two groups of male Wistar rats, applying 1 mg sterigmatocystin in 0.1 mℓ DMSO or 0.15 mℓ in acetone to the shaved dorsal skin surface twice weekly for 70 weeks. Skin papillomas were seen in four of ten, and skin carcinomas were seen in six of ten rats given sterigmatocystin in DMSO; among rats given sterigmatocystin in acetone, three of ten had papillomas, and seven of ten had skin carcinomas. Liver cell carcinomas were also observed in five out of ten rats given sterigmatocystin in DMSO and seven rats given sterigmatocystin in acetone. No liver or skin tumors occurred in 30 controls.

Orally administered sterigmatocystin clearly plays a carcinogenic role in the development of lung and liver tumors. When applied to the skin, sterigmatocystin has produced skin and liver tumors, and it has produced sarcomas at the site of s.c. injection.

V. ZEARALENONE

Zearalenone is an estrogenic metabolite produced by several species of *Fusarium*, particularly *Fusarium roseum*. It is not a toxin in the true sense of the word, since it produces no known acute effects. Its hyperestrogenic effects, however, affect food production by decreasing the production of swine; in this manner, it has an impact on human populations. As noted in an earlier chapter, zearalenone is a beta-resorcylic acid-lactone. There are at least five naturally occurring derivatives of the compound, and its action apparently derives from its regulation of perithecia formation in the sexual stage of the life cycle.[142] Although no incidences of toxicity in human populations have been reported, zearalenone has been found in corn destined for human use in the U.S. in amounts known to cause problems in swine. Because zearalenone has been identified mainly as a contaminant in corn, it may find its way into the food chain of some populations.

Stoloff and co-workers[143] conducted a survey in 1974 of corn stored in country elevators in the U.S. A total of 315 marketable and 57 damaged corn samples were collected at 116 different farms. There was a 10% incidence of zearalenone in marketable corn, with an average concentration of 117 ppb. Grossly damaged corn, which constituted 7% of the total lot, contained an average level of 2100 ppb. Zearalenone was found together with aflatoxin in three samples of marketable corn and twice in visibly damaged corn. Thus, there is a potential for public health problems from zearalenone, either singly or in combination with other mycotoxins.

Concentrations of zearalenone greater than 1 ppm are considered physiologically significant, based on responses observed in test animals. Table 13 shows effects of pure zearalenone in ration of bred sows.

Swine appear to be the animals most sensitive to zearalenone, and it was in this species that hyperestrogenism was first observed.[144] If zearalenone is present at 1 to 5 ppm, the animal will consume about 1 to 4 mg of it per day. Pure zearalenone (1 mg) administered daily to a 60- to 70-lb female pig (10 to 12 weeks old) will cause swelling of the vulva on the 4th day. A mature gilt ready for breeding weighs about 250 lb and consumes about 9 lb of feed per day. Thus, some problems in infertility and reproduction are of economic significance to the swine producers. Table 14 lists the occurrence of zearalenone in feed stuffs, the vehicle for the contaminant, and the diagnosis of the clinical problem.

Mirocha et al.[145] and Ueno et al.[146] have described effects of zearalenone on the reproductive tract of experimental and domestic animals. In general, a hyperestrogenism is produced; the weight of the uterus increases, and the vulva swells. There are mitosis and proliferation of uterine muscle, ovarian atrophy, and other signs of estrogenism.

In addition to hyperestrogenism, zearalenone has anabolic properties which result in weight gain of experimental animals. This discovery has resulted in the production of a commercially available product used in beef production.

Poultry are affected by zearalenone, but not to the same extent as swine. Turkey poults fed 10% corn heavily infected with *F. roseum* for 76 days ate only slightly less than those fed a ration containing good corn; body weight gain and feed efficiency of the two groups were nearly identical.[147] However, some of the turkeys eating the contaminated diet developed swollen vents, and three out of eight fed such a ration developed prolapsed cloacae and enlarged bursa of Fabricius. Other investigators have observed effects of zearalenone on poultry, indicating that it may be an economic problem that has not been adequately recognized.

The toxicity, teratogenicity, and effect on reproductive processes in rats have been evaluated by Bailey and co-workers.[148] Zearalenone, given at up to 10 mg/kg of body weight per day, depressed growth and food intake. Fertility was impaired and there was a significant number of resorptions and stillbirths. No teratogenic responses were observed, but the number of medullary trabeculae in the marrow cavity of the femur increased.

Table 1
LD$_{50}$ OF AFB$_1$ IN DIFFERENT SPECIES

Species	LD$_{50}$ (oral, mg/kg)	Ref.
Duckling	0.36—0.73	1, 7
Rat		
100-g, M, outbread	7.2	8
150-g, F, outbread	17.9	
100-g, M, Fischer	5.5	9
200-g, M, Fischer	1.16[a]	7
Mouse	60	3
Hamster	10.2	3
Guinea pig	1.4—2.0	10, 11
Monkey		
Macaca fascicularis, F	7.8	12
M. irus, M	2.2	11
Cereopithecus aethiops	3.7[b]	5
Baboon	2.0	13
Rainbow trout	0.81	14

[a] Given i.p.
[b] Mixed aflatoxin, 71% AFB$_1$.

Table 2
COMPARISON OF TOXICITY OF NATURALLY OCCURRING AFLATOXINS IN DUCKLINGS AND RATS[a]

Compound	Duckling (po)	Rat (i.p.)
AFB$_1$	0.73	1.16
AFG$_1$	1.18	1.5—2.0
AFB$_2$	1.76	>200
AFG$_2$	2.83	>200

[a] LD$_{50}$ (mg/kg).

From Wogan, *Cancer Res.*, 31, 136, 1971. With permission.

Table 3
TOXICITY TO RATS OF REPEATED DOSES OF AFLATOXINS

Compound	Daily dose[a]	Total dose (mg/rat)	Mortality (4 weeks)
AFB$_1$	25	0.5	0/10
	50	1.0	8/10
AFG$_1$	50	1.0	0/10
	100	2.0	4/10
AFB$_2$	3750	52.5	0/11

[a] Miligrams per rat per day, 5 days/week, 4 weeks by gastric intubation. AFB$_2$ given on alternate days.

From Wogan, G. N., Edwards, G. S., and Newberne, P. M., *Cancer Res.*, 31, 1936, 1971. With permission.

Table 4
AFB$_1$ TOXICITY IN 6-WEEK-OLD MALE RATS FED COMPLETE OR HIGH-FAT, LIPOTROPE-DEFICIENT DIETS

Diet	Total aflatoxin B$_1$ dose (mg/kg body weight)	Route	Mortality (2 weeks)
Sprague-Dawley			
Complete	7	i.g.	3/5
	9	i.g.	4/5
	7	i.p.	5/5
Deficient	7	i.g.	0/5
	9	i.g.	0/10
	7	i.p.	0/5
Fischer			
Complete	7	i.g.	10/10
Deficient	7	i.g.	0/10

From Rogers, A. E. and Newberne, P. M., *Toxicol. Appl. Pharmacol.*, 20, 113, 1971. With permission.

Table 5
ESTIMATED LIFETIME HEPATOCARCINOMA RISKS IN RATS EXPOSED TO AFLATOXIN IN DIETS

	Calculated incidence of hepatocarcinoma per 10^5 rats fed	
Study	Aflatoxin (0.1 mg/kg)	Aflatoxin (0.3 mg/kg)
1	30	160
2	70	360
3	140	650
4	320	1400
Combined	240	1100

From Food and Drug Administration, Bureau of Foods, *Assessment of Estimated Risk Resulting from Aflatoxins in Consumer Peanut Products and Other Food Commodities*, Food and Drug Administration, Bureau of Foods, Washington, D.C., January 19, 1978.

Table 6
INCIDENCE OF HEPATOCARCINOMA INDUCED BY AFLATOXIN B_1 GIVEN INTRAGASTRICALLY TO MALE FISCHER RATS FED A NUTRITIONALLY COMPLETE PURIFIED DIET

Schedule of AFB_1 administration	Total dose of AFB_1 per rat	Incidence of hepatocarcinoma (%)	Ref.
25 μg/day, 5 l days/week	350	11	26
15 μg/day, 3—5 l days/week	375	11—15	36, 59
70 μg/day, 5 l days/week	630	50	7
25 μg/day, 5 l days/week	1000	100	7

Table 7
TOXICOLOGICAL RESPONSES TO TRICHOTHECENE TOXINS

1. Anorexia and/or failure to eat
2. Lowered white cell counts and platelets
3. Depression of the hematopoetic tissues
4. CNS disorders
5. Brain hemorrhage involving the meninges
6. Vomiting and diarrhea
7. Edema, hemorrhage, and necrosis of dermal tissues
8. Hemorrhage into the mucosa of the stomach and intestines

Table 8
LONG-TERM EFFECTS OF TRICHOTHECENES[126]

Animal and treatment	Route of administration	Number of animals	Neoplasms observed
Rats, male			
Control	Oral	10	0
F. nivale (moldy rice)	Oral	30	0
F. graminearum (moldy rice)	Oral	18	1 Lymphosarcoma
Fusarenon-X	Intragastric	20	1 Hepatoma
Mice, male and female			
Control	Oral	36	0
F. nivale (moldy rice)	Oral	32	1 Hepatoma 2 Leukemia 1 Intestinal carcinoma
F. graminearum (moldy rice)	Oral	36	2 Hemangiomas

Table 9
CHRONIC FEEDING OF FUSARENON-X TO RATS[126]

	Dosage group and tumors observed			
Tumor type	Control	50 μg/day, 2 years	100 μg/day, 1 year	100 μg/day, 2 years
Hypophyseal adenoma	1/33	1/35	2/19	1/14
Adrenal adenoma	0/33	0/35	1/19	0/14
Thyroid carcinoma	3/33	3/35	1/19	0/14
Liver adenoma	1/33	0/35	0/19	0/14
Stomach carcinoma	1/33	1/35	0/19	0/14
Bladder carcinoma	0/33	1/35	0/19	0/14
All tumors	6/33	6/35	4/19	1/14

Table 10
FORESTOMACH LESIONS AND TUMORS IN MICE FED T-2 TOXIN[126]

T-2 Toxin (ppm)	Number of animals	15-month survival	Forestomach lesions	Tumors
15	16	5	17	2 hepatocellular carcinoma; 1 adenocarcinoma stomach
10	19	5	12	1 thymoma, 1 uterine angiosarcoma
0	22	0	0	1 leukemia, 1 reticulum cell sarcoma

Table 11
TOXICITY OF STERIGMATOCYSTIN IN VERVET MONKEYS

Body weight (kg)	Dose (mg/kg body weight)	Survival time (days)
5.2	15	10
4.4	15	10
3.8	32	10
3.3	32	4
5.1	70	7
4.9	70	2
2.5	150	6
4.2	150	2

From van der Watt, J. J. and Purchase, I. F. H., *Br. J. Exp. Pathol.,* 51, 183, 1970. With permission.

Table 12
LONG-TERM FEEDING STERIGMATOCYSTIN TO ICR MICE[139]

Dosage (ppm) Culture	Pure toxin	Mice with tumors Male	Female
0	0	2/16	2/21
0	0	10/34	11/43
0	5	12/15	9/10
0	5	10/13	7/13
5	0	14/22	19/33
5	0	16/26	21/33

Table 13
ZEARALENONE AND SWINE REPRODUCTION

Zearalenone in diet	Animal number	Litter size	Average piglet weight at birth (lb)
25	1	4	2.5
	2	8	2.6
	3	0	0
50	1	9	3.3
	2	7	2.3
	3	1	2.5
100	1	0	0
	2	0	0
	3	0	0
	4	0	0
Control	1	9	3.5

From Chang, K., Ph.D. thesis, University of Minnesota, Minneapolis, 1977.

Table 14
OCCURRENCE OF ZEARALENONE IN FEEDSTUFF

Source	Diagnosis	Zearalenone
Commercial ration	Porcine abortion	10.3
Whole kernel corn	Porcine estrogenism	35.6
Sorghum	Bovine abortion	12
Maize	Porcine infertility	0
Commercial ration[a]	Porcine estrogenism	5
		50
	Porcine infertility	0
	Porcine estrogenism	0
Whole kernel corn	Porcine estrogenism	2.7
Ground maize	Porcine abortion	32

[a] Diethylstilbestrol found in sample.

FIGURE 1. Liver of duckling following exposure to a single LD$_{50}$ of aflatoxin B$_1$ (AFB). Liver on left is a control; that on right is aflatoxin poisoned; it is shrunken and greenish brown from bile retention.

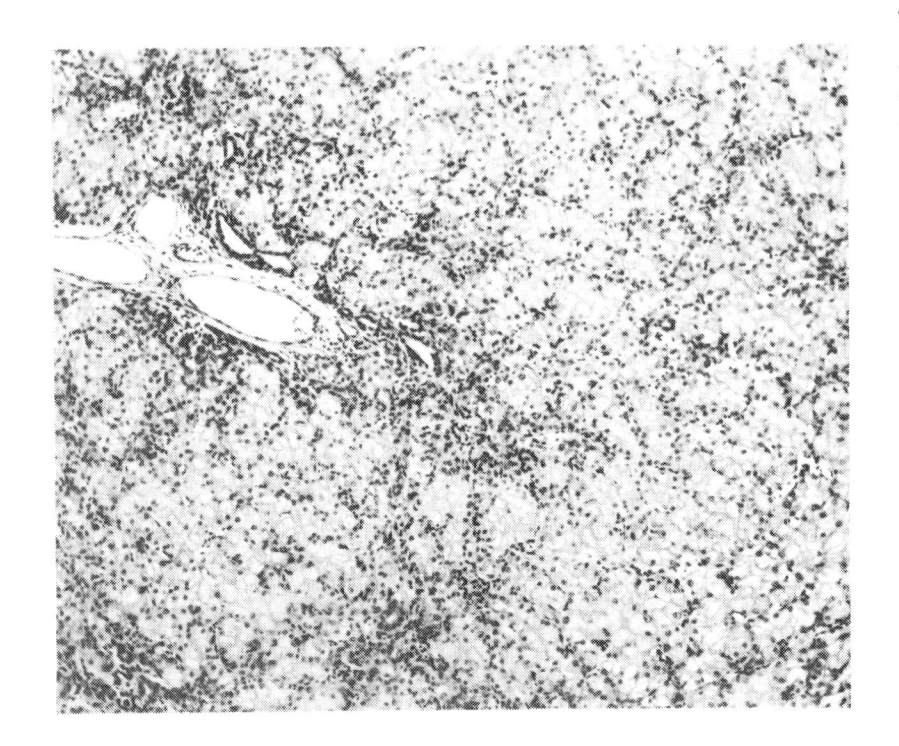

FIGURE 3. Bile duct proliferation, 72 hr after exposure to single dose of AFB$_1$. Note fatty appearance in addition to bile duct proliferation.

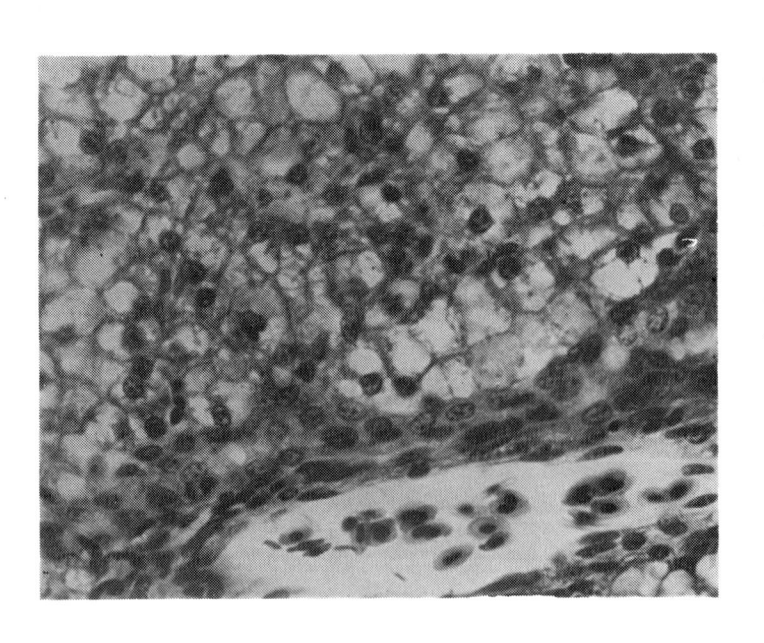

FIGURE 2. Fatty liver of 3-day-old duckling exposed to aflatoxin about 48 hr earlier. Note large amount of fat distending hepatocyte cytoplasm. (The newly hatched duckling has large amounts of liver fat, but this is normally decreased by the age of 3 days.) In the liver of an aflatoxin-poisoned duckling, the mechanism for lipid removal is deranged.

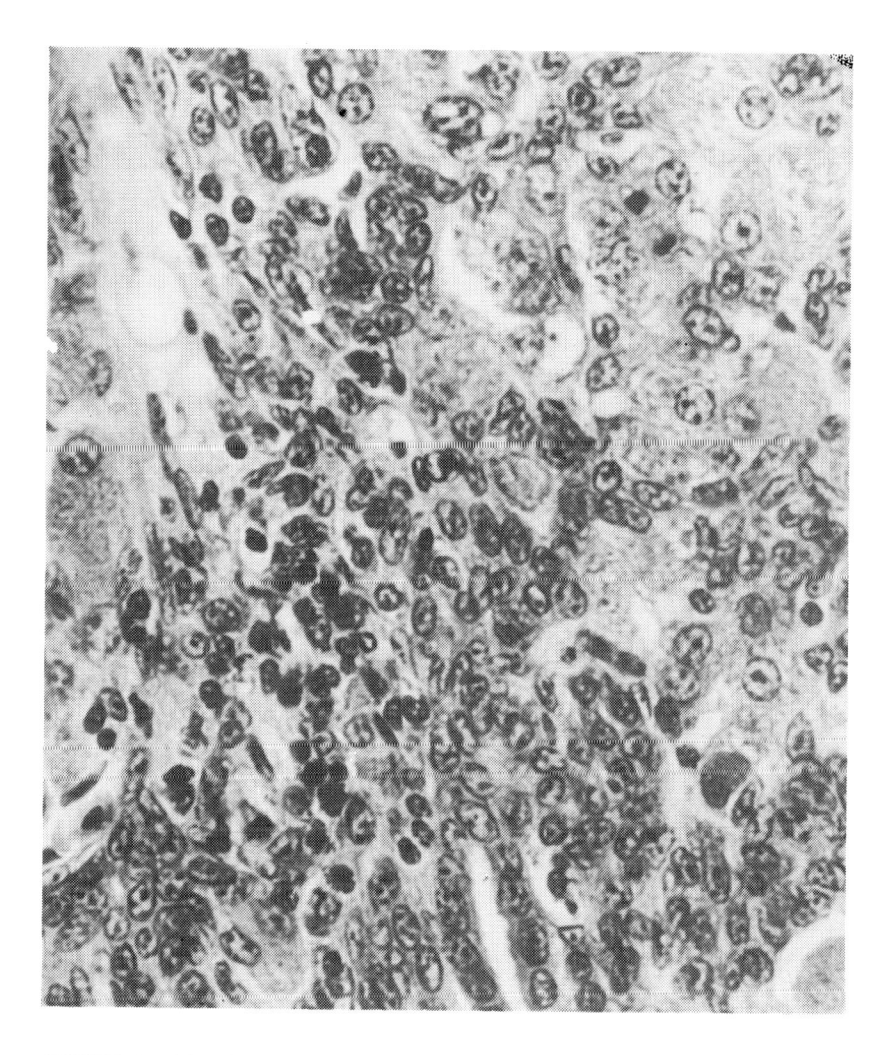

FIGURE 4. Liver from duckling exposed to an LD$_{50}$ of aflatoxin. Note mitotic figure in hepatocyte 2 weeks after exposure to toxin, a common observation in poisoned liver attempting to replace injured and dead cells.

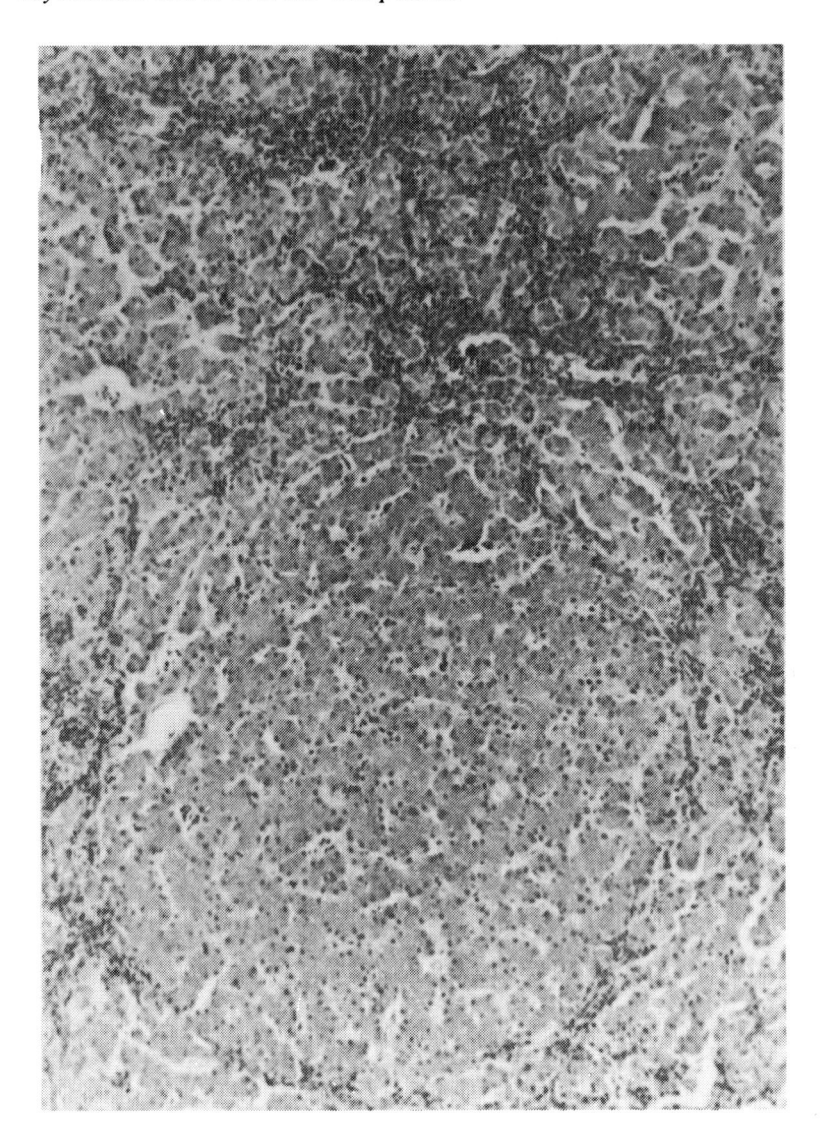

FIGURE 5. Liver from duckling exposed to dietary aflatoxin, 1 ppm, for 3 weeks. Nodular hyperplasia and bile duct proliferation are prominent.

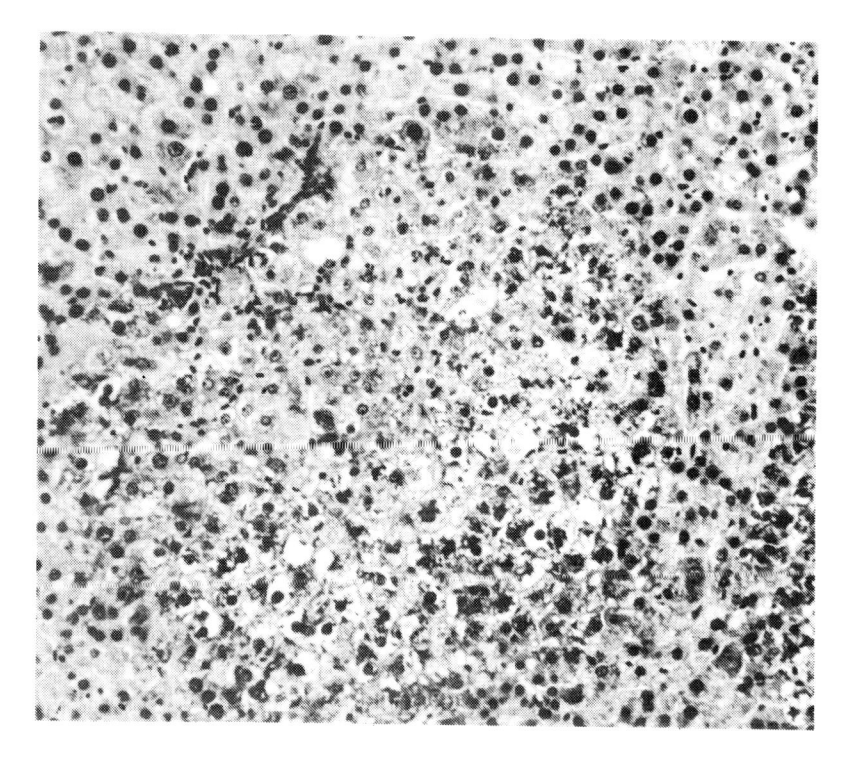

FIGURE 7. Periportal hepatic necrosis of liver from rat exposed to AFB$_1$. Note fatty accumulation and liver cell necrosis of periportal hepatocytes.

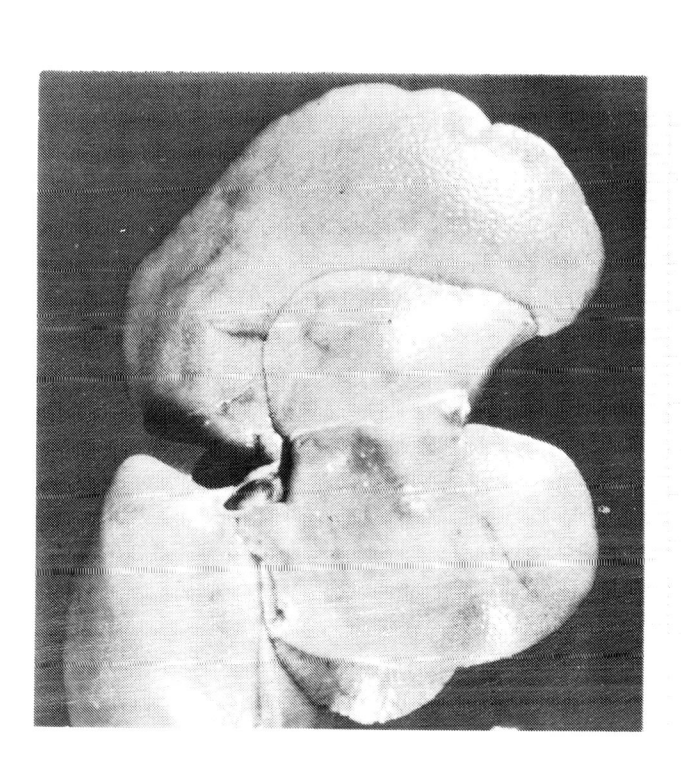

FIGURE 6. Pale, soft, and enlarged fatty liver of rat exposed to an LD$_{50}$ of aflatoxin.

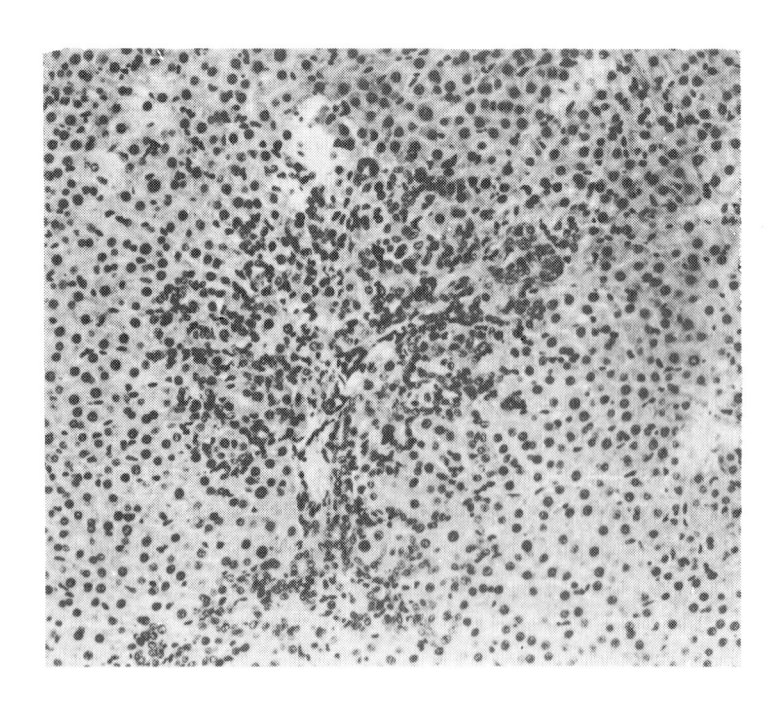

76 *Mycotoxins and N-Nitroso Compounds*

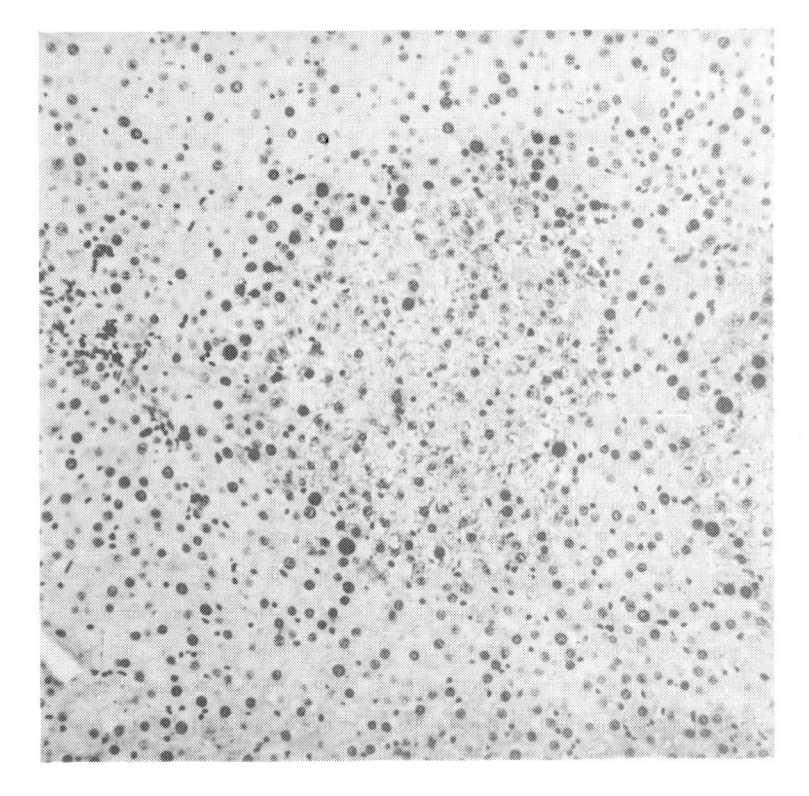

FIGURE 9. Areas of early focal necrosis and hepatocyte regeneration in rat exposed to aflatoxin and to a single dose of tritiated thymidine (1 µCi/g body weight) just prior to sacrifice. Dark cells at periphery of lesion are synthesizing DNA therefore and incorporating the radioactive nucleotide into newly synthesized nucleic acid.

FIGURE 8. Periportal bile duct proliferation in rat exposed to aflatoxin several weeks before sacrifice. Note that repair of hepatocytes has been largely completed in this zone.

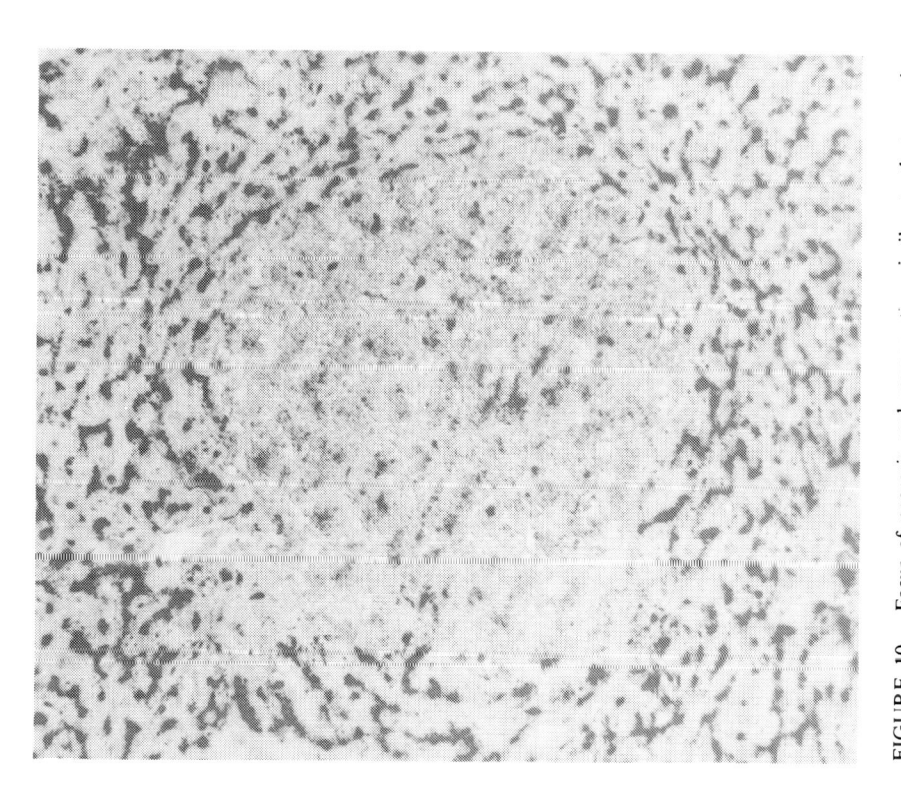

FIGURE 10. Foci of necrosis and regeneration similar to that seen in Figure 9; cells have lost the enzyme, glucose-6-phosphatase.

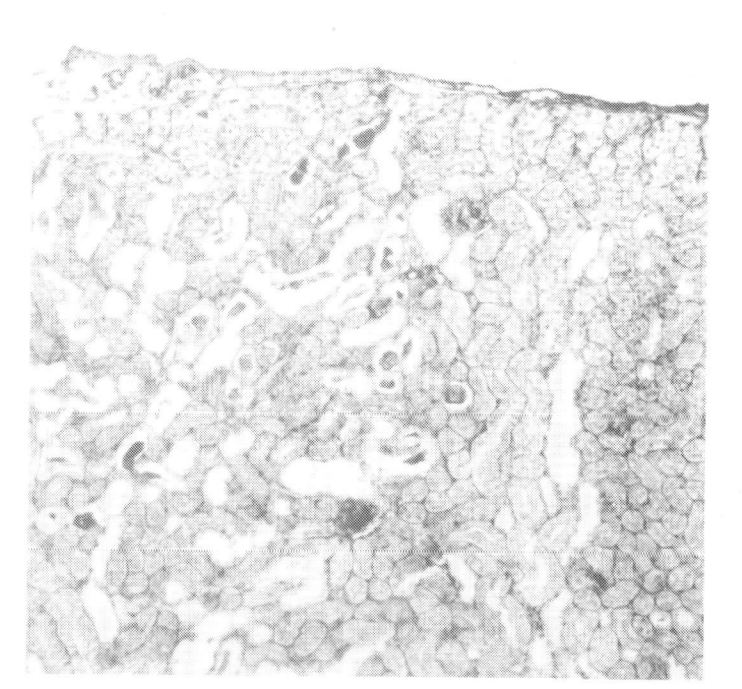

FIGURE 11. Kidney from rat exposed to an acute toxic level of aflatoxin, illustrating renal tubular necrosis often seen with such exposures.

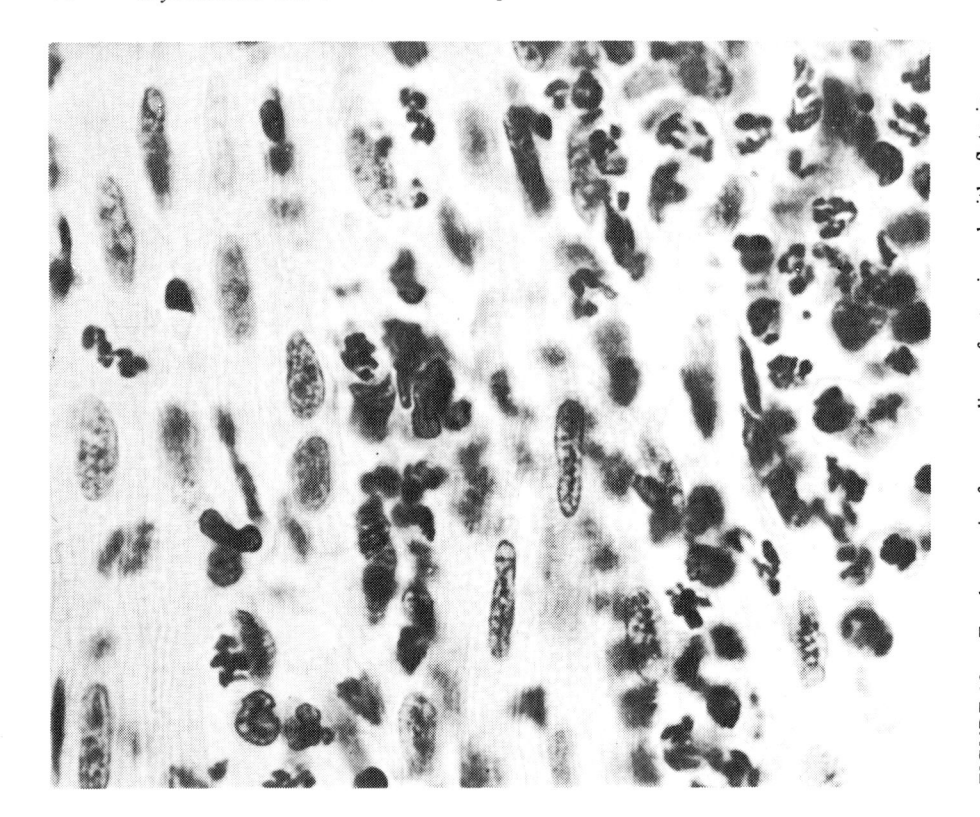

FIGURE 13. Focal necrosis of myocardium of rat poisoned with aflatoxin.

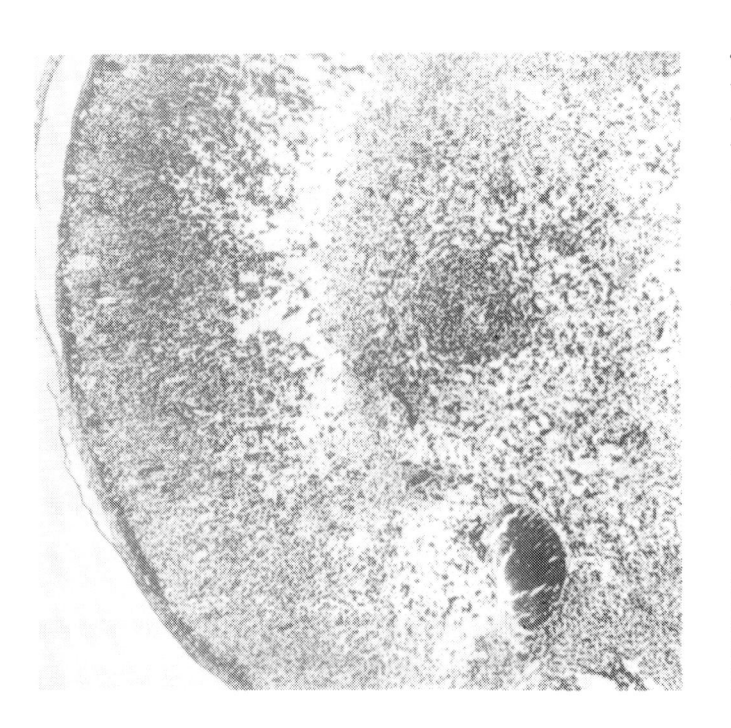

FIGURE 12. Adrenal from rat exposed to an acute toxic level of aflatoxin, illustrating cortical necrosis often seen with such exposures.

FIGURE 15. Microscopic appearance of liver shown in Figure 14. Note large amount of accumulated centrilobular fat and areas of necrosis.

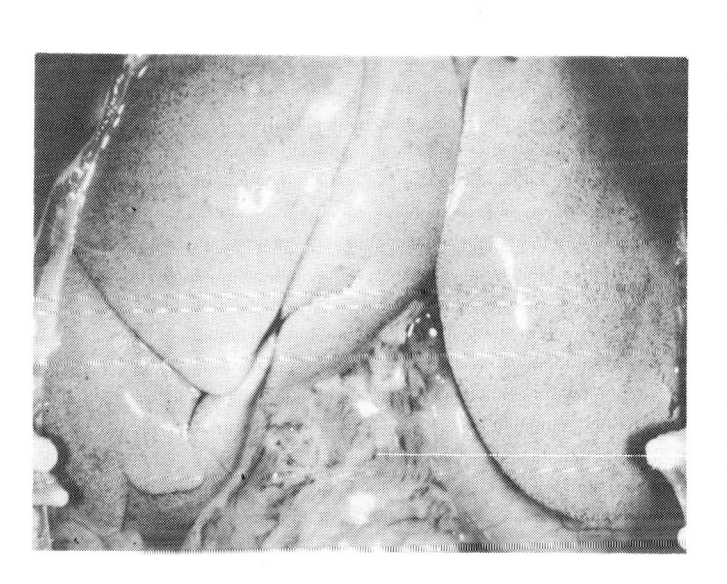

FIGURE 14. Gross appearance of liver from dog given toxic dose of AFB_1. The liver is swollen and fatty; the adjacent mesentery is markedly edematous

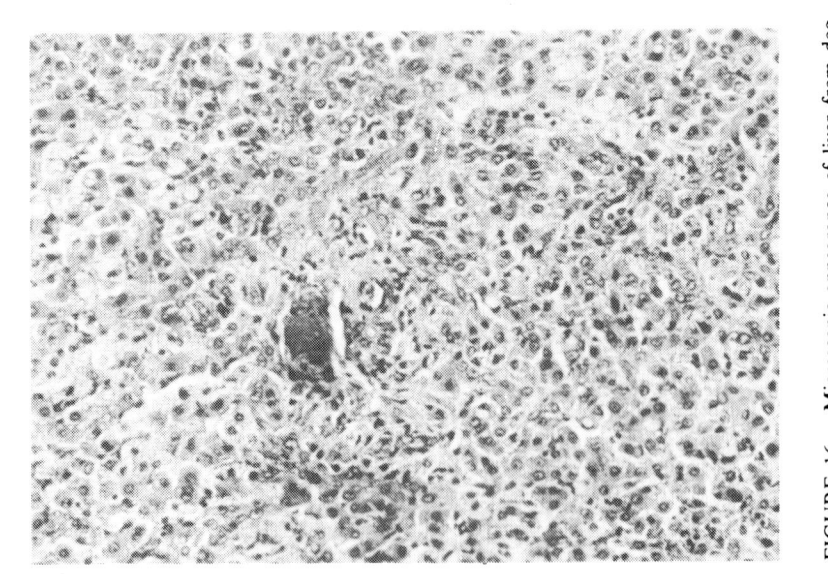

FIGURE 16. Microscopic appearance of liver from dog with subchronic poisoning with AFB$_1$. Bile stasis, bile duct hyperplasia, and regeneration of hepatocytes are characteristic of this less acute but more prolonged type of poisoning.

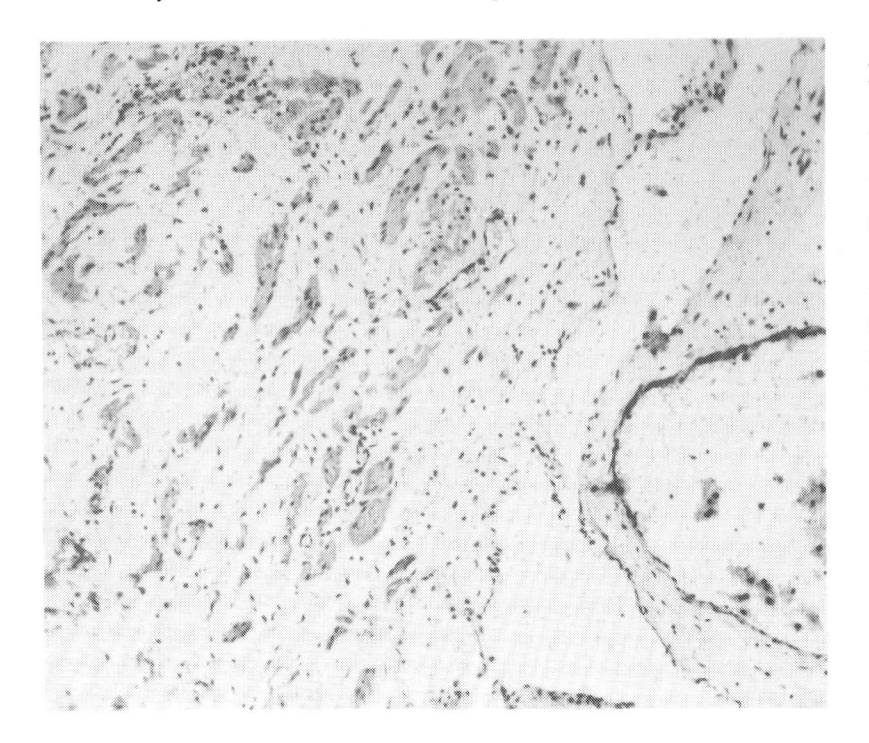

FIGURE 17. Severe edema of gall bladder from aflatoxin-poisoned dog. Note wide separation of muscle fibers by edema fluid.

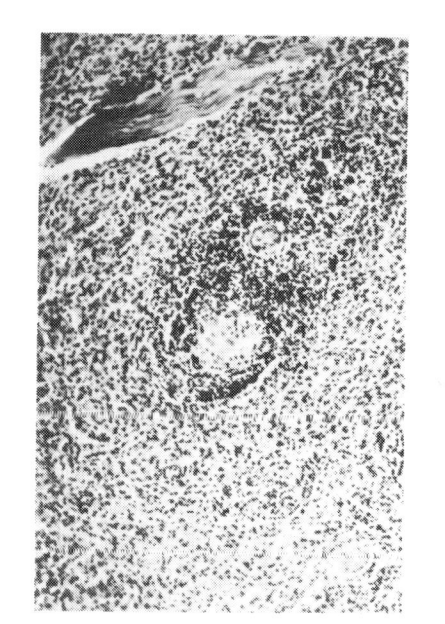

FIGURE 19. Follicle from spleen of dog exposed to aflatoxin; note depletion of lymphocytes.

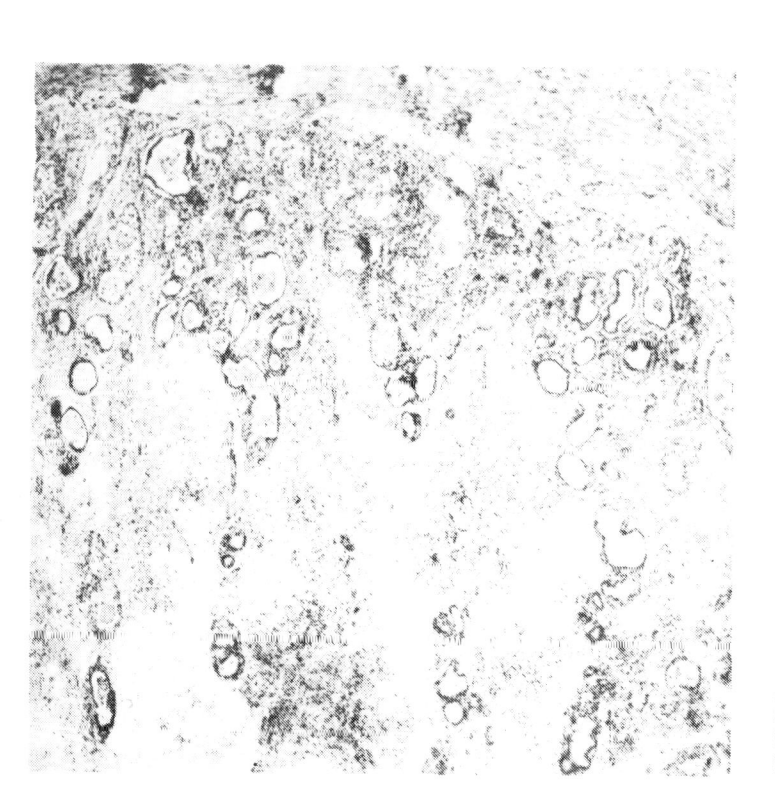

FIGURE 18. Edema and hemorrhage in mucosa of gall bladder from dog given acute dose of aflatoxin. Lesions such as this and the one shown in Figure 17 are seen in as short a time as 12 to 15 hr after a single LD_{50} of AFB_1.

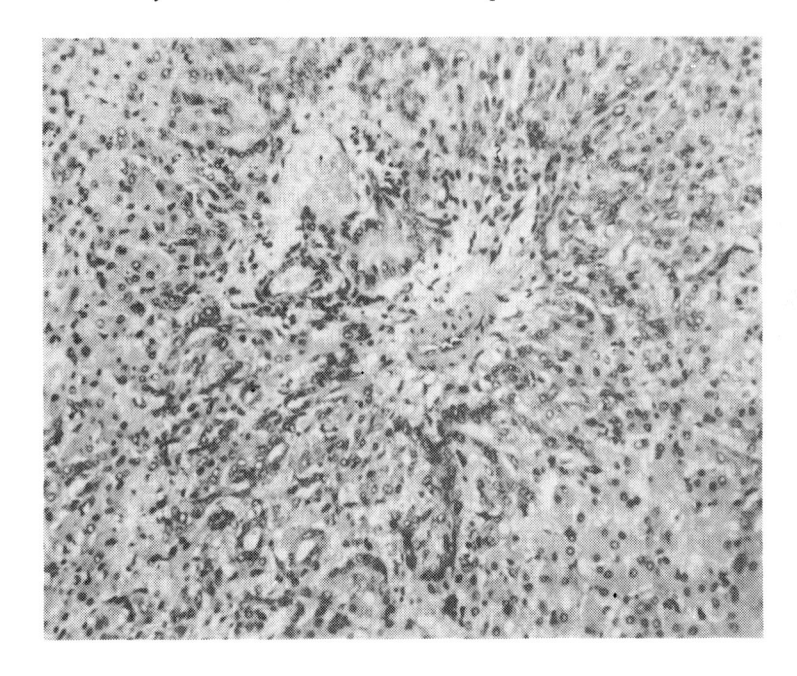

FIGURE 21. Guinea pig liver with marked bile duct hyperplasia following exposure to aflatoxin at subchronic concentration.

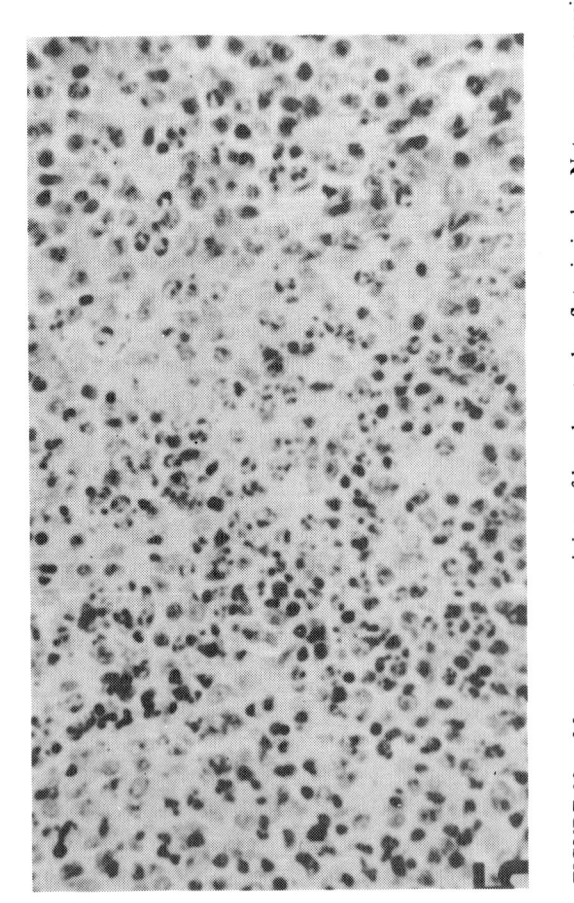

FIGURE 20. More severe, acute injury of lymphocytes by aflatoxin in dog. Note severe necrosis and lysis.

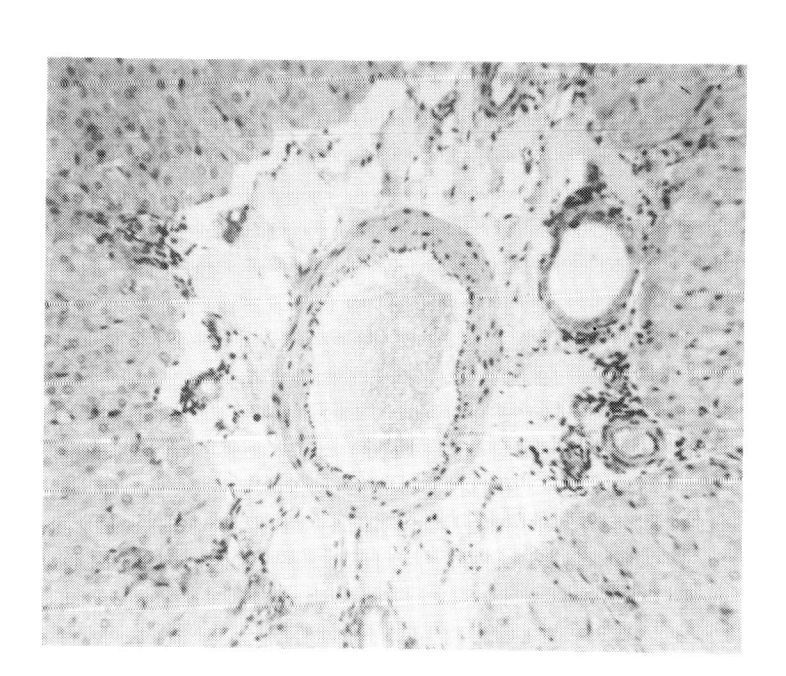

FIGURE 22. Dilatation of periportal lymphatics of guinea pig exposed to acute dose of aflatoxin.

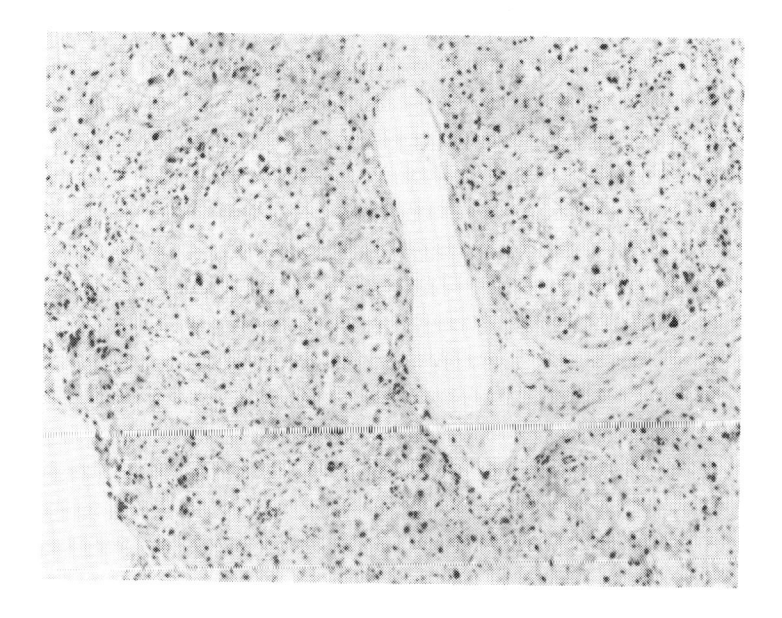

FIGURE 23. Necrosis, fibrosis, and fatty change in liver from monkey given nonlethal dose of AFB_1.

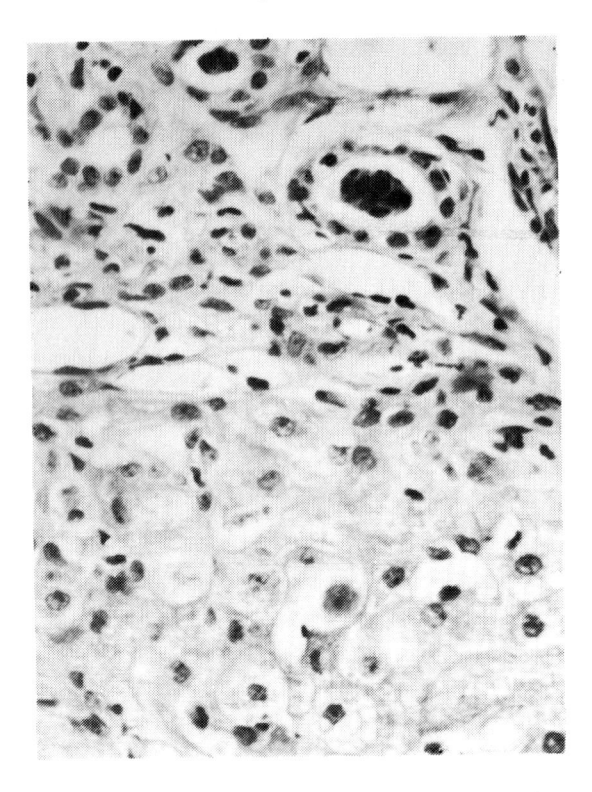

FIGURE 24. Bile duct stasis and pleomorphic nuclei in liver of aflatoxin-poisoned monkey. Bile ducts and ductules are occluded with bile, and there is mild fibrosis in the portal triad area.

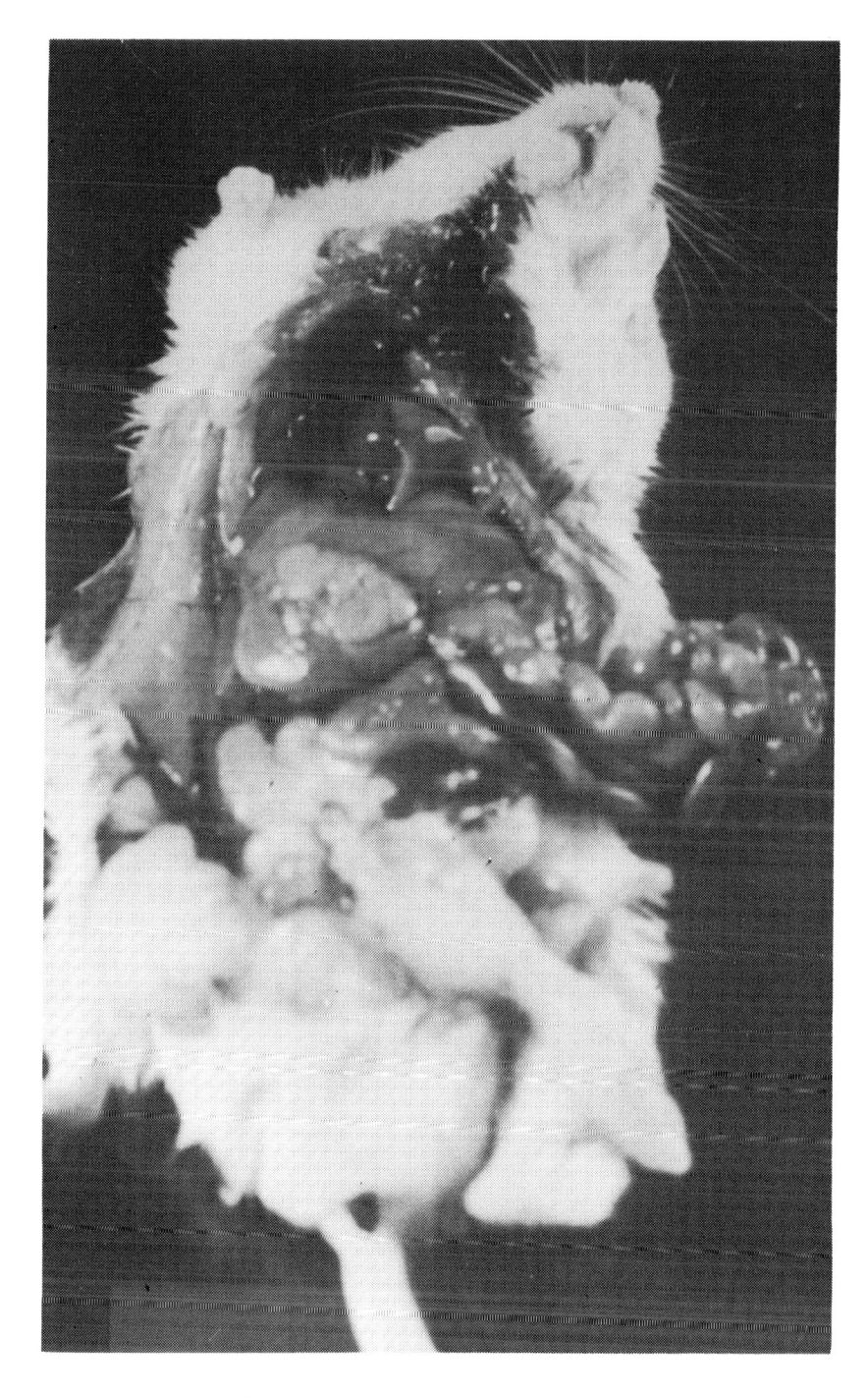

FIGURE 25. Tumor in mouse fed aflatoxin-contaminated diet. There was some question about the etiology and malignancy of tumors observed in this series of studies.

86 *Mycotoxins and N-Nitroso Compounds*

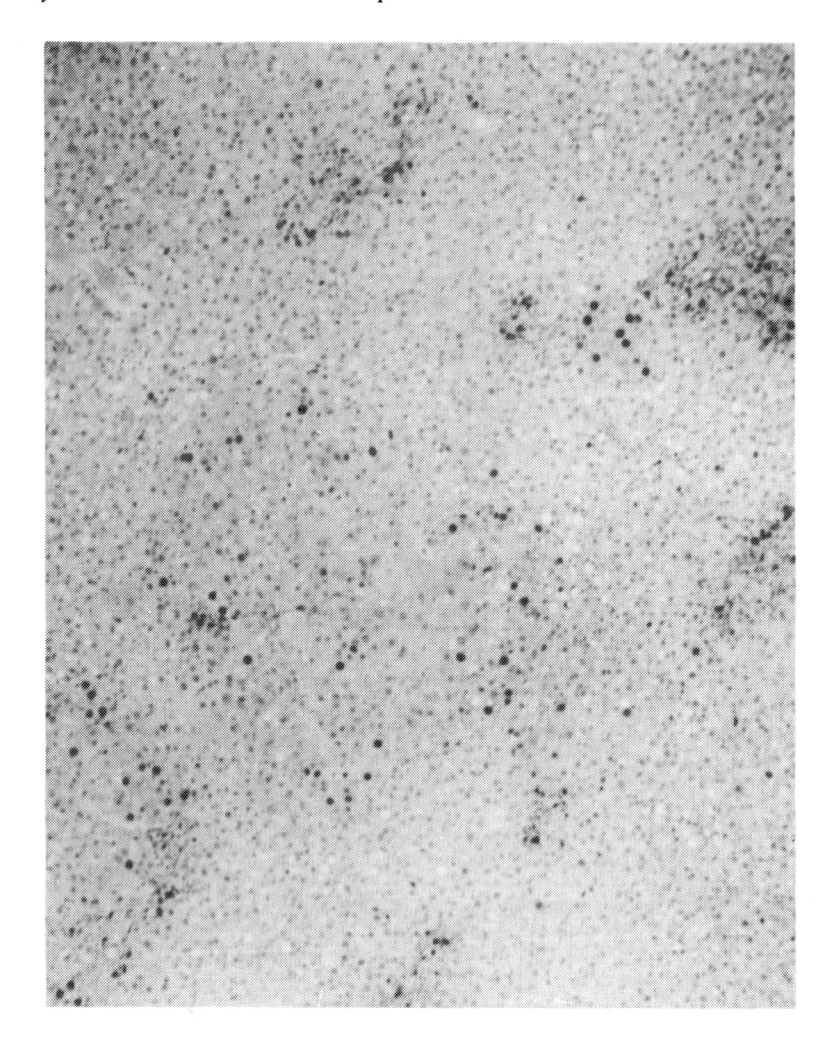

FIGURE 26. Liver from rat exposed to a carcinogenic regimen of AFB_1, injected with tritiated thymidine, and sacrificed for histologic preparations. Note labeling of early and more mature nodules, some of which would have developed into hepatocarcinoma.

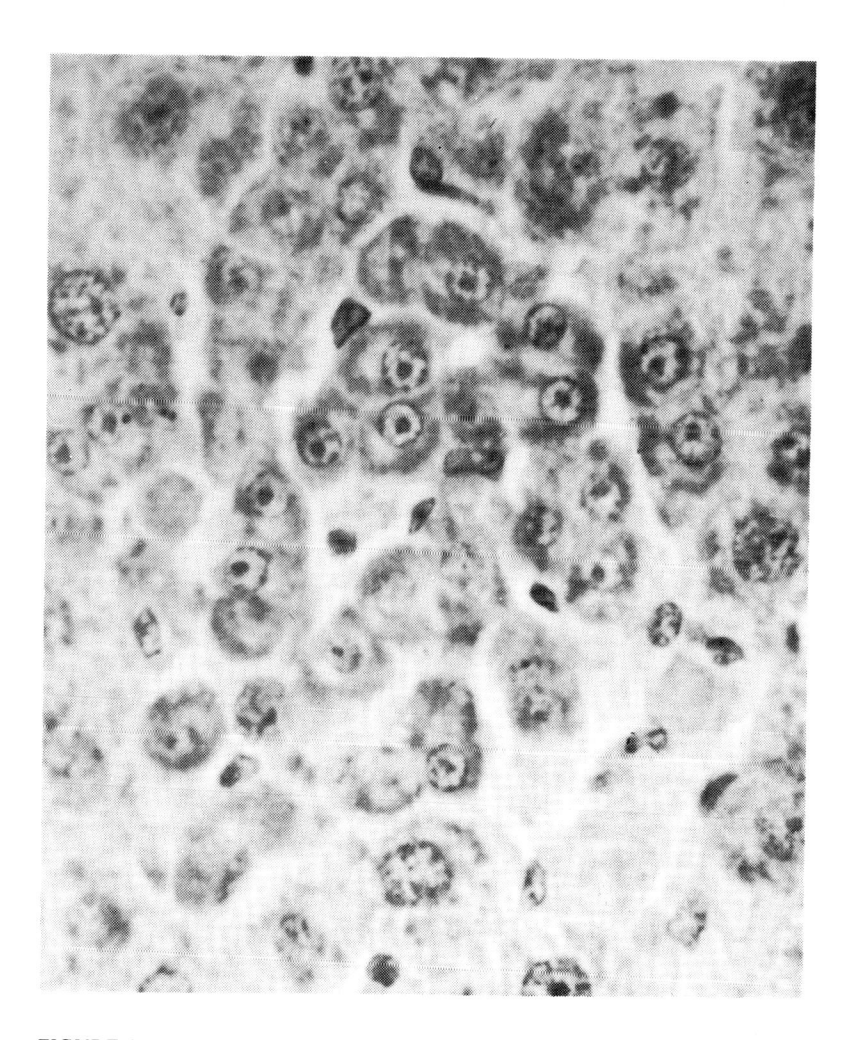

FIGURE 27. The earliest histologic change observed in the liver of a rat exposed to aflatoxin. Small clusters of hepatocytes that stain deeply with hematoxylin and label heavily with tritiated thymidine (see Figure 26) are observed in the lobule. They are precursors of nodules, some of which under favorable circumstances proceed to hepatocellular carcinoma.

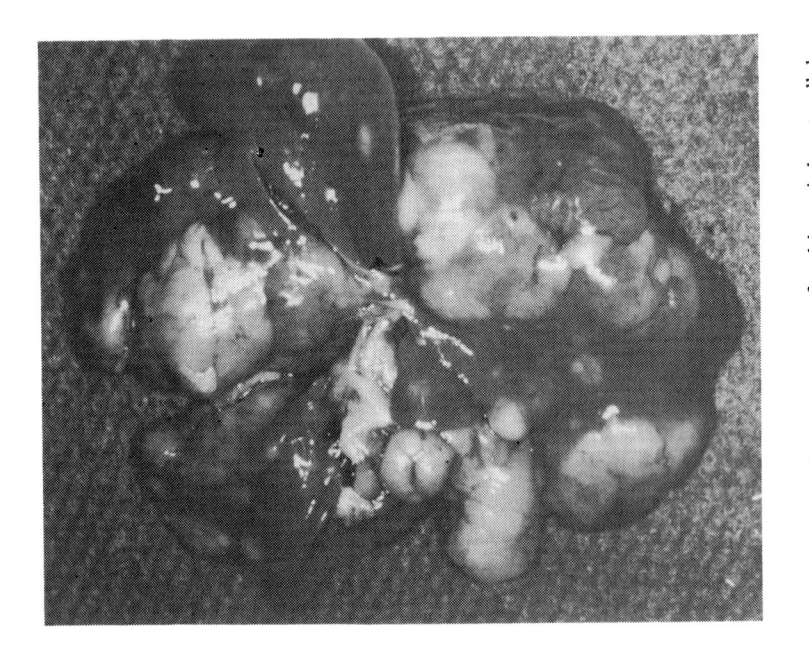

FIGURE 29. Microscopic appearance of liver cell carcinoma induced by aflatoxin. These nodules exhibit the trabecular form of tumor.

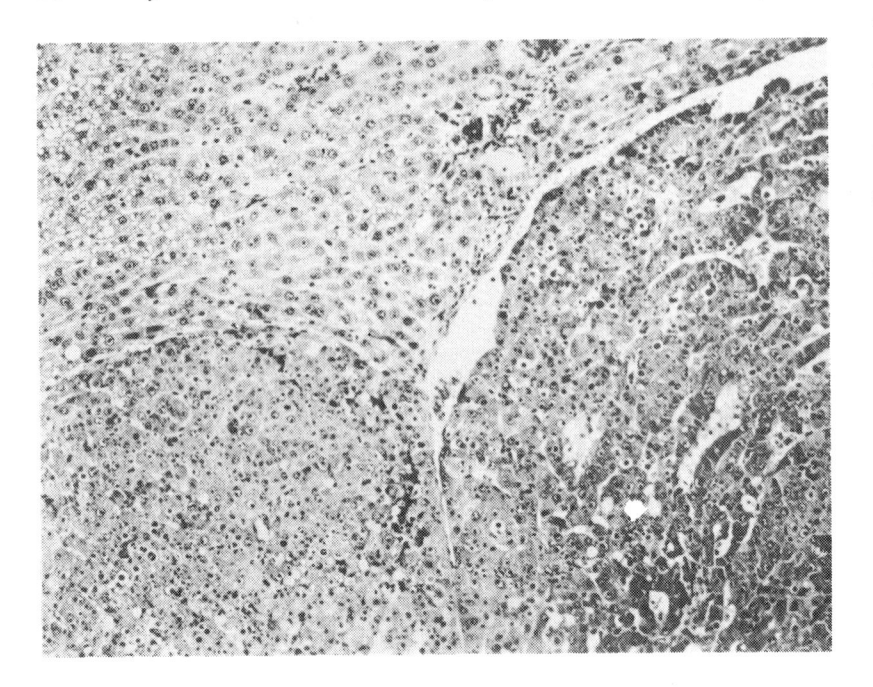

FIGURE 28. Gross appearance of multicentric hepatocellular carcinoma induced by aflatoxin.

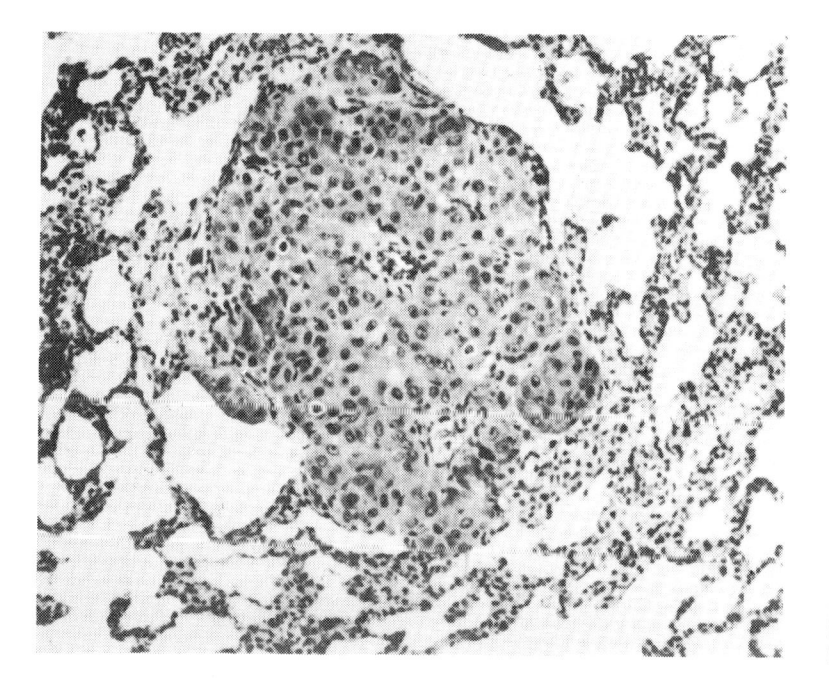

FIGURE 31. A metastatic lesion of the liver cell carcinoma shown in Figures 28 and 29. Such lung lesions are common with aflatoxin-induced tumors.

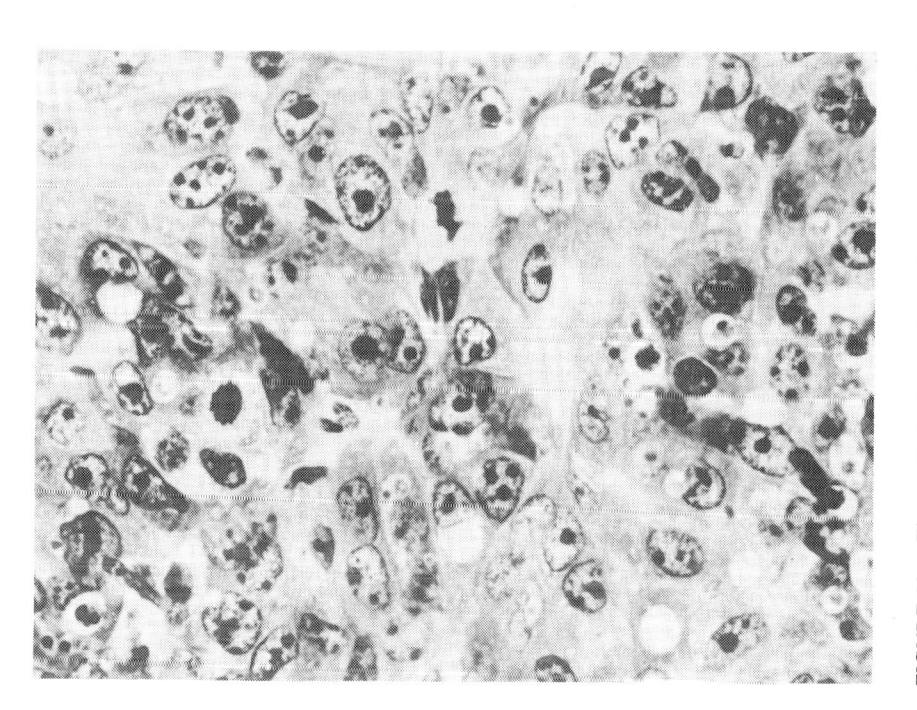

FIGURE 30. The anaplastic form of liver cell carcinoma induced by aflatoxin. Note mitotic figures.

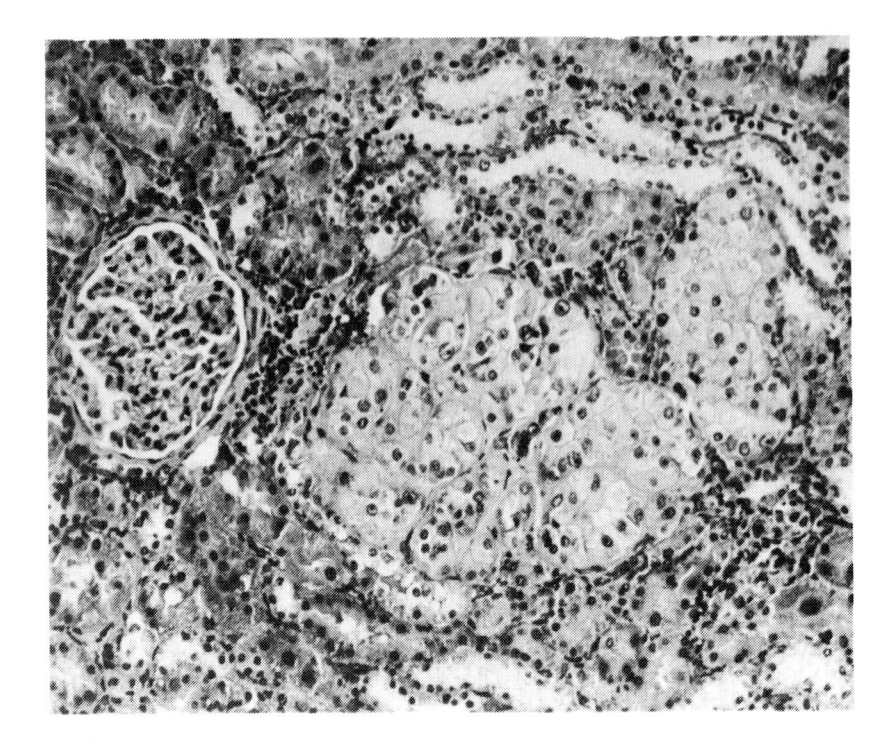

FIGURE 32. Renal adenoma, typical of tumors found in rats treated with aflatoxin. Aflatoxin G_1 (AFG_1) is apparently the most potent fraction for inducing renal tumors, some of which progress to carcinoma.

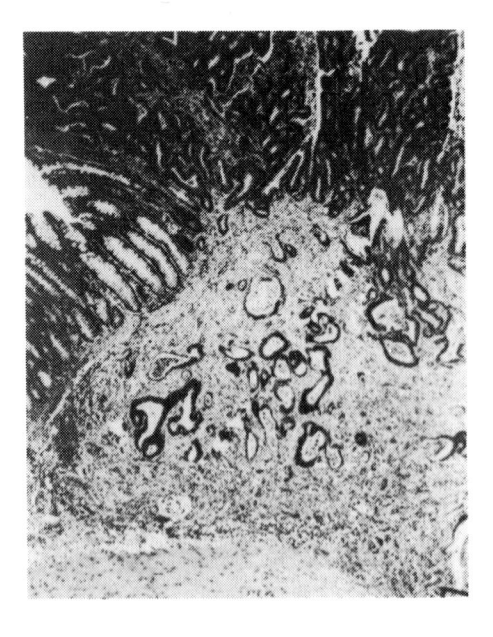

FIGURE 33. Colon carcinoma observed in rats fed diets contaminated with aflatoxin or in animals intubated with pure AFB_1. Note invasion of polypoid stalk by carcinoma.

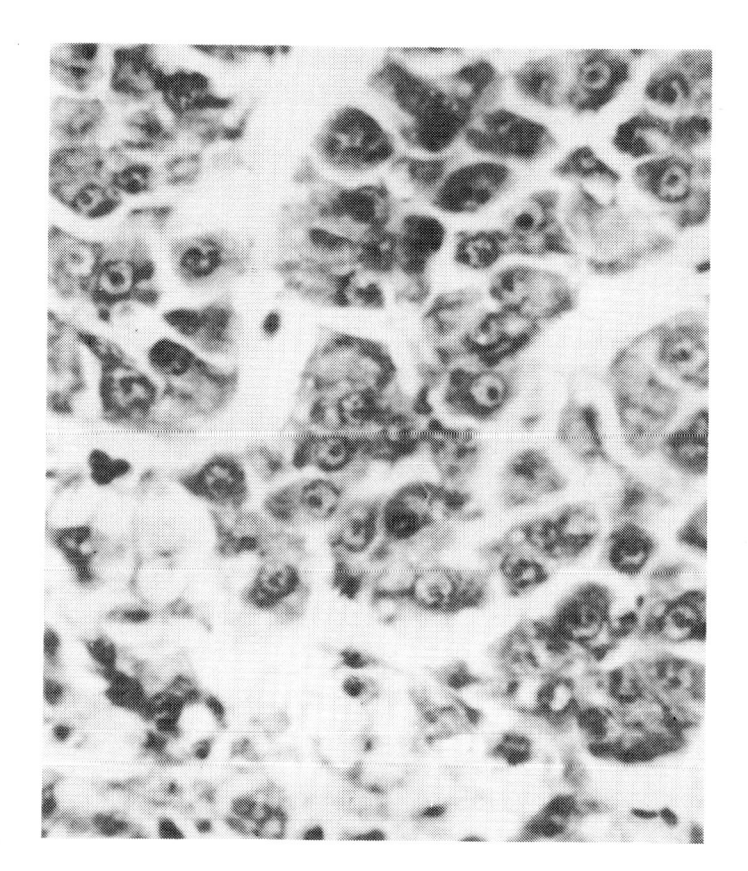

FIGURE 35. Focus of hepatocellular carcinoma from trout exposed to very low concentrations of aflatoxin (4 µg/kg diet).

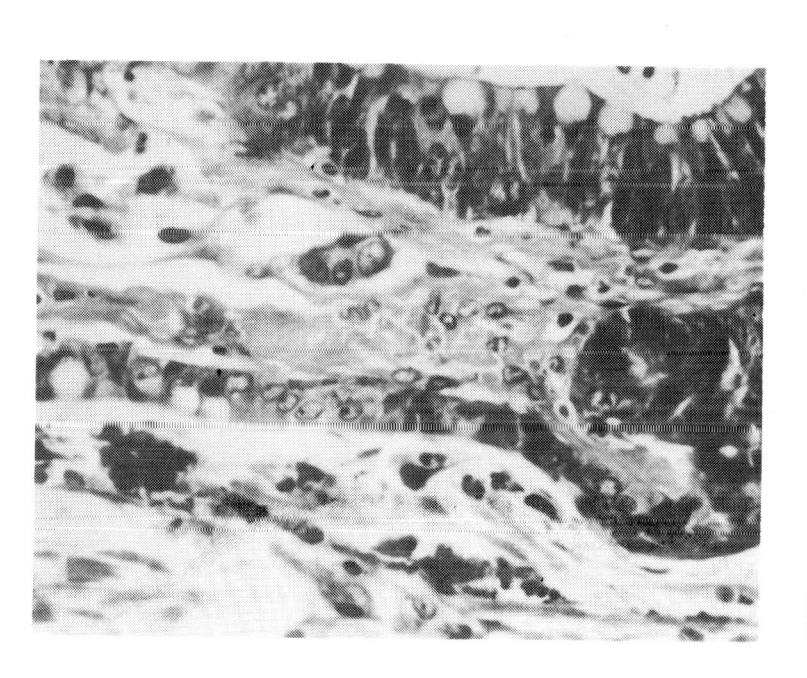

FIGURE 34. Higher magnification of colon carcinoma in aflatoxin-treated rat. Note invasion of subepithelial stroma by tumor cells.

92 *Mycotoxins and N-Nitroso Compounds*

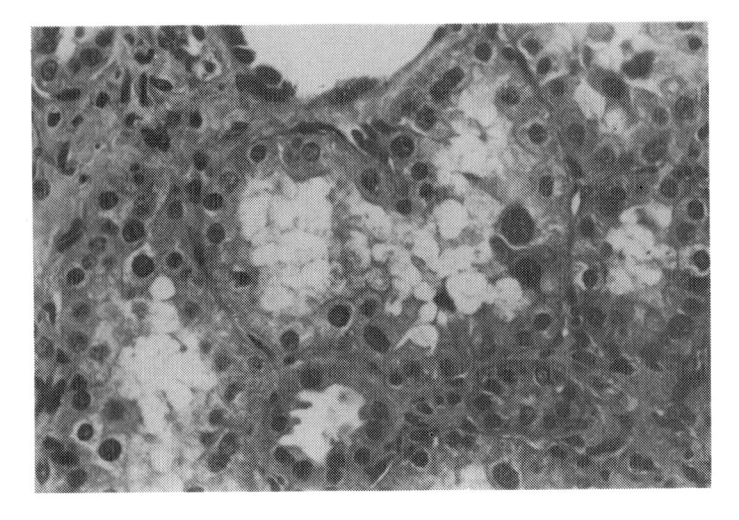

FIGURE 37. Dilated tubules and degeneration of tubular epithelium in kidney from pig exposed to ochratoxin. (From Krogh, P., Axelson, N. H., Elling, F., Gyrd-Hansen, N., Hald, B., Hyldgarrd-Jenson, J., Larsen, A. E., Madsen, A., Mortensen, H. P., Peterson, O. K., Ravnskow, U., Rostgaard, M., and Aulund, O., *Acta Pathol. Microbiol. Scand. Sect. A.* Suppl. No. 246, 1974.)

FIGURE 36. Kidneys from control pig and pigs exposed to ochratoxin. The top and middle kidneys were from poisoned pigs; the lower is a control. (From Krogh, P., Axelson, N. H., Elling, F., Gyrd-Hansen, N., Hald, B., Hyldgaard-Jenson, J., Larsen, A. E., Madsen, A., Mortensen, H. P., Peterson, O. K., Ravnskow, U., Rostgaard, M., and Aulund, O., *Acta Pathol. Microbiol. Scand. Sect. A.* Suppl. No. 246, 1974.)

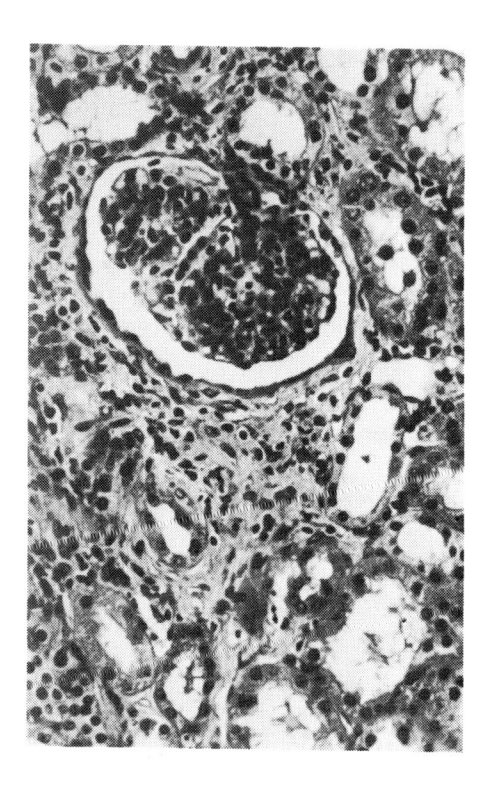

FIGURE 38. Fibrosis and sclerosis of interstitium, and atrophy of glomerulus in kidney from pig fed ochratoxin. (From Krogh, P., Axelson, N. H., Elling, F., Gyrd-Hansen, N., Hald, B., Hyldgaard-Jenson, J., Larsen, A. E., Madsen, A., Mortensen, H. P., Peterson, O. K., Ravnskow, U., Rostgaard, M., and Aulund, O., *Acta Pathol. Microbiol. Scand. Sect. A*, Supplementum No. 246, 1974.)

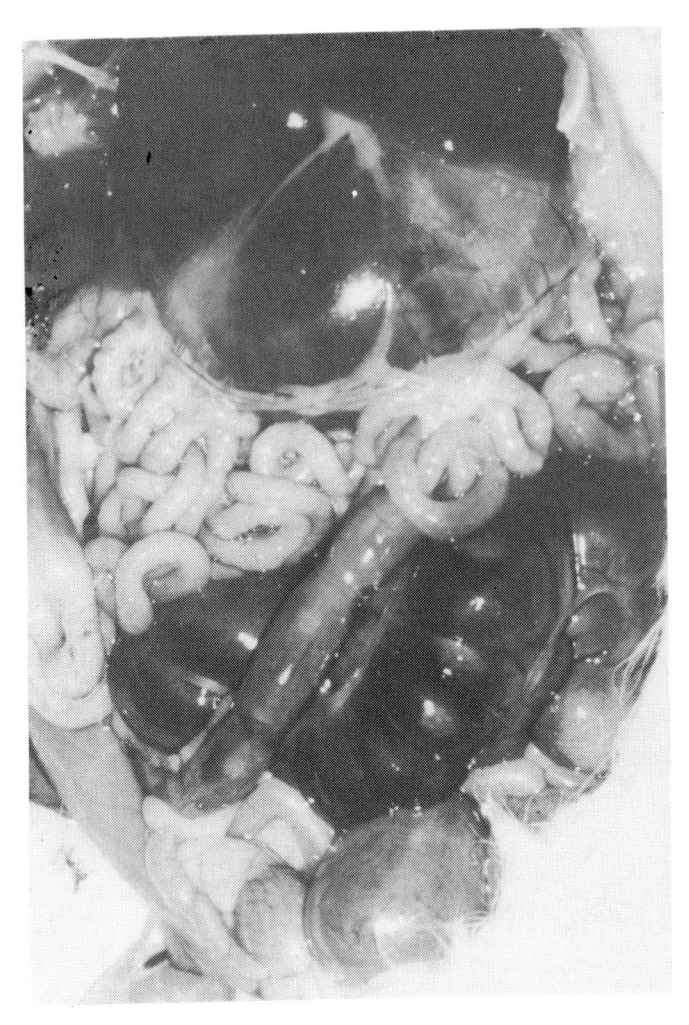

FIGURE 39. Viscera from guinea pig exposed to T-2 toxin, 5 mg/kg, several hours after exposure. Note stomach and cecal hemorrhages and edema. (Courtesy of Dr. W. W. Carlton.)

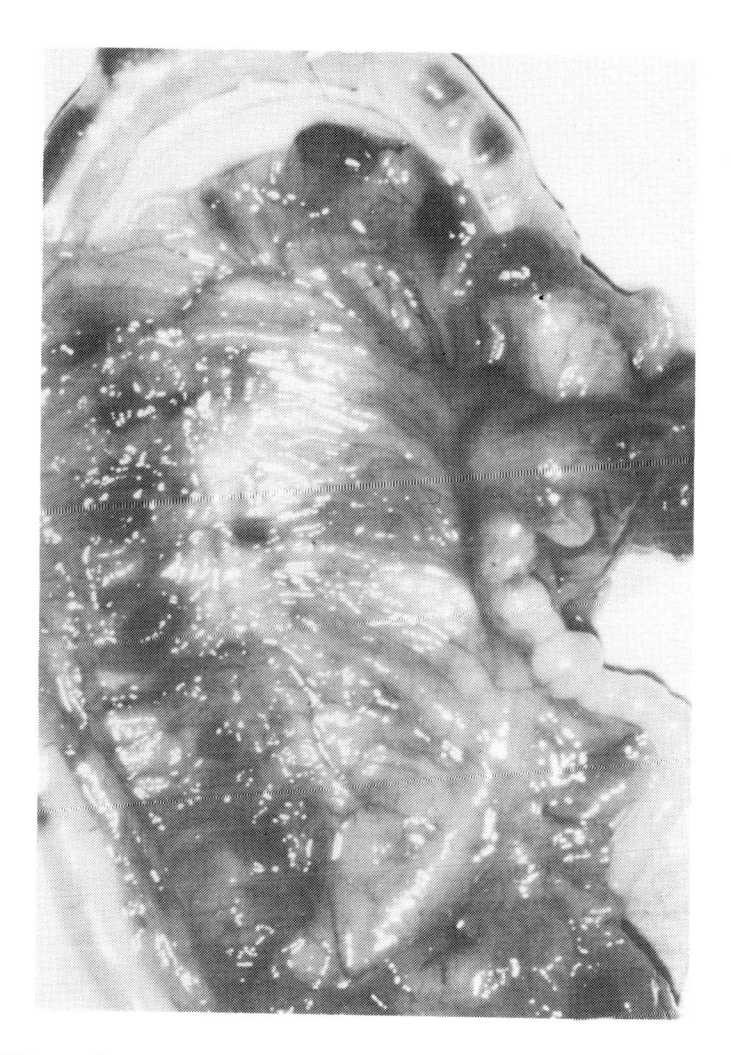

FIGURE 40. Cecal mucosal hemorrhages and edema in guinea pig exposed to T-2 toxin 17 hr prior to sacrifice. Peyers patches were hemorrhagic also. (Courtesy of Dr. W. W. Carlton.)

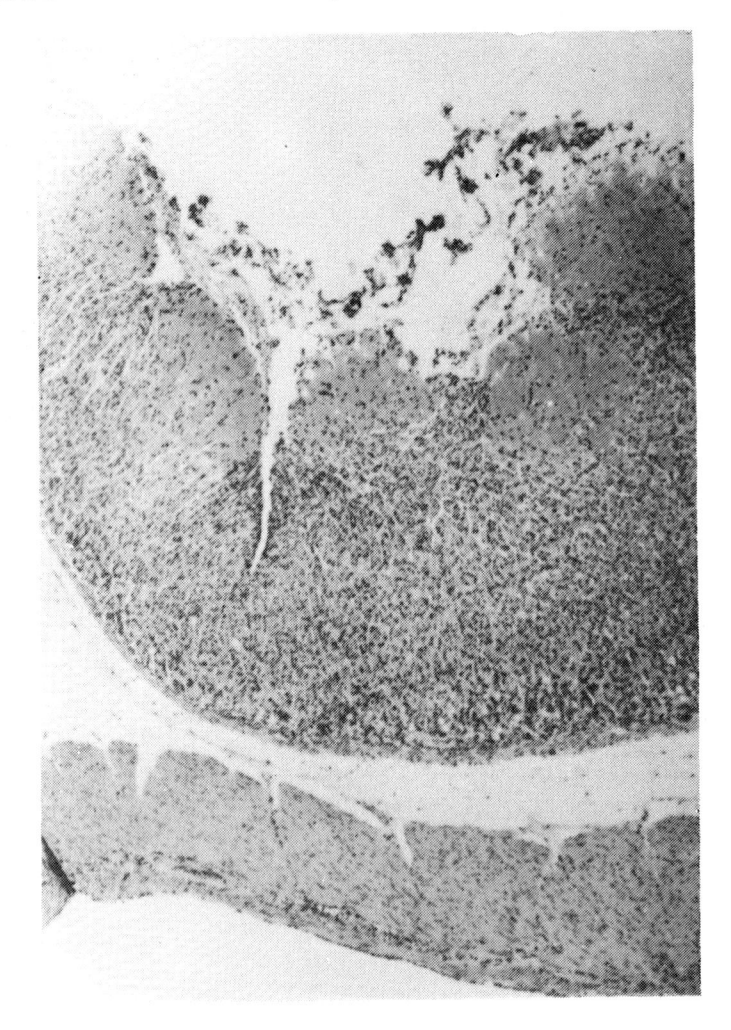

FIGURE 41. Hemorrhagic necrosis, fundus of stomach of guinea pig exposed to T-2 toxin 17 hr prior to sacrifice. (Courtesy of Dr. W. W. Carlton.)

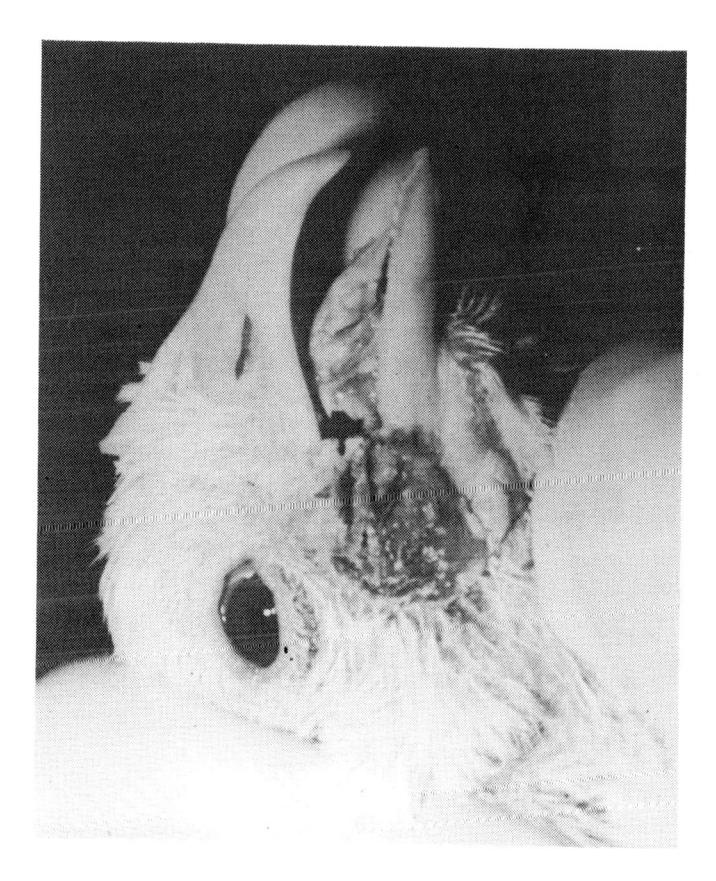

FIGURE 43. Severe oral lesions in chicken, caused by indigestion of T-2 toxin. (Courtesy of Dr. W. W. Carlton.)

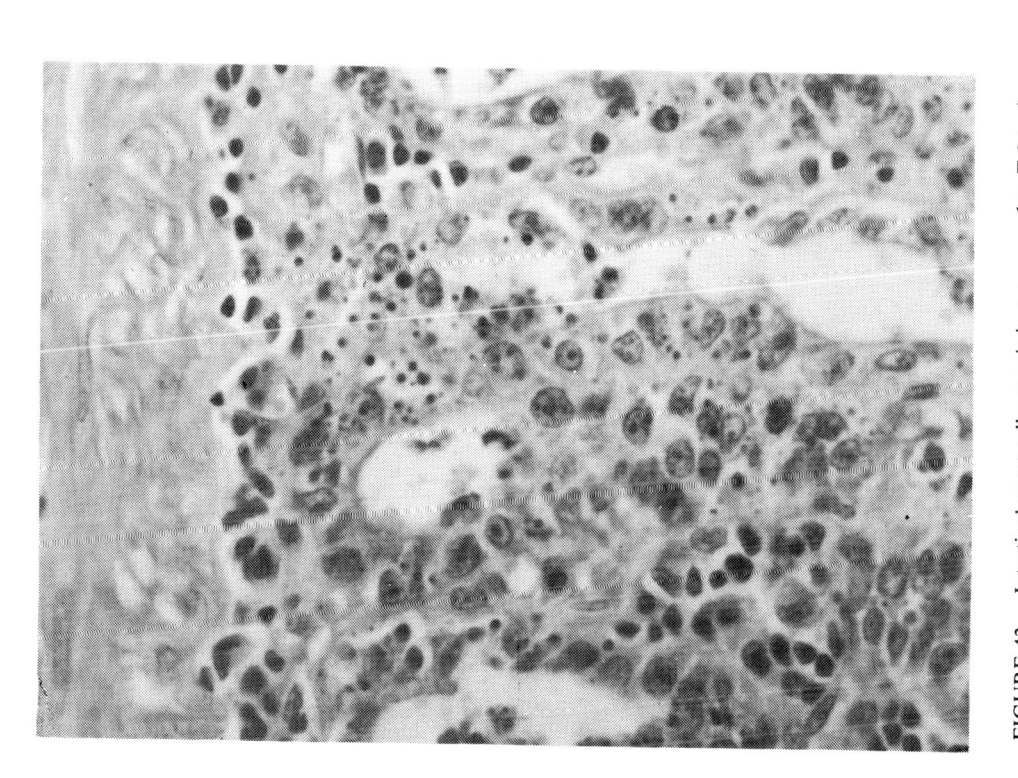

FIGURE 42. Intestinal crypt cell necrosis in cat exposed to T-2 toxin.

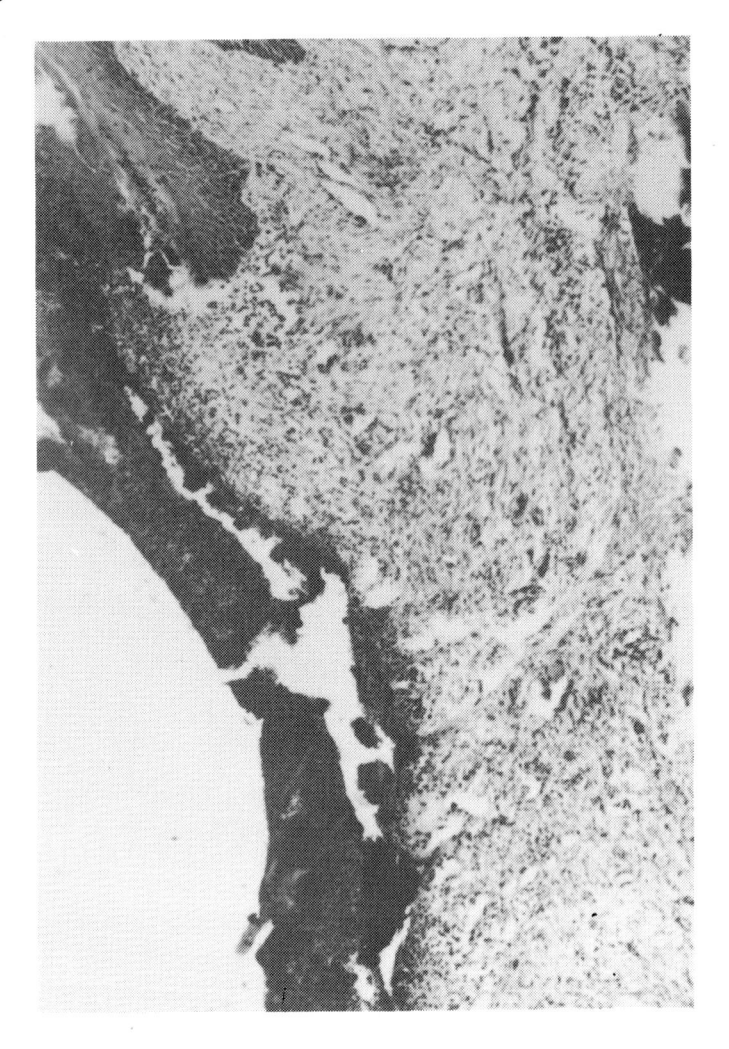

FIGURE 44. Microscopic appearance of oral cavity of bird shown in Figure 43. Severe necrosis and hemorrhage are prominent. (Courtesy of Dr. W. W. Carlton.)

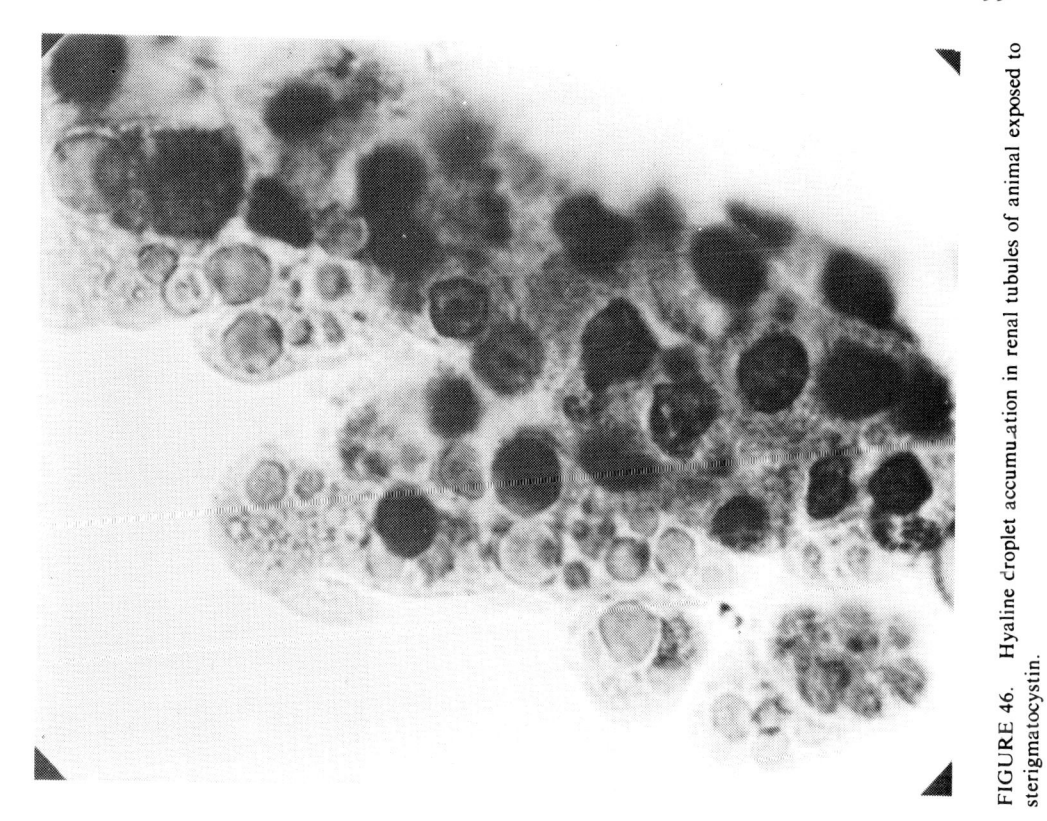

FIGURE 46. Hyaline droplet accumuation in renal tubules of animal exposed to sterigmatocystin.

FIGURE 45. Focal, hemorrhagic necrosis of liver.

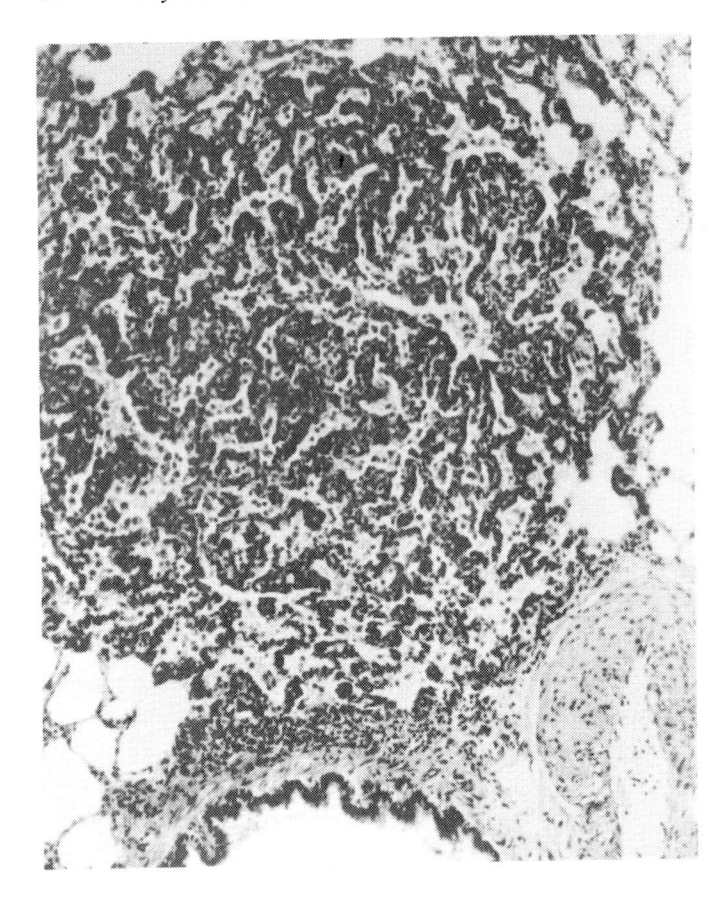

FIGURE 48. Pulmonary adenomas in mouse fed pure sterigmatocystin.

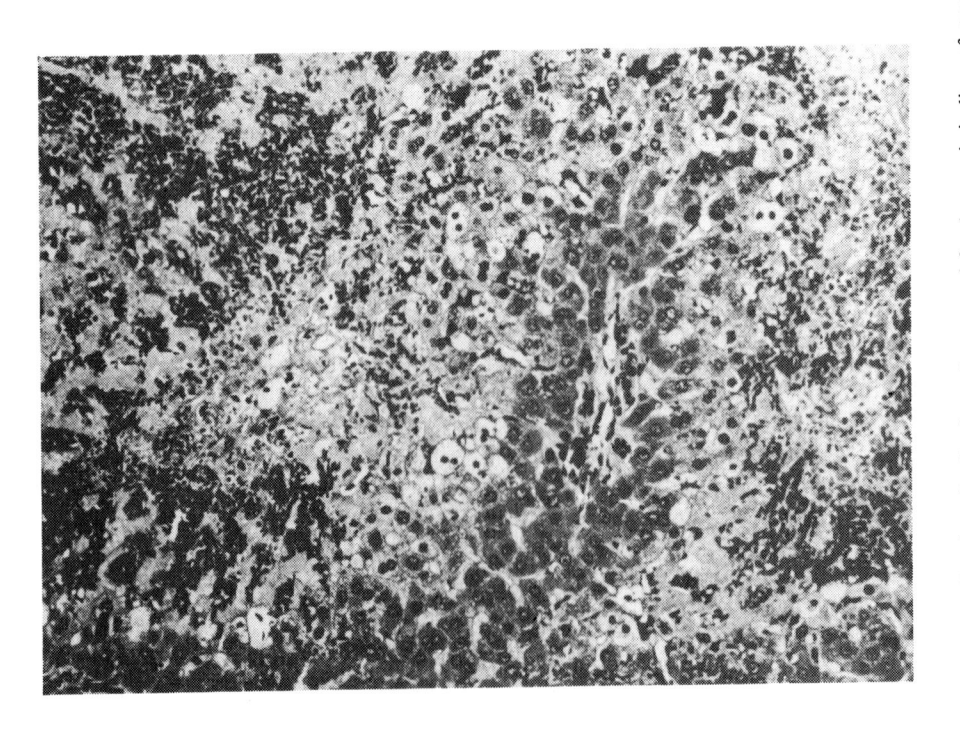

FIGURE 47. Ballooning of cells, fatty change, and focal necrosis in liver of monkey given sterigmatocystin.

REFERENCES

1. **Newberne, P. M. and Butler, W. H.**, Acute and chronic effects of aflatoxin on the liver of domestic and laboratory animals: a review, *Cancer Res.*, 29, 236, 1969.
2. **Butler, W. H.**, Aflatoxin, in *Mycotoxins*, Purchase, I. F. H., Ed., Elsevier, Amsterdam, 1974, 1.
3. **Wogan, G. N.**, Aflatoxin carcinogenesis, in *Methods in Cancer Research*, Bush, M., Ed., Academic Press, New York, 1973, 309.
4. **Bourgeois, C. H., Shank, R. C., Grossman, R. A., Johnsen, D. O., Wooding, W. L., and Chandavimol, P.**, Acute aflatoxin B_1 toxicity in the Macaque and similarities to Reye's Syndrome, *Lab. Invest.*, 24, 206, 1971.
5. **van der Watt, J. J. and Purchase, I. F. H.**, The acute toxicity of retrorsine, aflatoxin and sterigmatocystin in vervet monkeys, *Br. J. Exp. Pathol.*, 51, 183, 1970.
6. **Bababunmi, E. A. and Bassir, O.**, Species differences in the anticoagulant activities of aflatoxin B_1 and 4-hydroxycoumarin, *Afr. J. Med. Sci.*, 3, 97, 1972.
7. **Wogan, G. N., Edwards, G. S., and Newberne, P. M.**, Structure-activity relationships in toxicity and carcinogenicity of aflatoxins and analogs, *Cancer Res.*, 31, 1936, 1971.
8. **Butler, W. H.**, Acute toxicity of aflatoxin B_1 in rats, *Br. J. Cancer*, 18, 756, 1964.
9. **Wogan, G. N.**, Chemical nature and biological effects of the aflatoxins, *Bacteriol. Rev.*, 30, 460, 1966.
10. **Butler, W. H.**, Acute toxicity of aflatoxin B_1 in guinea pigs, *J. Pathol. Bacteriol.*, 91, 277, 1966.
11. **Rao, K. S. and Gehring, P. J.**, Acute toxicity of aflatoxin B_1 in monkeys, *Toxicol. Appl. Pharmacol.*, 19, 169, 1971.
12. **Shank, R. C., Johnsen, D. O., Tanticharocnyoz, P., Wooding, W. L., and Bourgeois, C. H.**, Acute toxicity of aflatoxin B_1 in the macaque monkey, *Toxicol. Appl. Pharmacol.*, 20, 277, 1971.
13. **Peers, F. G. and Linsell, C. A.**, Acute toxicity of aflatoxin B_1 for baboons, *Food Cosmet. Toxicol.*, 14, 227, 1976.
14. **Bauer, D. H., Lee, D. J., and Sinnhuber, R. D.**, Acute toxicity of aflatoxins B_1 and G_1 in the rainbow trout (*Salmo girdneri*), *Toxicol. Appl. Pharmacol.*, 15, 415, 1969.
15. **Allcroft, R. and Lewis, G.**, Groundnut toxicity in cattle: experimental poisoning of calves and a report on clinical effects in older cattle, *Vet. Rec.*, 75, 489, 1963.
16. **Newberne, P. M.**, Carcinogenicity of aflatoxin-contaminated peanut meals, in *Mycotoxins in Foodstuffs*, Wogan, G. N., Ed., MIT Press, Cambridge, Mass., 1965, 187.
17. **Carnaghan, R. B. A., Lewis, G., Patterson, D. S. P., and Allcroft, R.**, Biochemical and pathological aspects of groundnut poisoning in chickens, *Pathol. Vet.*, 3, 601, 1966.
18. **Cuthbertson, W. F. J., Laursen, A. C., and Pratt, D. A. H.**, Effect of groundnut meal containing aflatoxin on Cynomolgus monkeys, *Br. J. Nutr.*, 21, 893, 1967.
19. **Krogh, P., Hald, B., Hasselager, E., Madsen, A., Mortensen, H. P., Larsen, A. E., and Campbell, A. D.**, Aflatoxin residues in bacon pigs, *Pure Appl. Chem.*, 35, 275, 1973.
20. **Madhavan, T. V., Tulpule, P. G., and Gopalan, C.**, Aflatoxin-induced hepatic fibrosis in rhesus monkeys, *Arch. Pathol.*, 79, 466, 1965.
21. **Magwood, W. E., Annau, E., and Corner, A. H.**, Induced tolerance in turkeys to aflatoxin poisoning, *Can. J. Comp. Med. Vet. Sci.*, 30, 17, 1966.
22. **Wilson, B. J., Teer, P. A., Barney, G. H., and Blood, F. R.**, Relationship of aflatoxin to epizootics of toxic hepatitis among animals in southern United States, *Am. J. Vet. Res.*, 283, 1217, 1967.
23. **Carnaghan, R. B. A., Hartley, R. D., and O'Kelly, J.**, Toxcity and fluoresence properties of the aflatoxins, *Nature (London)*, 200, 1101, 1963.
24. **Newberne, P. M., Carlton, W. W., and Wogan, G. N.**, Hepatomas in rats and hepatorenal injury in ducklings fed peanut meal or *Aspergillus flavus* extract, *Pathol. Vet.*, 1, 105, 1964.
25. **Gumbmann, M. R., Williams, S. N., Booth, A. N., Vohra, P., Ernst, R. A., and Bethard, M.**, Aflatoxin susceptibility in various breeds of poultry, *Proc. Soc. Exp. Biol. Med.*, 134, 683, 1970.
26. **Rogers, A. E. and Newberne, P. M.**, Diet and aflatoxin B_1 toxicity in rats, *Toxicol. Appl. Pharmacol.*, 20, 113, 1971.
27. **Rogers, A. E., Kula, N. S., and Newberne, P. M.**, Absence of an effect of partial hepatectomy on aflatoxin B_1 carcinogenesis, *Cancer Res.*, 31, 491, 1971.
28. **Tilak, T. B. G. and Krishnamurthi, D.**, Liver regeneration following partial hepatectomy, *Arch. Pathol.*, 96, 18, 1973.
29. **Newberne, P. M., Russo, R., and Wogan, G. N.**, Acute toxicity of aflatoxin B_1 in the dog, *Pathol. Vet.*, 3, 331, 1966.
30. **Tupule, P. G., Madhavan, T. V., and Gopalan, C.**, Effect of feeding aflatoxin to young monkeys, *Lancet*, 1, 962, 1964.
31. **Newberne, P. M.**, Chronic aflatoxicosis, *J. Am. Vet. Med. Assoc.*, 163, 1262, 1973.
32. **Armbrecht, B. H., Geleta, J. N., Shalkop, W. T., and Durbin, C. G.**, A subacute exposure of beagle dogs to aflatoxin, *Toxicol. Appl. Pharmacol.*, 18, 579, 1971.

33. **Allcroft, R.,** Aflatoxicosis in farm animals, in *Aflatoxin — Scientific Background, Control and Implications,* Goldblatt, L. A., Ed., Academic Press, New York, 1969, 237.
34. **Patterson, D. S. P.,** Metabolism as a factor in determining the toxic action of the aflatoxin in different animal species, *Food Cosmet. Toxicol.,* 11, 287, 1973.
35. **Campbell, T. C. and Hayes, J. R.,** The role of aflatoxin metabolism in its toxic lesion, *Toxicol. Appl. Pharmacol.,* 35, 199, 1976.
36. **Rogers, A. E.,** Variable effects of a lipotrope-deficient, high fat diet on chemical carcinogenesis in rats, *Cancer Res.,* 35, 2469, 1975.
37. **Smith, J. W., Hill, C. H., and Hamilton, P. B.,** The effect of dietary modifications on aflatoxicosis in broiler chicken, *Poultry Sci.,* 50, 768, 1971.
38. **Newberne, P. M. and Conner, M. W.,** Effect of selenium on acute response to aflatoxin B_1, in *Trace Substances in Environmental Health — VIII. A Symposium,* Hemphill, D. D., Ed., University of Missouri, Columbia, 1974, 323.
39. **Newberne, P. M., Chan, W. -Ch. M., and Rogers, A. E.,** Influence of light, riboflavin, and carotene on the response of rats to the acute toxicity of aflatoxin and monocrotaline, *Toxicol. Appl. Pharmacol.,* 28, 200, 1974.
40. **Reddy, G. S., Tilak, T. B. G., Krishnamurthi, D.,** Susceptibility of vitamin A-deficient rats to aflatoxin, *Food Cosmet. Toxicol.,* 11, 467, 1973.
41. **Roebuck, B. D. and Wogan, G. N.,** Species comparison *in vitro* metabolism of aflatoxin B_1, *Cancer Res.,* 37, 1649, 1977.
42. **Ueno, Y., Kubota, K., Ito, T., and Nakamura, Y.,** Mutagenicity of carcinogenic mycotoxins in *Salmonella typhimurium, Cancer Res.,* 38, 536, 1978.
43. **Suit, J. L., Rogers, A. E., Jetten, M. E. R., and Luria, S. E.,** Effects of diet on conversion of aflatoxin B_1 to bacterial mutagen(s) by rats in vivo and by rat hepatic microsomes in vitro, *Mutat. Res.,* 46, 313, 1977.
44. **Elis, J. and Di Paolo, J. A.,** Aflatoxin B_1, *Arch. Pathol.,* 83, 53, 1967.
45. **Hendricks, J. D., Sinnhuber, R. O., Nixon, J. E., Wales, J. H., Putnam, G. B., Loveland, P. M., Masri, M. S., and Hsieh, D. P.,** Carcinogenicity of aflatoxin Q_1 to rainbow trout and its potentiation by cyclopropene fatty acids, *Fed. Proc. Fed. Am. Soc. Exp. Biol.,* 37, 451, 1978.
46. **Ayres, J. L., Lee, D. J., Wales, J. H., and Sinnhuber, R. O.,** Aflatoxin structure and hepatocarcinogenicity in rainbow trout (*Salmo gairdneri*), *J. Natl. Cancer Inst.,* 46, 561, 1971.
47. **Wogan, G. N. and Paglialunga, S.** Carcinogenicity of synthetic aflatoxin M, in rats, *Food Cosmet. Toxicol.,* 12, 381, 1974.
48. **Sinnhuber, R. O., Wales, J. H., Ayres, J. L., Engebrecht, R. H., and Amed, D. L.,** Dietary factors and hepatoma in rainbow trout (*Salmo gairdneri*). I. Aflatoxins in vegetable protein feedstuffs, *J. Natl. Cancer Inst.,* 41, 711, 1968.
49. **Wogan, G. N., Paglialunga, S., and Newberne, P. M.,** Carcinogenic effects of low dietary levels of aflatoxin B_1 in rats, *Food Cosmet. Toxicol.,* 12, 681, 1974.
50. Food and Drug Administration, Bureau of Foods, Assessment of Estimated Risk Resulting from Aflatoxins in Consumer Peanut Products and Other Food Commodities, Food and Drug Administration, Washington, D.C., January 19, 1978.
51. **Vesselinovitch, S. D., Milhailovich, N., Wogan, G. N., Lombard, L. S., and Rao, K. V. N.,** Aflatoxin B_1, a hepatocarcinogen in the infant mouse, *Cancer Res.,* 32, 2289, 1972.
52. **Halver, J. E.,** Crystalline aflatoxin and other vectors for trout hepatoma, in Trout Hepatoma Research Conference Papers, Research Report 70, Haver, J. E. and Mitchell, I. A., Eds., Department of Health, Education and Welfare, Public Health Service and the Department of the Interior, Fish and Wildlife Service, Bureau of Sport Fisheries and Wildlife, 1967.
53. **Wales, J. H., Sinnhuber, R. O., Hendricks, J. D., Nixon, J. E., and Eisele, T. A.,** Aflatoxin B_1 induction of hepatocellular carcinoma in the embryos of rainbow trout (*Salmo gairdneri*), *J. Natl. Cancer Inst.,* 60, 1133, 1978.
54. **Newberne, P. M. and Rogers, A. E.,** Rat colon carcinomas associated with aflatoxin and marginal vitamin A., *J. Natl. Cancer Inst.,* 50, 439, 1973.
55. **Nixon, J. E., Sinnhuber, R. O., Lee, D. J. Landers, M. K., and Harr, J. R.,** Effect of cyclopropenoid compounds on the carinogenic activity of diethylnitrosamine and aflatoxin B_1 in rats, *J. Natl. Cancer Inst.,* 53, 453, 1974.
56. **Epstein, S. M., Bartus, B., and Farber, E.,** Renal epithelial neoplasms induced in male Wistar rats by oral aflatoxin B_1, *Cancer Res.,* 29, 1045, 1969.
57. **Merkow, L. P., Epstein, S. M., Slifkin, M., and Pardo, M.,** The ultrastructure of renal neoplasms induced by aflatoxin B_1, *Cancer Res.,* 33, 1608, 1973.
58. **Butler, W. H. and Barnes, J. M.,** Carcinogenic action of groundnut meal containing aflatoxin in rats, *Food Cosmet. Toxicol.,* 6, 135, 1968.

59. **Rogers, A. E.,** Dietary effects on chemical carcinogenesis in the livers of rats, in *Rat Hepatic Neoplasia,* Newberne, P. M. and Butler, W. H., Eds., MIT Press, Cambridge, Mass., 1978, chap. 11.
60. **Carnaghan, R. B. A.,** Hepatic tumors and other chronic liver changes in rats following a single oral administration of aflatoxin, *Br. J. Cancer,* 21, 811, 1967.
61. **Wogan, G. N. and Newberne, P. M.,** Dose-response characteristics of aflatoxin B_1 carcinogenesis in the rat, *Cancer Res.,* 28, 770, 1968.
62. **Newberne, P. M. and Wogan, G. N.,** Sequential morphologic changes in aflatoxin B_1 carcinogenesis in the rat, *Cancer Res.,* 28, 770, 1968.
63. **Rogers, A. E. and Newberne, P. M.,** Aflatoxin B_1 carcinogenesis in lipotrope-deficient rats, *Cancer Res.,* 29, 1965, 1969.
64. **Kalengayi, M. M. R., Ronchi, G., and Desment, V. J.,** Histochemistry of gamma-glutamyl transpeptidase in rat liver during aflatoxin B_1-induced carcinogenesis, *J. Natl. Cancer Inst.,* 55, 579, 1975.
65. **Novi, A. M.,** Liver carcinogenesis in rats after aflatoxin B_1 administration, *Current Topics in Pathology,* Vol. 65, Grundmann, E. and Kirsten, W. H., Eds., Springer-Verlag, Basel, 1977, 115.
66. **Newberne, P. M. and Suphakarn, V.,** Preventive role of vitamin A in colon carcinogenesis in rats, *Cancer,* 40, 2553, 1977.
67. **Ward, J. M., Sontag, J. M., Weisburger, E. K., and Brown, C. A.,** Effect of lifetime exposure to aflatoxin B_1 in rats, *J. Natl. Cancer Inst.,* 55, 107, 1975.
68. **Dickens, F., Jones, H. E. H., and Waynforth, H. B.,** Oral subcutaneous and intratracheal administration of carcinogenic lactones and related substances: the intratracheal administration of cigarette tar in the rat, *Br. J. Cancer,* 20, 134, 1966.
69. **Rogers, A. E., Lenhart, G., and Morrison, G.,** Influence of dietary lipotrope and lipid content on aflatoxin B, N-2-fluorenylacetamide, and 1,2-dimethylhydrazine carcinogenesis in rats, *Cancer Res.,* 40, 2802, 1980.
70. **Newberne, P. M. and Rogers, A. E.,** Nutrition, monocrotaline, and aflatoxin B_1 in liver carcinogenesis, *Plant Foods Man,* 1, 23, 1973.
71. **Newberne, P. M., Harrington, D. H., and Wogan, G. N.,** Effects of cirrhosis and other liver insults on induction of liver tumors by aflatoxin, *Lab. Invest.,* 15, 962, 1966.
72. **Newberne, P. M., Rogers, A. E., and Wogan, G. N.,** Hepatorenal lesions in rats fed a low lipotrope diet and exposed to aflatoxin, *J. Nutr.,* 94, 1968.
73. **Salmon, W. D. and Newberne, P. M.,** Occurrence of hepatomas in rats fed diets containing peanut meal as a major source of protein, *Cancer Res.,* 23, 571, 1963.
74. **Temcharoen, P., Anukarahanonta, T., and Bhamarapravati, N.,** Influence of dietary protein and vitamin B_{12} on the toxicity and carcinogenicity of aflatoxins in rat liver, *Cancer Res.,* 38, 2185, 1978.
75. **Newberne, P. M. and Williams, G.,** Inhibition of aflatoxin carcinogenesis by diethylstilbesterol in male rats, *Arch. Environ. Health,* 19, 489, 1969.
76. **Goodal, C. M. and Butler, W. H.,** Aflatoxin carcinogenesis: inhibition of liver cancer induction in hypophysectomized rats, *Int. J. Cancer,* 4, 422, 1969.
77. **Chedid, A., Bundeally, A. E., and Mendenhall, C. L.,** Inhibition of hepatocarcinogenesis by adrenocorticotropin in aflatoxin B_1-treated rats, *J. Natl. Cancer Inst.,* 58, 339, 1977.
78. **Ghittino, P.,** Nutritional factors in trout hepatoma, *Prog. Exp. Tumor Res.,* 20, 317, 1976.
79. **Halver, J. E., Ashley, L. M., and Smith, R. R.,** Aflatoxicosis in Coho salmon, National Cancer Institute Monograph 31, 141, 1969.
80. **Scarpelli, D. G., Greider, M. G., and Frajola, W. J.,** Observations on hepatic cell hyperplasia, adenoma, and hepatoma of rainbow trout (*Salmo gairdnerii*), *Cancer Res.,* 23, 848, 1963.
81. **Campbell, T. C., Sinnhuber, R. O., Lee, D. J., Wales, J. H., and Salamat, L.,** Hepatocarcinogenic material in urine specimens from humans consuming aflatoxin, *J. Natl. Cancer Inst.,* 52, 1647, 1974.
82. **Sinnhuber, R. O., Lee, D. J., Wales, J. H., and Ayres, J. L.,** Dietary factors and hepatoma in rainbow trout (*Salmo gairdneri*). II. Cocarcinogenesis by cyclopropenoid fatty acids and the effect of gossypol and altered lipids on aflatoxin-induced liver cancer, *J. Natl. Cancer Inst.,* 41, 1293, 1968.
83. **Wales, J. H. and Sinnhuber, R. O.,** Hepatomas induced by aflatoxin in the sockeye salmon (*Oncorhynchus nerka*), *J. Natl. Cancer Inst.,* 48, 1529, 1972.
84. **Hendricks, J. D., Putnam, T. P., Bills, D. D., and Sinnhuber, R. O.,** Inhibitory effect of a polychlorinated biphenyl (Aroclor 1254) on aflatoxin B_1 carcinogenesis in rainbow trout (*Salmo gairdneri*), *J. Natl. Cancer Inst.,* 59, 1545, 1977.
85. **Scarpelli, D. G.,** Drug metabolism and aflatoxin-induced hepatoma in rainbow trout (*Salmo gairdneri*), *Prog. Exp. Tumor Res.,* 25, 339, 1976.
86. **Sato, S., Matsushima, T., Tanaka, N., Sugimura, T., and Takashima, F.,** Hepatic tumors in the guppy (*Lebistes reticulatus*) induced by aflatoxin B_1, dimethyl-nitrosamine, and 2-acetylamino-fluorene, *J. Natl. Cancer Inst.,* 50, 765, 1973.
87. **Gopalan, C., Tulpule, P. G., and Krishnamurthi, D.,** Induction of hepatic carcinoma with aflatoxin in the rhesus monkey, *Food Cosmet. Toxicol.,* 10, 519, 1972.

88. **Tilak, T. B.,** Induction of cholangiocarcinoma following treatment of a rhesus monkey with aflatoxin, *Food Cosmet. Toxicol.,* 13, 247, 1975.
89. **Adamson, R. H., Correa, P., Sieber, S. M., McIntire, K. R., and Dalgard, D. W.,** Carcinogenicity of aflatoxin B_1 in rhesus monkeys: two additional cases of primary liver cancer, *J. Natl. Cancer Inst.,* 57, 67, 1976.
90. **Lin, J. J., Liu, C., and Svoboda, D. J.,** Long-term effects of aflatoxin B_1 and viral hepatitis on marmoset liver, *Lab. Invest.,* 30, 267, 1974.
91. **Reddy, J. K., Svoboda, D. J., and Rao, M.S.,** Induction of liver tumors by aflatoxin B_1 in the tree shrew (*Tupaia glis*), a nonhuman primate, *Cancer Res.,* 36, 151, 1976.
92. **Carnaghan, R. B. A.,** Hepatic tumors in ducks fed on low level of toxic ground-nut meal, *Nature (London),* 208, 308, 1965.
93. **Miller, R. W.,** Cancer epidemics in the People's Republic of China, *J. Natl. Cancer Inst.,* 60, 1195, 1978.
94. **Krogh, P., Axelsen, N. H., Elling, F., Gyrd-Hansen, N., Hald, B., Hyldgaard-Jensen, J., Larsen, A. E., Madsen, A., Mortensen, H. P., Miller, T., Petersen, O. K., Ravnskow, U., Rostgaard, M., and Aulund, O.,** Experimental porcine nephropathy: changes of renal function and structure induced by ochratoxin A-contaminated feed, *Acta Pathol. Microbiol. Scand. Sect. A,* Suppl. No. 246, 1974.
95. **Krogh, P.,** Ochratoxins, in *Mycotoxins in Human and Animal Health,* Rodricks, J. V., Hesseltine, C. W., and Mehlman, M. A., Eds., Pathotox, Park Forest South, Ill., 1977, 489.
96. **Krogh, P.,** Causal associations of mycotoxic nephropathy, *Acta Pathol. Microbiol. Scand. Sect. A,* Suppl. No. 269, 1978.
97. **Suzuki, S., Kozuka, Y., Satoh, T., and Yamazaki, M.,** Studies on the nephrotoxicity of ochratoxin A in rats, *Toxicol. Appl. Pharmacol.* 34, 479, 1975.
98. **Elling, F.,** Mycotoxic nephropathy, in *Mycotoxins in Human and Animal Health,* Rodricks, J. V., Hesseltine, C. W., and Mehlman, M. A., Eds., Pathotox, Park Forest South, Ill., 1977, 499.
99. **Szczech, G. M., Carlton, W. W., and Tuite, J.,** Ochratoxicosis in beagle dogs. II. Pathology, *Vet. Pathol.,* 10, 219, 1973.
100. **Kitchen, D. N., Carlton, W. W., and Hinsman, E. J.,** Ochratoxin A and citrinin induced nephrosis in beagle dogs. III. Terminal renal ultrastructural alterations, *Vet. Pathol.,* 14, 392, 1977.
101. **Munro, I. C., Moodie, C. A., Kuiper-Goodman, T., Scott, P. M., and Grice, H. C.,** Toxicologic changes in rats fed graded dietary levels of ochratoxin A., *Toxicol. Appl. Pharmacol.,* 28, 180, 1974.
102. **Peckham, J. C., Doupnik, B., and Jones, O. H., Jr.,** Acute toxicity of ochratoxins A and B in chicks, *Appl. Microbiol.,* 21, 492, 1971.
103. **Galtier, P., More, J., and Alvinerie, M.,** Acute and short-term toxicity of ochratoxin A in 10-day-old chicks, *Food Cosmet. Toxicol.,* 14, 129, 1976.
104. **Prior, M. G., Sisodia, C. S., and O'Neil, J. B.,** Acute oral ochratoxicosis in day-old white leghorns, turkeys and Japanese quail, *Poult. Sci.,* 55, 786, 1976.
105. **Doster, R. C., Sinnhuber, R. O., and Wales, J. H.,** Acute intraperitoneal toxicity of ochratoxins A and B in rainbow trout (*Salmo gairdneri*), *Food Cosmet. Toxicol.,* 10, 85, 1972.
106. **Hayes, A. W., Hood, R. D., and Lee, H. L.,** Teratogenic effects of ochratoxin A in mice, *Teratology,* 9, 93, 1974.
107. **Brown, M. H., Szczech, G. M., and Purmalis, B. P.,** Teratogenic and toxic effects of ochratoxin A in rats, *Toxicol. Appl. Pharmacol.,* 37, 331, 1976.
108. **Hood, R. D., Naughton, M. J., and Hayes, A. W.,** Prenatal effects of ochratoxin A in hamsters, *Teratology,* 13, 11, 1976.
109. **Ueno, Y., Sato, N., Ishii, K., Sakai, K., Tsunoda, H., and Enomoto, M.,** Biological and chemical detection of tricothecene mycotoxins of *Fusarium* species, *Appl. Microbiol.,* 25, 699, 1973.
110. **Ueno, Y.,** Mode of action of trichothecenes, *Pure Appl. Chem.,* 49, 1737, 1977.
111. **Ueno, Y., Hosoya, M., Morita, Y., Ueno, I., and Tatsuno, T.,** Inhibition of protein synthesis in rabbit reticulocyte by nivalenol, a toxic principle isolated from *Fusarium nivale*-growing rice, *J. Biochem. (Tokyo),* 64, 479, 1968.
112. **Ueno, Y., Hosoya, M., and Ishikawa, Y.,** Inhibitory effects of mycotoxins on protein synthesis in rabbit reticulocytes, *J. Biochem. (Tokyo),* 66, 419, 1969.
113. **Ueno, Y. and Fukushima, K.,** Inhibition of protein and DNA synthesis in Erlich Ascites tumor by nivalenol, a toxic principle of *Fusarium nivale*-growing rice, *Experientia,* 24, 1032, 1969.
114. **Ueno, Y., Nakajima, M., Sakai, K., Ishii, K., Sato, N., and Shimade, N.,** Comparative toxicology of trichothecene mycotoxins: inhibition of protein synthesis in animal cells, *J. Biochem. (Tokyo),* 74, 285, 1973.
115. **Ueno, Y., Ueno, I., Iitoi, Y., Tsunoda, H., Enomoto, M., and Ohtsubo, K.,** Toxicological approaches to the metabolites of *Fusaria*. III. Acute toxicity of Fusarion-X, *Jpn. J. Exp. Med.,* 41, 521, 1971.
116. **Ueno, Y., Ishii, K., Sato, N., and Ohtsubo, K.,** Toxicological approaches to the metabolites of *Fusaria*. VI. Vomiting factor from moldy corn infected with *Fusarium* spp., *Jpn. J. Exp. Med.,* 44, 123, 1974.

117. **Ueno, Y., Ishikawa, Y., Amakai, K., Nakajima, M., Saito, M., Enomoto, M., and Ohtsubo, K.,** Comparative study on skin-necrotizing effect of scirpene metabolites of *Fusaria, Jpn. J. Exp. Med.,* 40, 33, 1970.
118. **Sato, N., Ueno, Y., and Enomoto, M.,** Toxicological approaches to the toxic metabolites of *Fusaria.* VIII. Acute and subacute toxicities of the T-2 toxin in cats, *Jpn. J. Pharmacol.,* 25, 263, 1975.
119. **Joffe, A. Z.,** Toxicity of *Fusarium poae* and *F. sporotrichioides* and its relation to alimentary toxic aleukia, in *Mycotoxins,* Purchase, I. F. H., Ed., Elsevier, Amsterdam, 1974, 419.
120. **Smalley, E. B. and Strong, F. M.,** Toxic trichothecenes, in *Mycotoxins,* Purchase, I. F. H., Ed., Elsevier, Amsterdam, 1974, 199.
121. **Hsu, I., Smalley, E. B., Strong, F. M., and Ribelin, W. E.,** Identification of T-2 toxin in moldy corn associated with a lethal toxicosis in dairy cattle, *Appl. Microbiol.,* 24, 684, 1972.
122. **Kosuri, N. R.,** Toxicity of *Fusarium tricinctum* in Rats and Cattle, Ph.D. thesis, University of Wisconsin, Madison, University Microfilms, Ann Arbor, Mich., 1969.
123. **Wyatt, R. D., Weeks, B. A., Hamilton, P. D., and Burmeister, H. R.,** Severe oral lesions in chickens caused by ingestion of dietary fusaritoxin T-2, *Appl. Microbiol.,* 24, 251, 1972.
124. **Ohtsubo, K. and Saito, M.,** Cytotoxic effects of scirpene compounds, fusarenon-X produced by *Fusarium nivale,* dihydronivalenol and dihydrofusarenon-X, on HeLa cells, *Jpn. J. Med. Sci. Biol.,* 23, 217, 1970.
125. **Saito, M. and Ohtsubo, K.,** Trichothecene toxins of Fusarium species, in *Mycotoxins,* Purchase, I. F. H., Ed., Elsevier, Amsterdam, 1974, 263.
126. **Ohtusbo, K. and Saito, M.,** Chronic effects of trichothecene toxins, in *Mycotoxins in Human and Animal Health,* Rodricks, J. V., Hesseltine, C. W., and Mehlman, M. A., Eds., Pathotox, Park Forest South, Ill., 1977.
127. **Marasas, W. F. O., Bamburg, J. R., Smalley, E. B., Strong, F. M., Ragland, W. L., and Degurse, P. E.,** Toxic effect on trout, rats, and mice of T-2 toxin produced by the fungus *Fusarium tricinctum* (Cd.) Snyd. et Hans., *Toxicol. Appl. Pharmacol.,* 15, 471, 1969.
128. **Hatsuda, Y. and Kuyama, S.,** Studies on metabolic products of *Aspergillus versicolor.* I. Cultivation of *Aspergillus versicolor,* isolation and purification of metabolic products, *J. Agric. Chem. Soc. Jpn.,* 28, 989, 1954.
129. **Davies, J. E., Kirkaldy, D., and Roberts, J. C.,** Studies in mycological chemistry. VII. Sterigmatocystin, a metabolite of *Aspergillus versicolor* (Vuillemin) Tiraboschi, *J. Chem. Soc.,* 2169, 1960.
130. **Ballantine, J. A., Hassall, C. H., and Jones, G.,** The biosynthesis of phenols. IX. Asperugin, a metabolic product of *Aspergillus rugulosis, J. Chem. Soc.,* 4672, 1965.
131. **Holzapfel, C. W., Purchase, I. F. H., Steyn, P. S., and Gouws, L.,** The toxicity and chemical assay of sterigmatocystin, a carcinogenic mycotoxin, and its isolation from two new fungal sources, *S. Afr. Med. J.,* 40, 1100, 1966.
132. **Purchase, I. F. H. and van der Watt, J. J.,** Carcinogenicity of sterigmatocystin, *Food Cosmet. Toxicol.,* 8, 289, 1970.
133. **Lillehoj, E. B. and Ciegler, A.,** Biological activity of sterigmatocystin, *Mycopathologia,* 35, 373, 1968.
134. **van der Watt, J. J.,** Sterigmatocystin, in *Mycotoxins,* Purchase, I. F. H., Ed., Elsevier, Amsterdam, 377, 1974.
135. **Thiel, P. G. and Steyn, M.,** Urinary excretion of the mycotoxin, sterigmatocystin by vervet monkeys, *Biochem. Pharmacol.,* 22, 3267, 1973.
136. **Rodricks, J. V., Henery-Logan, K. R., Campbell, A. D., Stoloff, L., and Verret, M. J.,** Isolation of a new toxin from cultures of *Aspergillus flavus, Nature (London),* 217, 668, 1968.
137. **Newberne, P. M.,** unpublished observations.
138. **Terao, K., Takano, N., and Yamazaki, Y.,** The effects of o-acetylsterigmatocystin and related compounds on rat liver and cultured chicken embryonal liver cells, *Chem. Biol. Interact.,* 11, 507, 1975.
139. **Zwicker, G. M., Carlton, W. W., and Tuite, J.,** Long-term administration of sterigmatocystin and *Penicillium viridicatum* to mice, *Food Cosmet. Toxicol.,* 12, 491, 1974.
140. **Fujii, K., Kurata, H., Odashima, S., and Hatsuda, Y.,** Tumor induction by a single subcutaneous injection of sterigmatocystin in newborn mice, *Cancer Res.,* 36, 1615, 1976.
141. **Purchase, I. F. H. and van der Watt, J. J.,** Carcinogenicity of sterigmatocystin to rat skin, *Toxicol. Appl. Pharmacol.,* 26, 274, 1973.
142. **Mirocha, C. J, and Christensen, C. M.,** Oestrogenic mycotoxins synthesized by *Fusarium,* in *Mycotoxins,* Purchase, I. F. H., Ed., Elsevier, Amsterdam, 1974.
143. **Stoloff, L., Henry, S., and Francis, O. J., Jr.,** Survey for aflatoxins and zearalenone in 1973 crop corn stored on farms in country elevators, *J. Assoc. Off. Anal. Chem.,* 59, 118, 1976.
144. **Christensen, C. M., Nelson, G. H., and Mirocha, C. J.,** Estrogenic metabolite produced by *Fusarium graminearum* in stored corn, *Appl. Microbiol.,* 13, 653, 1965.
145. **Mirocha, C. J., Christensen, C. M., and Nelson, G. H.,** Physiologic activity of some fungal estrogens produced by *Fusarium, Cancer Res.,* 28, 2319, 1968.
146. **Ueno, Y., Shimada, N., Yagasaki, S., et al.,** Toxicological approaches to the metabolites of Fusaria. VII. Effects of zearalenone on the uteri of mice and rats, *Chem. Pharm. Bull. (Tokyo),* 22, 2830, 1974.

147. **Meronuck, R. A., Garren, K. H., Christensen, C. M., Nelson, G. H. and Bates, F.**, Effects on turkey poults and chicks of rations containing corn invaded by *Penicillium* and *Fusarium* species, *Am. J. Vet. Res.*, 31, 551, 1970.
148. **Bailey, D. E., Cox, G. E., Morgareide, K., and Taylor, J.**, Acute and subacute toxicity of zearalenone in the rat, *Toxicol. Appl. Pharmacol.*, 37, 144, 1976.

Chapter 3

ENVIRONMENTAL TOXICOSES IN HUMANS

Ronald C. Shank

TABLE OF CONTENTS

I.	Introduction	108
II.	Ergot Alkaloids	108
	A. History and Chemistry	108
	B. Signs and Symptoms of Ergot Poisoning	109
	C. Control	111
III.	Alimentary Toxic Aleukia	112
	A. History and Chemistry	112
	B. Signs and Symptoms	113
	1. First Stage	113
	2. Second Stage	113
	3. Third Stage	113
	4. Fourth Stage	114
IV.	Yellowed Rice Syndrome	114
	A. History and Chemistry	114
	B. Signs and Symptoms	114
V.	Balkan (Endemic) Nephropathy	115
	A. History and Chemistry	115
	B. Signs and Symptoms	116
VI.	Bishydroxycoumarin (Dicumarol)	116
VII.	Pink Rot Disease	117
VIII.	Aflatoxins and Human Liver Cancer	120
	A. Geographic Distribution of Human Liver Cancer	120
	B. Epidemiological Evidence Associating Aflatoxin with Human Liver Cancer	121
	1. Uganda	121
	2. Philippines	122
	3. Thailand	123
	4. Kenya	125
	5. Mozambique	126
	6. Swaziland	127
	C. Other Factors	127
	1. Nutrition	127
	2. Infection	129
	3. Cointoxication	129

IX. Aflatoxins and Acute Poisonings ... 130
   A. Taiwan Outbreak ... 130
   B. Ugandan Case ... 130
   C. Reye's Syndrome (Thailand) ... 131
      1. Characteristics of the Disease ... 131
      2. Implication of Aflatoxin ... 131
      3. Associations Outside Thailand ... 132
      4. Contributing Factors ... 132
   D. Indian Childhood Cirrhosis ... 134
   E. Indian Hepatitis Outbreak ... 134
   F. Other Possible Cases ... 134

References ... 135

I. INTRODUCTION

There have been several episodes of human poisoning by fungal metabolites, in almost every case resulting from the consumption of heavily contaminated foodstuffs. The preceding chapters have discussed the environmental occurrence of those mycotoxins important to human health, the toxic effects these components exert on domestic and laboratory animals, and the mechanisms of toxic action for these important environmental contaminants. The present chapter reviews the evidence associating acute and chronic human poisonings with the individual mycotoxins.

II. ERGOT ALKALOIDS

A. History and Chemistry

Two major monographs have been written on ergot alkaloids in human poisonings, that of Barger[1] in 1931 and the more recent study by Bové.[2] An important review has been written by Groger.[3]

The biological effects of the ergot alkaloids have been known since ancient times; episodes of acute poisoning may date back as far as 430 B.C. in Greece, and sporadic outbreaks appear to have occurred ever since, one of the most recent being in France in 1951. It is not certain when the first epidemic clearly because of ergot consumption occurred, but according to Bové,[2] the outbreak of ignis plaga or fire plague in Paris in 945 seems to be the best documented case. In writing the history of the cathedral in Reims, the archivist Frodoard described this early Paris epidemic as attacking the limbs of the body, slowly destroying them by fire, with death the only relief for most victims. If some victims left their homes and visited certain holy places, the disease subsided, only to flare again upon returning home. It was described that hundreds of victims in Paris were cured by going to a particular chapel in which the food was supplied by Count Hugo of Paris; most historians ascribe the relief as because of leaving home and changing the diet. Such outbreaks occurred with little pattern, sometimes involved hundreds and even up to tens of thousands of people in Europe, but were never definitely shown to result from a dietary component, let alone ergot alkaloids.

The importance of changing the diet to bring about relief from this poisoning began to come into focus near the end of the 11th century. The remains of St. Anthony, who died in Egypt in 356 as a hermit famed for his healing powers, were placed in a shrine in Dauphiné in France in 1070. The fame of St. Anthony's instant cures attracted to the shrine many pilgrims, including those afflicted with ergot poisoning. In approximately 1090, a nobleman's son suffering from "the fire" was said to have been cured by a visit to the shrine; in gratitude the nobleman financed a hospital to help other victims of the disease. This concentrated medical attention at one specialized center, and the physicians began to record in detail the development of the illness. Gradually associations between the holy fire and certain foods and agricultural crops led to efforts to determine the cause of the disease.

The pharmacological properties of ergot were recognized in the latter part of the 16th century, and about 100 years later, cereal grains containing ergot were shown to be the cause of these epidemics in Europe.

Chemical studies on ergot probably began in the early 18th century, but it was not until 1906 that Barger et al.[4] and Kraft[5] crystallized the first pharmacologically active substance from ergot, and finally Stoll[6] in 1920 identified the ergot alkaloid, ergotamine.

The last outbreak of purported ergot poisoning was described by Gabbai et al.[7] Apparently moldy rye was sold illegally (to avoid a grain tax) in central France to a miller, who ground the rye and mixed it with flour which he then sold to a baker 300 mi away in Pont St. Esprit, near Avignon. On August 15, 1951, many people in Pont St. Esprit ate small amounts of bread which may have been made from this contaminated flour. About 200 people became ill, as did many domestic animals fed the same bread. A total of 25 people suffered severe delirium, and four people died, one a previously healthy young man and three old people in poor health. Scientific evidence for the presence of ergot alkaloids was not given, and Bové[2] claims that subsequently toxicologists stated the cause of the outbreak was an organic phosphorous preparation that had been used to spray the grain. While organo-phosphorus poisoning would be consistent with most of the clinical symptoms in this mass poisoning, it would not explain all the reported conditions of the victims, and cotoxicity resulting from pesticide and moldy grain cannot be ruled out.

The ergot alkaloids of pharmacological and toxicological importance are derivatives of lysergic acid (see Figure 1A); a smaller group of ergot alkaloids, the clavine alkaloids, which are somewhat less complex, has not been as well characterized biologically. The simple lysergic acid derivatives (see Figure 1B) are summarized in Table 1. These compounds yield lysergic acid and an amine upon hydrolysis and therefore are often referred to as amine alkaloids of ergot. Pharmacologically these compounds are rapid acting, powerful oxytocics, i.e., they stimulate the smooth muscle of the uterus. They are also weak vasoconstrictors.

The other, larger group of ergot alkaloids derived from lysergic acid are the amino acid alkaloids (see Figure 1C); upon hydrolysis they yield lysergic acid and proline and another amino acid. These alkaloids, listed in Table 2, are potent vasoconstrictors and are highly active oxytocics if given intravenously, but not orally. The amino acid alkaloids also inhibit the activity of nerves stimulated by epinephrine (adrenaline), norepinephrine, and other sympathomimetic amines.

The clavine alkaloids (see Figure 1D) which are not derived from lysergic acid are also found in ergot, but their role, if any, in human ergotism, has not been identified; members of this group are listed in Table 3.

B. Signs and Symptoms of Ergot Poisoning

Descriptions of ergot poisoning written in the Middle Ages were quite vivid. Accounts were given of epidemics of a holy fire in which victims suffered awfully with a severe internal feeling of heat and intense thirst, multiple ulcerations of the skin, a burning

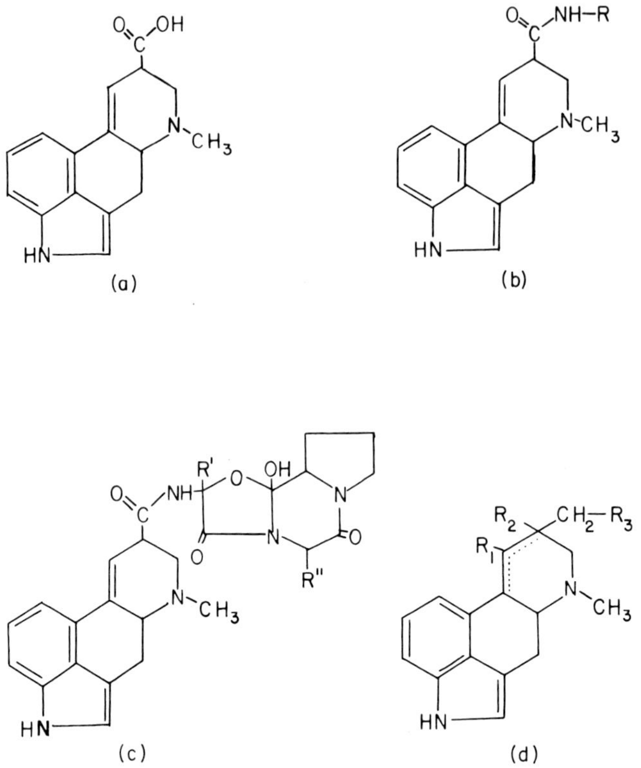

FIGURE 1. (a) Lysergic acid, (b) amine alkaloids of ergot (see Table 1), (c) amino acid alkaloids of ergot (see Table 2), (d) clavine alkaloids of ergot (see Table 3).

Table 1
AMINE ALKALOIDS DERIVED FROM LYSERGIC ACID

Alkaloid	R
Ergine	H
Lysergic acid methylcarbinolamide	$CH_3CH(OH)-$
Ergometrine (ergobasine, ergonovine)	$CH_2(OH)CH(CH_3)-$
Lysergic acid-L-valinemethylester	$(CH_3)_2CHCH(COOCH_3)-$

Note: Radicals refer to Figure 1B.

sensation of the limbs, the feeling of ants and mice crawling underneath the skin, the drying and turning black of hands, arms, feet, and legs, blindness, dementia, and mental degeneration. In modern terms, a clinical description is readily understandable from the known mechanisms of action of the ergot alkaloids.

Acute ergot poisoning today is essentially a problem only in chemotherapy and rarely has it occurred recently as a result of eating a contaminated foodstuff. The symptoms of acute ergotamine poisoning include vomiting, diarrhea, intense thirst, a tingling, itchy and cold skin, a rapid, weak pulse, confusion, and unconsciousness.

Table 2
AMINO ACID ALKALOIDS OF ERGOT

Alkaloid	R'	R''
Ergotamine	$-CH_3$	$-CH_2-\bigcirc$
Ergosine	$-CH_3$	$-CH_2CH(CH_3)_2$
Ergocristine	$-CHCH(CH_3)_2$	$-CH_2-\bigcirc$
α-Ergokryptine	$-CHCH(CH_3)_2$	$-CH_2CH(CH_3)_2$
β-Ergokryptine	$-CHCH(CH_3)_2$	$-CH(CH_3)CH_2CH_3$
Ergocornine	$-CHCH(CH_3)_2$	$-CHCH(CH_3)_2$
Ergostine	$-CH_2CH_3$	$-CH_2-\bigcirc$

Note: Radicals refer to Figure 1C.

Table 3
CLAVINE ALKALOIDS OF ERGOT

Alkaloids	R_1	R_2	R_3	D-ring double bond
Argoclavine	H	H	H	8,9
Elymoclavine	H	H	OH	8,9
Setoclavine	H	OH	H	9,10
Penniclavine	H	OH	OH	9,10
Festuclavine	2H	H	H	None
Lysergol	H	H	OH	9,10
Fumigoclavine A	H, $-OCOCH_3$	H	H	None

Note: Radicals refer to Figure 1D.

Chronic ergot poisoning today seems limited to therapeutic accidents in treating patients for migraine headaches. The extremities, especially the feet and legs, become cold, pale, and numb because of the constriction of the local blood vessels and result in diminished blood flow. Walking becomes painful and eventually gangrene develops. The impaired blood flow is worsened by necrotic destruction of the peripheral capillary epithelium which can result in thrombi which block small arterioles. These changes in the circulatory system can produce anginal pain, a speeding or slowing of heart rate, and a raising or lowering of blood pressure. Headache, nausea, vomiting, diarrhea, and dizziness are also common in chronic ergot poisoning, as are muscular weakness, intense tingling and itching of the skin, and coldness of the skin. Ergotism also produces central nervous system (CNS) changes, including decreased vasomotor activity, depressed respiration, reduced secretion of anterior pituitary hormones, confusion, drowsiness, and depression.

C. Control

Control of ergot production in human foodstuffs is an obvious success story as epidemics of ergotism are no longer seen. Recognizing that the disease was closely associated with the presence of compact black or purple masses (ergots) of fungal mycelia (of

Claviceps purpurea and *Claviceps paspali*) in cereal grains, especially rye, lead to ready methods to prevent human poisonings. When such infested grains were seen, they were not used as human foodstuffs. Rapid harvesting and drying of the grasses greatly reduced the incidence of ergot in these foodstuffs. Such control measures may seem obvious by today's standards, yet human mycotoxicoses because of mold infestations not related to ergot seem still to occur. Ergotism was probably the first instance of mycotoxin poisoning in man; unfortunately it was not the last.

III. ALIMENTARY TOXIC ALEUKIA

A. History and Chemistry

In spite of the thousands of people afflicted with this disease of the blood-forming system, surprisingly little progress has been made in identifying its causative agents. Mayer[8,9] and Joffe[10] have written comprehensive reviews in English, but the bulk of the literature is in Russian, for it is in the Soviet Union that the disease has been a problem.

Perhaps as early as 1913, a food-borne disease, possibly from eating fusarium-contaminated bread, occurred in far eastern Siberia and the Amur region. Larger outbreaks began to occur more frequently in the 1930s, involving several districts in Western Siberia and European Soviet Russia. During the war years of 1941 through 1945, the spread of alimentary toxic aleukia was especially rapid.

The outbreaks of disease followed a definite pattern, beginning early in May, reaching a peak in June, and ending usually by late autumn or early winter. Agricultural practices at the time seem to explain the seasonal nature of these annual outbreaks. The crops of cereal grasses were not harvested in the fall, but allowed to stand in the fields throughout winter; after the spring thaw the crops were harvested, and the grains were milled for immediate consumption in April. It took only a few weeks for the disease to manifest itself. If the winter had been marked by heavy snowfalls and was followed by an early spring with a slow progressive thawing, the cereal seed absorbed much moisture and provided a suitable medium for the growth of cryophilic fungi, especially the mold *Fusarium*. Russian scientists associated the disease with crops of rye with high toxicity. Soon the toxicity of the rye was related to the growth of toxigenic strains of *Fusarium* on the rye in the fields during the spring thaws.

In 1957 Olifson reported on the work of his doctoral research in which he isolated and identified two toxic glycosides from prosomillet. The prosomillet had been overgrown with *Fusarium sporotrichiodes* and *F. poae* which had been isolated previously from Russian overwintered grain. These molds produced two compounds named sporofusarin and poaefusarin. Experimental studies in the cat showed that these metabolites could produce symptoms similar to human alimentary toxic aleukia. Olifson's reports have been reviewed in English by Joffe.[10]

The purity of the preparations used by Olifson in his studies with the cat is important. Several species of *Fusarium* produce the toxins 12,13-epoxytrichothecenes, several of which have been shown to be radiomimetic; hence they would produce changes in animals similar to alimentary toxic aleukia. Mirocha and Pathre[11] examined a sample of poaefusarin and found it contained the epoxytrichothecene, T-2 toxin (see Figure 2), in sufficient quantity to account for the toxicity of the poaefusarin sample; smaller amounts of other epoxytrichothecenes and zearalenone were also present.

It has been suggested that the epoxytrichothecenes played a role in Russian alimentary toxic aleukia.[12] Wyatt and colleagues[13,14] have shown that T-2 poisoning in chickens resembles the mycotoxin poisoning associated with Russian overwintered grain.

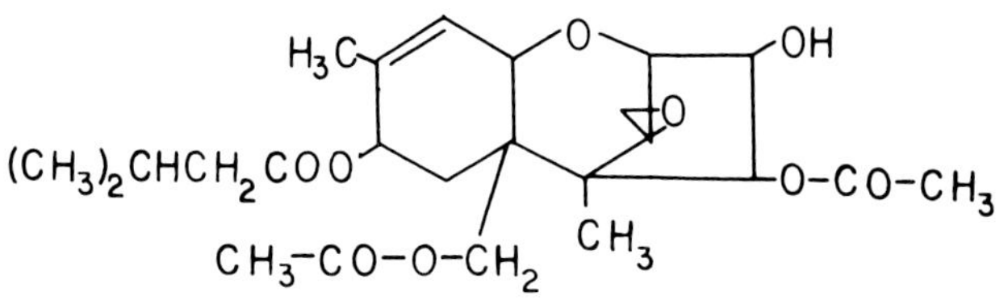

FIGURE 2. T-2 toxin, associated with alimentary toxic aleukia.

B. Signs and Symptoms

Mayer[8] summarized the Russian descriptions of the disease, dividing the clinical features into four stages; the disease seems to result from toxic injury to the hematopoietic, autonomic nervous, and endocrine systems.

1. First Stage

This is a rapid onset of irritation to the upper GI tract, beginning a few hours after ingestion of the toxic cereal product, often bread. The contaminated food would have a peppery taste and produce a burning sensation from the mouth to the stomach. Within a couple of days, the GI inflammation develops into acute gastroenteritis, with nausea, vomiting, and diarrhea. This local effect, without a change in body temperature, persists for 3 to 9 days and then spontaneously ends, even when the victim continues to eat the poisoned grain. Mayer[8] indicated such symptoms were observed in 92% of all cases in an outbreak.

2. Second Stage

The disappearance of the gastroenteritis is followed by a slow degeneration of the bone marrow; for some 2 to 9 weeks, the victim is often without outward symptoms. When the marrow changes are extensive enough, visible skin hemorrhages appear on what may appear an otherwise healthy individual. Hematologic examination reveals a progressive sharp decrease in the total number of leukocytes and a relative decrease in the percentage of granulocytes. Before the hemorrhages appear, some victims display nervous system problems: irritation, weakness, easy fatigue, vertigo, headache, palpitation, and slight asthma. The skin and mucous membranes may be pale, the pupils wide, heart sounds muffled, pulse soft and labile, and the blood pressure low. Hospitalization of victims through the second phase usually resulted in recovery.

3. Third Stage

The most serious stage comes about suddenly and has four essential features: hemorrhagic diathesis, necrosis, especially necrotic angina, toxic-infectious state, hematological changes, such as agranulocytosis and worsening of the blood abnormalities of the first two stages. The hemorrhagic syndrome begins with petechial hemorrhages on the skin of the trunk, axillary, and inguinal areas, arms, thighs, face, and head. Increased capillary fragility results in hemorrhages at the site of only slight traumas and also on the mucous membranes of the mouth, tongue, soft palate, tonsils, nose, stomach, and intestines. Necrosis begins in the throat and spreads throughout the mouth and into the larynx, vocal cords, lungs, stomach, bowels, and even the skin of the hands and face and in internal organs. The impaired hemotopoietic and reticuloendothelial systems permit widespread bacterial infection in the necrotic areas. Lymph nodes become enlarged.

Esophageal lesions involving the epiglottis may cause laryngeal edema and stenosis of the glottis, resulting in death because of asphyxiation.

4. Fourth Stage

If clinical help is provided in time, the recovery, or fourth stage, ends the course of the disease. Convalescence usually takes 2 months or more before the necrotic areas have healed and the bone marrow functions have returned to normal.

IV. YELLOWED RICE SYNDROME

A. History and Chemistry

Approximately a hundred years ago, epidemics of an acute heart disease broke out in rural Japan, the etiology of which was never determined. Retrospectively, Uraguchi[15,16] analyzed the records of cases of this disease called acute cardiac beriberi (shoshin-kakke) and concluded that the ailment was probably a human mycotoxicosis.

Acute cardiac beriberi was associated with the consumption of polished rice and was initially thought to be an avitaminosis. In 1910, however, the Japanese government took action to reduce the quantity of moldy rice reaching the markets and the incidence of acute cardiac beriberi began a sharp decrease in that same year, although polished rice remained prominent in the Japanese diet and other forms of beriberi did not decline.

Uraguchi[16] suggested that acute cardiac beriberi may have resulted from eating "yellowed rice"; such a foodstuff became pigmented — and toxic — after the growth of a variety of *Penicillium* species. *Penicillium citreo-viride* Biourge produces a dark yellow metabolite, citreoviridin (see Figure 3) which, when given to rats as a crude extract from moldy rice, causes paralysis and respiratory failure. The compound concentrates in the CNS, liver, kidneys, and the cortex of the adrenals; the acute toxic response in the rat resembles acute cardiac beriberi in the human.

Pencillium islandicum Sopp is another mold common to yellowed rice; this species produces at least nine toxic metabolites: the anthraquinones, islandicin, catenarin, luteoskyrin, rubroskyrin, iridoskyrin, and skyrin; the chlorinated cyclic polypeptides, cyclochlorotine and islanditoxin; and the polyene compound, erythroskyrin. The structures of all these compounds are given in a review by Shank.[17] The compounds are toxic to the visceral organs and their experimental toxicity does not resemble acute cardiac beriberi of Japan.

B. Signs and Symptoms

The clinical manifestations of acute cardiac beriberi, first outlined by Irisawa and colleagues in 1936,[16] began with palpitation, precordial distress, and tachypnea; nausea and vomiting followed, and it became more difficult to breathe. Within a few days, the patient suffered severe anguish, pain, severe restlessness, or sometimes violent mania. The right heart was dilated, heart sounds were abnormal, blood pressure was low, and the pulse was rapid, sometimes exceeding 120 beats/min, and faint. The dyspnea increased, the skin of the extremities became cold, dry, and cyanotic, and the voice became husky. Finally the pulse became feeble, the pupils dilated, consciousness was lost, and respiration failed.

The course of the disease could be sudden and rapid. Seemingly healthy people would die within a few days. Animals suffering from avitaminosis, in experimental attempts to reproduce the disease in a model, could recover fully when fed a diet supplemented with liver from victims of acute cardiac beriberi. Such observations make avitaminosis an unlikely explanation for this disease, and many today agree with Uraguchi[16] that this

Table 4
PREVALENCE OF DERMATITIS IN MICHIGAN CELERY WORKERS, 1959 (%)[39]

Celery handling procedures	Field workers	Shed workers	Total workers
Hand cutting	82	55	68
Mechanical cutting	10	10	10
Both	62	46	54

	Light complexion	Dark complexion	Total
Hand cutting	72	29	68
Mechanical cutting	11	7	10
Both	65	15	54

Patch tests were done on field and shed workers using normal celery stalks, crushed normal celery leaves, and pink rot celery. The patches were applied for 24 hr to both arms; one arm remained covered, while the second was freed of the patches and exposed for 24 hr. A third of the exposed pink rot patch areas developed the dermatitis, and all other patch areas were negative, including the covered pink rot patch areas. The field workers, in an effort to demonstrate the alleged association to the investigating medical team, rubbed pieces of pink rot celery on their arms, leaving the skin uncovered. In each instance, they developed the localized dermatitis within 48 hr.

In laboratory tests, patches were applied to five volunteers. In four volunteers, a purplish erythema developed at the site where pink rot celery was applied and then exposed to sunlight; three of the four (the fourth was black) developed dermatitis by 48 hr, with all covered areas negative. Only 30 sec of sunlight was necessary to photoactivate the patch areas sufficiently to produce erythema and edema; 5 min of sunlight were required before bulla would develop.

A controversy has proceeded for 15 years on whether the celery contains the psoralen compounds in sufficient quantity, without mold action, to cause the dermatitis, or whether the mold itself produces a photoallergic dermatitis, or whether the mold indeed produces the toxins itself, either as a fungal metabolite or by inducing the celery to produce a phytoallexin. Italian workers[39,40] claimed in 1954 that celery and parsley contained bergapten (5-methoxypsoralen), but this could not be confirmed in the U.S., where only diseased portions of celery stalks were shown to contain phototoxic compounds, namely 8-methoxypsoralen and trimethylpsoralen.[41,42] The Chinese[43] reported the 8-methoxypsoralen and trimethylpsoralen occur in celery infested with *S. sclerotiorum*, but not when the same mold is grown on potato dextrose agar or autoclaved celery extract. The same psoralens occur when celery is infected with *Sclerotinia rolfsii*, *Rhizoctonia solani*, and *Erwinia aroideae* or when celery is soaked in 5% NaCl. This would suggest then that these phototoxins are produced by the celery plant itself when under fungal or chemical attack and as such would be phytoallexins and not mycotoxins.

The mechanism of action of the psoralens in toxic dermatitis is not known. Cole and others[44] have shown that interstrand crosslinks are formed in DNA of mammalian cells treated with trimethylpsoralen after earlier treatment with light at 360 nm (long UV). This cross-linking may result from the reaction of a photoexcited psoralen molecule with pyrimidine bases in opposite strands of the DNA duplex. The psoralens intercalate between base pairs in DNA rotating thymidine moieties enough to bring the 5,6,-double bond of the pyrimidine adjacent to the reactive sites of the psoralen compound, resulting in cross-linking. How this cross-linking is related to the dermatitis is unknown, but it may be quite relevant to the putative carcinogenicity of these compounds. Attention should be called to the similarity between the unsaturated furano rings of both 8-methoxypsoralen and aflatoxin B_1 (see Figure 7) a highly potent carcinogen completely dependent upon this focus of unsaturation for its carcinogenicity (see Chapter 3). The importance of this group in 8-methoxypsoralen has not yet been determined, but it is tempting to speculate that it too may undergo epoxidation by cellular monooxygenases and subsequently break down to a highly reactive compound and ultimate carcinogen.

VIII. AFLATOXINS AND HUMAN LIVER CANCER

Demonstrating the association between aflatoxin consumption and human liver cancer was the first instance in which a compound was found to be carcinogenic in animals, and then a large-scale international effort was made to determine what role, if any, the compound played in the etiology of human cancer. Although evidence for a dose-response relationship for aflatoxin consumption and human liver cancer has been obtained in four separate field studies, in Thailand, Kenya, Mozambique, and Swaziland, it must be emphasized that such evidence is presumptive of a causative role for the mycotoxins and not scientific proof of such a relationship. It is unlikely that scientific proof of the role of the aflatoxins in the etiology of liver cancer in man will ever be obtained, for such proof would require unconscionable direct human testing.

A. Geographic Distribution of Human Liver Cancer

Primary cancer of the liver is not a common disease in most areas of the world; the usual incidence expressed in terms of an annual rate for males and females of all ages is approximately 2 cases per 100,000 people.[45,46] There are particular geographic areas, however, where the annual liver cancer rate is reported to be well above this level. Certain populations in Africa,[47-49] southern India, Japan,[50-51] and Southeast Asia[52-55] have unusually high incidences of liver cancer. Such areas also have agricultural practices in the harvesting, handling, and storage of food crops that favor mold growth and contamination by aflatoxins; the climates are also conducive to mycotoxin production. Not all areas with similar agricultural practices and climate have high liver cancer rates, for instance, certain portions of tropical South America. Also, all areas with high liver cancer rates do not share cultures and climates favoring mold growth and aflatoxin production, e.g., Geneva, Switzerland.[56] Nevertheless, the association seems to hold in the majority of cases, and in 1965 Oettlé[57] proposed that mycotoxins may play a part in the high incidence of liver cancer in Africa.

In these areas of high liver cancer rate, the sex ratio for hepatocellular carcinoma is elevated; in Uganda[58] and Kenya,[59] the male to female ratio is 3:1; Singapore, 3.5:1;[60] Swaziland, 5:1;[61,62] Mozambique, 2:1 to 5.4:1;[49,63] and in Thailand between 5 and 6:1.[64,65] In most other areas of the world, where liver cancer rates are low, the sex ratio is between 1 and 1.5:1.[46]

Shank[66] has briefly reviewed several proposed etiologic agents and important environmental factors in human liver cancer. Cirrhosis is often associated with liver cancer and has been considered in a causative role — so are viral hepatitis, parasitic diseases, and nutritional inadequacies. It may be that several of these diseases act together, even in concert, to induce primary liver cancer in man. Chemical carcinogenesis, especially aflatoxin-induced carcinogenesis in Africa and Southeast Asia, is now becoming more widely accepted as having an important, if not central, role in human liver cancer.

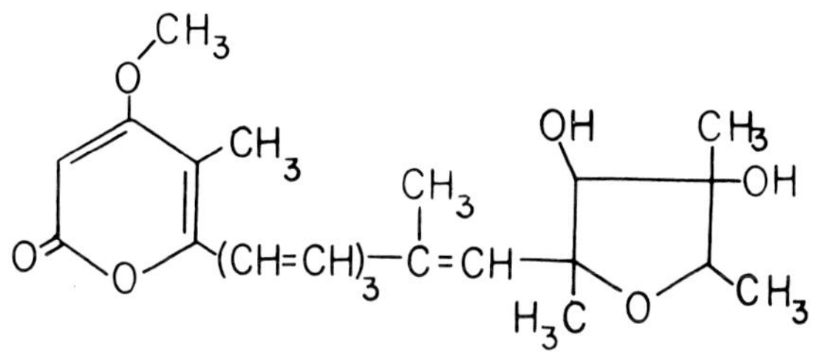

FIGURE 3. Citreoviridin, "yellowed rice" toxin.

disease probably was closely related to yellowed rice and that a mycotoxin, such as citreoviridin, could have had an important role in the etiology of that disease.

V. BALKAN (ENDEMIC) NEPHROPATHY

A. History and Chemistry

In 1957 to 1958, an unusual chronic disease of the kidney was recognized in contiguous areas of Yugoslavia, Rumania, and Bulgaria.[18] The endemic nephropathy occurred in limited districts within larger areas, on both sides of the Carpathian mountains, in villages, but not towns. Some villages were affected heavily, involving 59 to 75% of the households, while other villages were unaffected, and in still others 20% of the households were affected.

The geographical area is characterized by a Mediterranean climate with high humidity, elevations between 200 and 300 m, and rich fauna and flora. Affected villages were usually on valley floors as opposed to unaffected villages on hilltops above the same valleys.

Genetic factors may have a role in the etiology of Balkan nephropathy.[19] The strong familial character of the disease led to investigations of migrations in the 18th century of afflicted families into this basin of the Danube; the nephropathy followed the families and continued on in the children. All the facets of this disease cannot be explained solely on genetics, but it has been suggested repeatedly that perhaps there is a genetic predisposition to mishandle metabolically an environmental toxin; either sensitive people fail to detoxify the environmental agent, or they are especially efficient at transforming the agent from a nontoxic to a toxic intermediate (see the discussion following the paper of Bulic[19]).

Danilović and Stojimirovic[20] studied 12 families in a village hamlet in Serbia, who had a long history of this kidney disease. Their findings were unable to support an infection in the etiology of the disease, but they did find in the blood and urine of patients, the food-grade flour, well water, and in local soils levels of lead two to five times greater than normal. The lead in the flour came from deteriorated grinding stones, and the suggestion was made that only after many years of consuming flour from these mills and accumulating the lead would the levels of this heavy metal be sufficient to precipitate the nephropathy; indeed, the majority of patients are between the ages of 30 and 60 years.

A similar study in Bosnia[21] demonstrated a correlation between the nephropathy and the drinking of certain ground waters, but lead, cadmium, and several pesticides and fertilizers were excluded as possible contributory contaminants.

Barnes[22] suggested that plant toxins or mycotoxins may be an environmental factor in the causation of this human disease. Krogh and co-workers[23] presented preliminary evidence to associate the human disease with ingestion of ochratoxin A (see Figure 4), a dihydroisocoumarin derivative and metabolite of several species of *Aspergillus* and *Penicillium*. This nephrotoxic compound occurs in feeds and foodstuffs and is considered a major determinant of porcine nephropathy, a form of kidney damage strikingly similar to that seen in Balkan nephropathy cases. Krogh and colleagues found that ochratoxin A contamination of foodstuffs in Yugoslavia is more frequent in an area where the human nephropathy is prevalent compared to nonendemic areas. In 47 samples of cereals from an endemic area, 6 (13%) contained ochratoxin A (18 to 90 μg/kg in maize, 12 to 55 μg/kg wheat, and one sample of barley containing 5 μg/kg); in the nonendemic area, one sample of maize out of 64 cereal samples analyzed contained ochratoxin A (14 μg/kg). The study is still in progress, and any conclusions will have to be reserved until the final reports are issued.

B. Signs and Symptoms

The disease is of indefinite onset without acute manifestations. Among the earliest and most frequent complaints are headache, lassitude, easy fatigue, and anorexia. In some there is epigastric or diffuse abdominal pain, and in fewer cases, severe lumbar pains simulating nephrolithiotic crises accompanied by hematuria.[24] The typical syndrome includes a sallow, copper-colored skin, yellowing of the palms and soles, anemia in the preazotemic stage, and perhaps occasional profuse intermittent hematuria, due to tumors of the urinary passages; there is no hypertension or edema. Impaired renal function, including restricted concentrating power and reduced renal plasma flow are early changes, followed by a decrease in glomerular filtration rate probably due to compression by thickened capsules. In almost half of the victims, there is a slight depression of the excretory function of the liver.

Early in the nephropathy, dystrophy and necrosis, as well as tubular regeneration and acellular interstitial sclerosis, are seen. Later the kidneys become small and smooth, with one fibrous zone containing collagen, hyalinized glomeruli, and atrophied tubules and a second, inner zone of intact glomeruli, moderate interstitial round-cell infiltration and dystrophy, atrophy, and tubule epithelial regeneration. The glomerular changes follow the tubular changes. Severe anemia of the internal organs is common. The marrow of the flat bones is hypoplastic, while in long bones, regeneration is not observed. Death is due to azotemic uremia.

Balkan nephropathy is an adult disease, with almost all cases occuring between the ages of 30 and 60 years, and the disease seems to be more prevalent in the female. Although there is a familial character to the disease, several observations are inconsistent with a hereditary mechanism: inhabitants of affected areas, who migrated before the age of 15, were not affected; immigrants from unaffected areas were at high risk after living in affected areas for more than 15 years; a twin living in an endemic area would fall victim to the nephropathy, while the second twin, living in an unaffected area, would remain healthy.

VI. BISHYDROXYCOUMARIN (DICUMAROL)

Much of the foundation of our knowledge on mycotoxins stems from veterinary poisonings, especially those that occurred on large scale. One such instance was the severe bleeding seen in cattle in Alberta and North Dakota in the early 1920s. Herds grazing on sweet clover developed a hemorrhagic disease, and large numbers of animals became

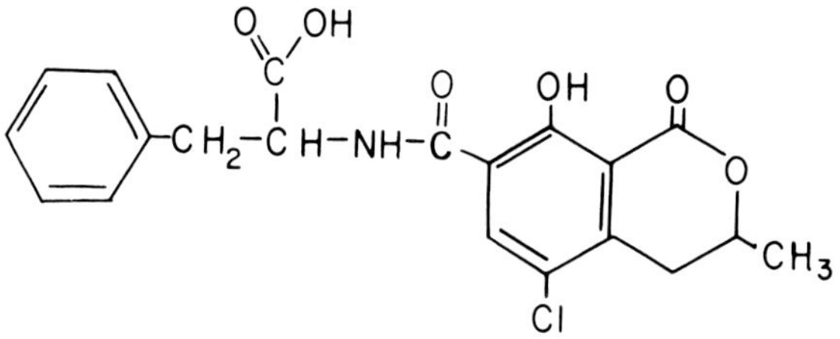

FIGURE 4. Ochratoxin A, associated with Balkan nephropathy.

involved. The Canadian veterinarian, Schofield,[25] associated sweet clover disease with clover overgrown with mold; Link and co-workers[26-28] demonstrated that the toxic agent was bishydroxycoumarin (see Figure 5). The compound has strong anticoagulant properties and hence is used clinically in the prevention and treatment of a variety of thromboembolic disorders. Almost all human poisonings due to this compound arise from therapeutic accidents and therefore have a small role in environmental toxicology.

Bishydroxycoumarin acts by inhibiting blood-clotting mechanisms through its interference with the synthesis of vitamin K-dependent clotting factors in the liver. The effects on clotting, then, are not immediate, but require up to 24 hr before prothrombin times (the times necessary for whole blood to clot under specified test conditions) increase. Bishydroxycoumarin is a competitive inhibitor of vitamin K. Cause of death in man from bishydroxycoumarin poisoning is usually due to massive hemorrhage from an unsuspected ulcer or neoplasm.

VII. PINK ROT DISEASE

Several psoralens, furanocoumarins, are found in a variety of plants and have been known for their pharmacologic value since ancient Egypt. Vitiligo is an acquired, progressive, localized achromia of the skin resulting from a functional abnormality of the melanocytes, apparently due to the loss of function of the tyrosinase system. Psoralens, especially 8 methoxypsoralen (see Figure 6A) and trimethylpsoralen (see Figure 6B), have been used successfully to treat this disorder and repigment the skin, as long as functional melanocytes still remain. Indeed, in 1958 a symposium on psoralens and their photoactivation in clinical studies was held in Michigan and the proceedings have been published.[29]

Becker[30] has described not only the use, but also the abuse of these medications. In apparent careless use of psoralens as suntanning agents to augment the tanning capabilities of the skin, excessive amounts of the agents were used, and after exposure to the sun, a severe dermatitis developed. Nausea, vomiting, vertigo, and mental excitation resulted from ingestion of 20 mg or more of 8-methoxypsoralen; whether liver damage resulted is still an unsolved question.[31-33]

Perhaps of greater importance, mice given 8-methoxypsoralen topically, by i.p. injection, and, to a much lesser extent, by the oral route, were found to develop carcinomas of the exposed skin if exposure to long UV radiation quickly followed toxin administration.[34-37] Although skin tumors normally develop in mice repeatedly exposed to UV light, psoralen-induced tumors appear in half the time (a halving of the induction period), and they are in greater number, if the animals are pretreated with the psoralen derivative

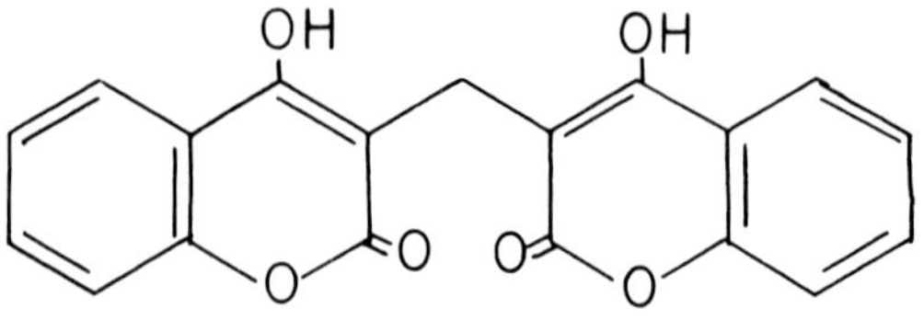

FIGURE 5. Bishydroxycoumarin, anticoagulant in sweet clover disease.

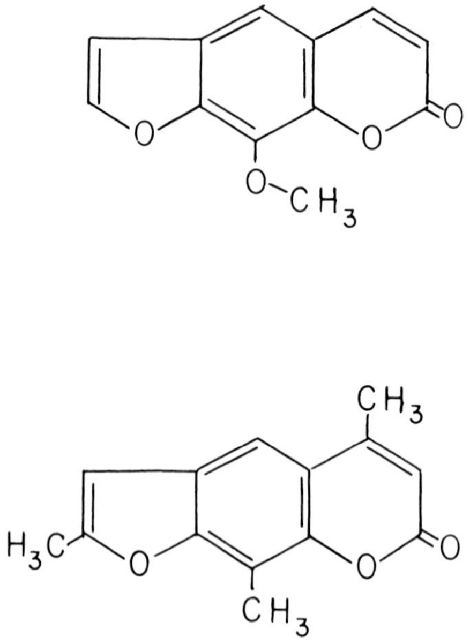

FIGURE 6. (a) 8-Methoxypsoralen, (b) trimethylpsoralen.

immediately before exposure to the light. This has not been reported in man, but then the high rate at which skin tumors are seen in humans may mask any effect psoralen treatment may have had in the past.

In 1961 Birmingham and colleagues[38] reported a phototoxic dermatitis which has been shown to be endemic among white harvesters of celery. The problem was first recognized in 1959 in Michigan. The Birmingham team studied 302 celery workers and found that 163 of them, 54%, had some degree of vesicular and bullous dermatitis, with areas of depigmentation and hyperpigmentation in areas from previous lesions (see Table 4).

The affected areas were the hands and forearms, but the lower legs, chest, abdomen, and back were also included if the worker wore shorts and no shirt. The lesions healed, with depigmentation lasting many months. The incidence was said to be greater after a rain and before the use of fungicides to control mold growth on the celery. The workers associated the dermatitis with pink rot disease, a fungal (*Sclerotinia sclerotiorum*) infection of celery.

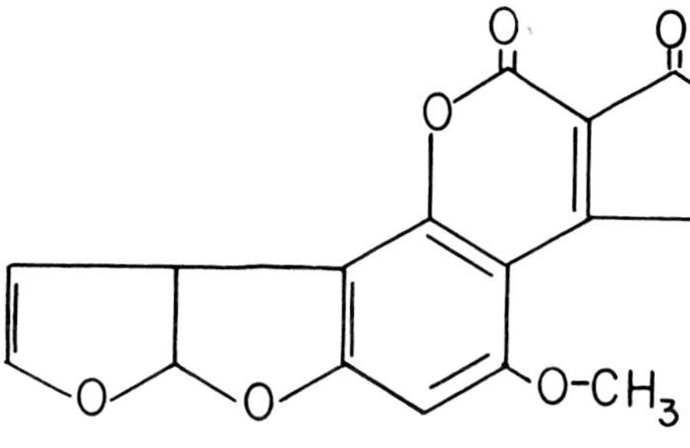

FIGURE 7. Aflatoxin B$_1$.

B. Epidemiological Evidence Associating Aflatoxins with Human Liver Cancer

The several field studies which have associated consumption of aflatoxins with human liver cancer have been recently reviewed.[67-68] The studies took place from 1966 to 1973 in Uganda, the Philippines, Thailand, Kenya, Mozambique, and Swaziland, in approximately that order. In most cases, a conscious effort was made to conduct the studies in such a manner as to facilitate comparing results between countries and test populations.

1. Uganda

The pioneering effort in the field associations was undertaken in 1966 by Alpert and colleagues at Harvard Medical School and the Massachusetts Institute of Technology.[69] Uganda provided an adequate site for the field study because of the high proportion of cereal grain and oil seeds in the diet, the traditional agricultural practices, and the varied climate: arid in the semideserts of the north, temperate in the southwestern mountains, and subtropical in the plains and marshes of the south.

Over a 10-month period, food samples were collected from village markets and home granaries throughout Uganda by staff and medical students on vacation leave from Kampala. Samples of maize, peanuts, cereal grains, beans, peas, and cassava were sealed upon collection and kept in cold storage until shipped by air freight to the U.S. for chemical assay for aflatoxins by the thin-layer chromatographic methods of Eppley[70] and Andrellos and Reid.[71]

The concentration of aflatoxins in the Uganda foodstuffs is summarized in Table 5. Approximately 30% of the 480 food samples analyzed contained aflatoxins, but most of these (61%) contained less than 100 μg of total aflatoxins per kilogram sample (100 ppb). Aflatoxins occurred most frequently in beans (72% of the samples), whereas peanuts (18%) and cassava (12%) were contaminated less frequently; the aflatoxin concentrations sometimes exceeded 1000 μg per sample (1 ppm).

The geographical distribution of aflatoxin contamination of foodstuffs was uneven. In the Toro district, 79% of the 29 samples analyzed contained alfatoxins; in Masaka district, 61% of 43 samples and in Karamoja 44% of 105 samples were contaminated.

At the time the aflatoxin survey was being made, local cancer registry records covering 1964 to 1966 were being used to estimate the geographical distribution of liver cancer in Uganda. The total number of liver cancer cases diagnosed during this period in the populaton of 6 million tribal people was 403, with histopathologic evidence on 310 of these cases. Davies and Owor[72] and Alpert et al.[73] claim that clinical diagnosis of hepatoma in Uganda is accurate in approximately 85% of the cases.

Table 6 gives the relation between the incidence of liver cancer and the aflatoxin contamination of foodstuffs in Uganda. Hepatoma occurred at an annual rate of 1.0 to

Table 5
AFLATOXIN CONCENTRATIONS IN UGANDAN FOODSTUFFS[60]

		Percent contaminated			
		Total aflatoxin concentration (μg/kg sample)			
Foodstuff	Number contaminated/ Number assayed	1—100	100—1000	>1000	All
Beans, peas	49/83	67	22	10	59
Cereal grains	40/162	75	15	15	25
Maize, peanuts	49/201	49	35	16	24
Cassava	4/34	0	50	50	12
Total	142/480	61	25	15	30

Table 6
HEPATOMA INCIDENCE AND FREQUENCY OF AFLATOXIN CONTAMINATION OF FOODSTUFFS IN UGANDA[60]

Area	Tribe	Hepatoma incidence	Number assayed	Percent contaminated	1—100	100—1000	>1000
Toro	Bwamba	—	29	79	10	31	38
Karamoja	Pokot(Suk) and Karamojang	6.8[135]	105	44	24	15	5
Buganda	Buganda Immigrants	2 3	149	29	23	4	1
West Nile	West Nile tribes	2.7	26	23	19	4	0
Acholi	Acholi	2.7	26	15	15	0	0
Busoga	Soga	2.4	39	10	5	5	0
Ankole	Ankole	1.4	37	11	11	0	0

2.7 cases per 100,000 people (all ages, both sexes) for most Ugandan tribes for which 10 to 23% of the food samples contained aflatoxins. In other areas where aflatoxin contamination was more frequent and more severe, the incidence of liver cancer was higher. For the Hutu and Tutsi immigrants to Buganda Province, aflatoxin contamination occurred in 29% of the samples, and liver cancer incidence was 3.0. For the Pokot (Suk) and Karamojan tribes, hepatoma incidence was 6.8 cases per 100,000 per year,[66,135] and the frequency of aflatoxin contamination was 44%. Aflatoxin appeared to be common (79% contamination frequency) in the foods of the Bwamba people; liver cancer data for these people, however, were not available.[69,135] This first field study on aflatoxin and human liver cancer can be summarized as providing evidence in support of a causal role for the mycotoxins in the human disease.

2. Philippines

Peanut butter and maize have been shown to be major contributors of aflatoxins to Philippine food products.[74] Aflatoxins were found in almost all of the 149 samples of peanut butter, with an average concentration of aflatoxin B$_1$ of 213 μg/kg peanut butter;

the most heavily contaminated sample of peanut butter contained 8600 μg/kg (8.6 ppm alfatoxin B$_1$). Analyses on whole peanuts showed that 80% of the samples were contaminated; the average concentration of aflatoxin B$_1$ was 98 μg/kg. Maize samples were contaminated to about the same extent, 95 out of 98 maize samples contained an average of 110 μg aflatoxin B$_1$ per kilogram.[67]

The incidence of liver cancer has not been measured directly in the Philippines. Estimates based on existing data for various areas of the country suggest a range of 0.15 to 1.17 cases per 100,000 people per year and for the country as a whole of 0.80 cases per 100,000 per year.[74] These estimates would appear quite low if indeed the contamination levels of peanut butter and maize are any indication of aflatoxin levels in the Philippine diet. According to Campbell and Salamat,[74] contaminated peanut butter is eaten principally by children; in fact, aflatoxin M$_1$, a metabolite of B$_1$, is found in the urine of these children. Maize is part of the diet apparently only of the people of the south central island of Cebu. As rice, the staple of most of the Philippines, seems to contain little aflatoxin, the general population may not be exposed to the toxins. This suggests that future epidemiological studies might focus on the inhabitants of Cebu and on unusual children's diseases which might have a toxicological etiology.

3. Thailand

With the appreciation that the aflatoxins were widespread in the dietaries of several populations in which liver cancer incidence seemed unusually high, several studies were begun to measure directly in defined population both aflatoxin consumption (not just frequency of contamination of raw commodities) and the incidence of histopathologically confirmed primary cancer of the liver. One of the first of these studies was done in Thailand from 1967 to 1970; the details of those studies are given in a series of reports by Shank and colleagues[65,75-78] and are summarized in a review.[66]

A report in 1966 by Bhamarapravati and Virranuvatti[55] showed that the frequency of primary cancer of the liver as revealed by liver biopsy on hospitalized patients was high; in 1,301 biopsies, the frequency of hepatocellular carcinoma was 251 (19% of all biopsies); cholangiocellular carcinoma was 61 cases (5% of all biopsies). A total of 86% of the liver cell carcinomas and 71% of the cholangiocarcinomas occurred in males. The high tumor frequency and the climate and agricultural practices which appeared to permit fungal invasion of foods and foodstuffs were principal factors in selecting Thailand for a major epidemiological survey to determine the role, if any, of aflatoxins in the etiology of human liver disease.

The initial phase of the study was designed to determine the extent of fungal invasion of Thai foods and foodstuffs and to determine which mycotoxins in addition to aflatoxin would have to be considered in the epidemiological study to follow. Over a 23-month period from September 1967 through July 1969, 2,179 samples of food and foodstuffs were examined in Bangkok for mold damage and aflatoxin content. Mycological studies on cereal grains, oil seeds, beans, cassava, dried fish, dried and fresh vegetables, spices, and prepared foods showed *Aspergillus* to be the most common contaminating fungus; in this genus, *Aspergillus flavus* predominated.[75] *Penicillium, Fusarium,* and *Rhizopus* fungi were also prevalent. Among 162 fungi isolated from the foodstuffs, 49 produced compounds other than aflatoxin, sterigmatocystin, and ochratoxins that were toxic to rats; three of the compounds have been identified and their structures elucidated: cytochalasin E[79] and tryptoquivaline and tryptoquivalone.[80]

Chemical assays for aflatoxins were carried out on each food sample.[76] The frequency and extent of contamination of Thai foodstuffs showed distinct seasonal and geographical distributions. The staple food in Thailand is rice, and analysis of 364 samples of uncooked milled rice revealed little contamination by aflatoxins (see Table 7). Peanuts were the

Table 7
CONCENTRATIONS OF AFLATOXINS IN THAI FOODSTUFFS[76]

Foodstuff	Number assayed	Percent contaminated	Total aflatoxin concentration (μg/kg sample) Mean (all samples)	Maximum
Peanuts	216	49	750	12,256
Chili peppers	106	11	14	966
Prepared foods	364	6	31	3,904
Dried fish/shrimp	139	5	8	772
Mung beans	140	5	1	112
Other beans	322	3	6	1,620
Sesame seeds	75	3	<1	<10
Cassava starch	65	3	5	294
Garlic/onions	58	3	2	60
Rice	364	2	<1	98

most frequently and highly contaminated with aflatoxins. Wheat, barley, Job's tear seeds, and dried chili peppers were also frequently contaminated, although, aflatoxin concentrations were considerably lower than those for peanuts.

The market survey of aflatoxins in foodstuffs provided a basis for the design of an epidemiological study to determine whether a relationship existed between human liver cancer and the daily rate of aflatoxin ingestion. In 1969, a 1-year pilot study was initiated to estimate simultaneously both dietary consumption of aflatoxins and incidence of liver cancer in three well-defined populations in Thailand.[77] The rationale used in selecting the survey villages included an estimation of high, intermediate, or low aflatoxin contamination of foods and proximity to provincial hospitals which could be used as centers for the liver cancer study.

The consumption of aflatoxin was determined by three separate surveys, each of 2-day duration, over a period of 1 year.[77] Within the three survey areas of Thailand (Singburi, Ratburi, and Songkhla), 144 households were selected at random in nine villages. Samples of food served were collected, and the amounts of each food eaten by the family were measured. Daily aflatoxin ingestion, expressed as nanograms of total aflatoxins consumed per kilogram body weight on a family rather than individual basis, was highest in Singburi (73 to 81), intermediate in Ratburi (45 to 77), and lowest in Songkhla (10 to 14). An important source of aflatoxins in the Thai diet was leftover cooked foods stored a day or longer without refrigeration. As noted in Table 8, one 72-year-old woman ingested relatively large amounts of aflatoxins; she lived alone and depended upon the village for her daily food, much of which was left over from meals served in other households the previous day.

Incidence of liver cancer was derived from investigation of all deaths occurring within the study population for the 1-year period coincident with the dietary survey. Persons aged 15 years or more who died during the study period were investigated by liver viscerotomy, irrespective of cause of death.[65] Incidence, as measured in this survey, was two new cases per 100,000 people per year in Songkhla and 6 new cases per 100,000 per year in Ratburi. National health records indicated that the incidence of primary liver cancer in the Singburi area was 14 deaths per 100,000 people per year, but this rate could not be measured directly as part of the aflatoxin study due to the unavailability of a key figure in the study. The association between aflatoxin consumption and human

Table 8
LIVER CANCER INCIDENCE AND AFLATOXIN CONSUMPTION IN THAILAND[65,77]

Area	Liver cancer incidence (cases/100,000/year)[a]	Average daily aflatoxin intake (ng/kg body weight) B_1	Average daily aflatoxin intake (ng/kg body weight) Total	Highest single day intake (ng/kg body weight) Total aflatoxins
Singburi	—	51—55	73—81	13,082
Ratburi	6.0	31—48	45—77	3,224
Songkhla	2.0	(5—6)[b]	(5—8)[b]	(1,072)[b]

[a] General population, all ages.
[b] Attributable mostly to 1 person, who consumed 1072 ng aflatoxins per kilogram body weight in 1 day and 380 ng/kg the following day, far exceeding daily intakes of all other subjects in the survey area; excluding this particular subject reduces average daily aflatoxin intakes to nil.

liver cancer is summarized in Table 8; the data were the first to support by direct measurement a causal role for the aflatoxins in this human disease.

4. Kenya

At the time of the Thailand study, another investigation was under way in Africa. The Murang'a district in central Kenya was chosen for the study primarily to make certain the general design and methods to be used were feasible and effective in demonstrating the association or lack of association of dietary aflatoxins with human liver cancer.[59] No evidence was available before the study to suggest that the incidence of liver cancer in Kenya was any different from that elsewhere in eastern Africa. Exposure to aflatoxins was considered likely due to climate and agricultural practices, although preliminary food surveys for the mycotoxins had not been done.

The topography of the Murang'a district is such that the survey populations were grouped as residents of the high altitude area with a population of nearly 40,000, the middle altitude area and population of over 160,000, and approximately 345,000 residents of the low altitude area. Differences between the three climates were predicted to result in differences in amounts of aflatoxin consumed and presumably the incidences of liver cancer.

The main evening meal was sampled over 2400 times in sample clusters of individuals distributed in 132 sublocations in the district. The collection period was 21 months, and samples were assayed for aflatoxins in Nairobi by standard chemical means. Estimation of the incidence of primary liver cancer in the district was based on data from the Kenya Cancer Registry; confirmation of most cases included histological diagnosis and a positive alpha-fetoprotein test and/or a rapid death following clinical diagnosis.

In the high-altitude area, 39 of 808 samples (5%) contained aflatoxins; mean concentration was 0.121 μg aflatoxin as B_1 per kilogram wet diet, including all negative samples; in the middle area, 7% (54 of 808) of the samples were contaminated with a mean concentration of 0.203 μg/kg; the low altitude area had the highest frequency, 78 of 816 (10%) of the samples, and highest mean concentration, 0.351 μg/kg, of aflatoxin contamination. The association between aflatoxin consumption and liver cancer in these survey populations is summarized in Table 9. A regression line was calculated for these data, expressing liver cancer incidence in arithmetic units and aflatoxin consumption in logarithmic units and combining males and females: cancer incidence, $y = 19.06 \log x - 10.16$, where x is the amount of aflatoxin B_1 consumed in nanograms per kilogram body weight per day. Even though this was a pilot study, as was the case in the inves-

Table 9
HEPATOMA INCIDENCE AND AFLATOXIN CONSUMPTION IN KENYA[59]

	Hepatoma incidence				Average daily aflatoxin B_1 intake (ng/kg body weight)	
	Cases/adult population[a] (1967—1970)		Cases/100,000 adults/year			
Altitude area	Male	Female	Male	Female	Male	Female
Low	16/30,949	9/41,375	12.9	5.4	14.81	10.03
Middle	13/30,105	6/45,693	10.8	3.3	7.84	5.86
High	1/8,027	0/10,885	3.1	<1/10,000/4 years	4.88	3.46

[a] 16 years of age and older.

tigation in Thailand, the results do show a positive correlation (correlation coefficient for the regression line was 0.87) between aflatoxin consumption and human liver cancer.

5. Mozambique

In 1965 Prates and Torres[49] reported the highest incidence of primary liver cancer in the world occurred in Mozambique; the estimated age-specific incidence rate of cancer of the liver and biliary passages in males was almost 110 new cases per 100,000 per year and for females was 29 new cases per 100,000 per year. Such cancer rates made consideration of mycotoxins as causal factors unattractive, for the amounts of aflatoxins that would have to be consumed would be far above any reasonable levels for the human food supply. For example, using the regression line of Peers and Linsell,[59] Mozambique males would have to consume approximately 2 mg of aflatoxin B_1 per kilogram per day to account for an incidence of 110 per 100,000 per year; this rate of consumption is approximately 100,000 times greater than the rates of aflatoxin consumption measured in Kenya.

In 1974 Van Rensburg and co-workers[63] reported results of a study they undertook to measure aflatoxin consumption in Mozambique, in particular the Inhambane district, the area of unusually high liver cancer incidence. Aflatoxin contamination of prepared foods consumed by the study population was measured by chemical assay of 880 meals. Approximately 9% of all samples contained the mycotoxin, with a mean concentration of 7.8 µg total aflatoxins per kilogram wet food for all samples. The mean daily per capita consumption of aflatoxins was calculated to be 222.4 ng/kg body weight, an order of magnitude greater than seen in either Thailand or Kenya.

Liver cancer rates were measured through a hospital registration program and from health records of goldminers originating in the study area. The hospital rate was 16 new cases per 100,000 people per year and was based on 460 histologically confirmed cases, however, field investigations suggested this was an underestimate. To the hospital rate was added the data from the goldminer's health records from 1969 to 1971, bringing the incidence of liver cancer to 25.4 cases per 100,000 people per year; for all males, the annual rate was 35 per 100,000 and for females it was 15.7 per 100,000. These data are in good agreement with the results from Thailand and Kenya and lend further support to a causal role of aflatoxins in human liver cancer in selected areas of the world.

6. Swaziland

Two studies on aflatoxins and human liver cancer have been done in Swaziland. In the first, reported by Keen and Martin[61] in 1971, cancer registry data for 1964 through 1968 indicated the crude annual rate of primary liver cancer was 8.6 cases per 100,000 males (all ages) and 1.6 cases per 100,000 females (all ages); actual cases recorded listed 75 males and 15 females over that 5-year period. Liver cancer appeared to have an uneven geographical distribution and to be more prevalent in Shangaan immigrants to Swaziland than it was for native Swazis.

Keen and Martin[61] found a geographical distribution for aflatoxins in peanut samples from the low, middle, and high velds which correlated with the distribution of liver cancer cases (see Table 10). Interviews with tribal groups indicated that Shangaans ate more peanuts (stored in powder form) more often and for longer periods than did the Swazis, and it was the Shangaans who had the greater liver cancer rates. Even the Swazis, who developed liver cancer, apparently lived in areas with Shangaans and ate more peanuts than Swazis living elsewhere; this apparently is due to Shangaan influence on Swazi eating habits.

In 1972 the International Agency for Research on Cancer and the Tropical Products Institute of London initiated a study in Swaziland which was modeled on their earlier study in Murang'a district of Kenya.[62] The incidence of liver cancer was estimated from the cancer registry data used by Keen and Martin.[61] Aflatoxin determinations were made from 1056 samples of the main meal and 455 samples of beer, sour porridge, sour drink, and peanuts not eaten with the main meal; the food samples were collected 6 times during a 1-year period in 11 areas from two randomly selected subchiefs per area and seven households living progressively further from the subchief. The results, summarized in Table 11, show a clear correlation between estimated aflatoxin consumption and liver cancer rates. The principal sources of aflatoxins were the main meals and peanut snacks. Peers and co-workers[62] calculated regression lines for aflatoxin ingestion correlated with primary liver cancer rates; for males, $y = 23.84 \log x - 13.66$, where y is cancer incidence per 100,000 adults per year, and x is the mean daily aflatoxin B_1 ingestion in nanograms per kilogram body weight; for females the line is $y = 4.68 \log x - 2.35$.

The impact that these studies in Swaziland, Mozambique, Kenya, and Thailand have on assessing the human cancer risk posed by the aflatoxins is discussed in detail in Chapter 4. The important conclusion here is that collectively these studies provide strong epidemiological evidence to support Oettlé's hypothesis that mycotoxins have a causative role in human liver cancer in Africa (and we can add Southeast Asia).[57]

C. Other Factors

1. Nutrition

The human diet is complex and variable. In laboratory studies, animals are maintained on a controlled diet, usually nutritionally adequate and much less variable in content than human diets. This distinction is important because it is now realized that the nutritional status of the animal can have profound effects on both the incidence of tumors and the different organs that may respond to the carcinogen.

The aflatoxins, as with most other environmental carcinogens, require metabolic activation to the ultimate carcinogenic form, a reactive electrophile (see Volume II, Chapter 2 for a complete discussion of metabolic activation of aflatoxin); the nutritional adequacy of an animal's diet may greatly influence the animal's response to aflatoxins.[81-84] Use of diets which depress or inhibit the metabolic activation of aflatoxin would be expected to result in lower tumor incidences; on the other hand, should the diet contain compounds which stimulate activation, the tumor incidence in aflatoxin-treated animals on such diets would be elevated. Rogers and Newberne[85] demonstrated that rats main-

Table 10
HEPATOMA RATE AND AFLATOXIN CONTAMINATION OF PEANUTS IN SWAZILAND[61]

Area	Number of liver cancer cases	Crude annual rate (cases/100,000, all ages)	Aflatoxin contamination of peanuts (%)
Lowveld	44	9.7	60
Middleveld	34	4.0	57
Highveld	11	2.2	20

TABLE 11
RELATION BETWEEN AFLATOXIN CONSUMPTION AND LIVER CANCER IN SWAZILAND[62]

Geographic area	Average daily aflatoxin B_1 intake (ng/kg body weight) Male	Female	Cases/adult population (1964—1968)[a] Male	Female	Cases/100,000 adults/year Male	Female	Cases/100,000/year, all ages Male	Female
Lowveld	53.34	43.14	35/26,266	7/24,907	26.65	5.62	15.3	3.1
Lebombo	19.89	15.40	4/8,713	0/10,034	18.65	<1/10,000	9.2	<1/10,000
Middleveld	14.43	8.89	24/32,464	5/45,225	14.79	2.21	6.9	1.2
Highveld	8.34	5.11	9/25,658	2/28,203	7.02	1.42	3.7	0.8

[a] 15 years of age and older.

tained on diet marginally deficient in lipotropes protected the animals against a single large dose of aflatoxin B_1, but made the animals more sensitive to the acute toxicity of repeated small doses of the mycotoxin. With repeated administration of aflatoxin, rats in the marginal lipotrope group developed hepatocarcinomas sooner and to a greater extent than did control animals (see Table 12).[86] Lee et al.[87] have demonstrated a variable liver cancer incidence in trout fed diets of varying protein content, but equal aflatoxin content.

Not only can cancer incidence be influenced by nutrition, but so also can cancer site be changed. Newberne and Rogers[88] fed rats diet containing vitamin A at nutritionally excessive, adequate, or marginally deficient levels. The frequency of liver cancer was little affected by varying the dietary vitamin A levels, however, rats maintained on the vitamin A-deficient diet also developed colon carcinomas. It is suggested that epithelial cells in the intestinal mucosa require vitamin A for the synthesis of a glycopeptide involved in absorption and metabolism of compounds in the gut,[89,90] and lowering the vitamin A level of the diet deprives the colon of normal protective mechanisms as yet uncharacterized. The impact of such findings is that although aflatoxin B_1 is generally regarded as a liver carcinogen, it can, under conditions not uncommon in the human environment, produce colon carinomas. It is unfortunate that the influence of vitamin A on aflatoxin carcinogenesis was not known before the epidemiological studies in Thailand and Africa were designed, for it might have been possible to associate aflatoxin ingestion not only with liver cancer, but also with colon cancer in vitamin A deficient populations. More work needs to be done on the role of nutrition in aflatoxin carcinogenicity and in carcinogenesis in general. Perhaps nutritional differences could explain

Table 12
EFFECT OF MARGINAL LIPOTROPE DIET ON AFLATOXIN CARCINOGENESIS IN THE RAT[86][a]

	Tumor incidence (%)			
	Months after initial exposure			
Diet	6	9	12	18
Control	0	0	30	60
Marginal lipotrope	25	50	60	85

[a] Aflatoxin B_1 dose: 25 μg/rat/day for 15 days.

why liver cancer appears to be a rare disease in Central and South America although aflatoxins seem to occur frequently and in high concentration in the food supplies for these areas.

2. Infection

Infection may influence carcinogenesis. In Thailand a retrospective study of deaths from liver cancer indicated that the northeastern section of that country had the highest death rate due to liver cancer and that most of the adult population of the northeast suffered to some extent from opisthorciasis, a liver fluke disease; this is also an area of high exposure to aflatoxins.[65,76] An analysis of liver biopsies from 1,301 patients admitted to a Bangkok hospital found 251 cases of liver cell carcinoma, 2% of which also had opisthorciasis; 61 cases of cholangiocarcinoma were found, 16% of which were associated with the liver fluke disease; the peak frequency of cholangiocarcinoma cases occurred during the 6th decade, while for liver cell carcinoma cases, the peak occurred 10 years earlier.[55] Recently, one of these investigators[91] has observed in an animal model, the hamster exposed to dimethylnitrosamine, a compound that causes liver cell carcinomas in normal animals, that the nitrosamine induces cholangiocarcinomas in animals infected with *Opisthorchis viverrini,* the liver fluke common to Thailand. Such results suggest that aflatoxin may induce cholangiocarcinomas in humans with opisthorciasis, in which case the infection would seem responsible for the cancers involving the new cell type.

Viral hepatitis can lead to liver cirrhosis, and because this infection occurs in many liver cancer areas, it has been claimed to have a causative role in the cancer. Increased frequencies in the appearance of hepatitis-associated antigen have been reported many times in hepatoma patients,[92-95] but this association does not always hold.[60,96-98] Unfortunately, the epidemiological question usually asked is "given a cluster of liver cancer cases, how many have had a history of viral hepatitis?", the converse is not asked, "given a cluster of viral hepatitis cases, how many develop liver cancer compared to a matched population free of viral hepatitis?". Nevertheless, even though viral hepatitis may not appear to be the sole etiological factor in human liver cancer, the frequent association between the infection and the tumor is undeniable and an influence by the infection on the development of the cancer is suggested.

3. Cointoxication

Epidemiological studies have associated ingestion of aflatoxins with human liver cancer in certain populations, but this does not necessarily indicate that the aflatoxins are the sole causative agents. Diets that can contain appreciable quantities of aflatoxins would appear capable of containing other mycotoxins as well. Because of the extraordinary potency of aflatoxin B_1 as a heptocarcinogen, only limited effort has been made to assess the role other mold metabolites might play in the etiology of human liver cancer. Purchase and Goncalves[99] surveyed selected African diets, already known to contain aflatoxin, for the carcinogenic mycotoxin, sterigmatocystin, but was unable to demonstrate a significant level of contamination by sterigmatocystin. Ochratoxin and penicillic acid can occur simultaneously in the same culture of *Aspergillus ochraceus*.[100] The investigation of the death of a Thai boy[101] demonstrated that 3 days before becoming ill the boy had eaten leftover cooked rice heavily contaminated with four toxigenic molds; the rice contained high concentrations of the aflatoxins and in addition a strain of *Aspergillus clavatus*[102] which produced the highly toxic cytochalasin E[79,103,104] and two quinazolone tremorgens.[80] Whether the cytochalasin E and tremorgens were also present in the rice could not be determined, and the cotoxicity of these agents has not yet been studied.

Lindenfelser et al.[105] demonstrated a synergism in the acute toxicity of aflatoxin B_1 coadministered with the trichothecene *Fusarium* mycotoxin, T-2 toxin.

Several reports from Asia and Africa have indicated that carcinogenic N-nitroso compounds may exist in dietaries (see Chapter 6) which are already known to contain aflatoxins and may also contain plant toxins such as the pyrrolizidine alkaloids. It is not yet known how all these food-borne toxins may interact, and much research is needed on this important environmental problem.

IX. AFLATOXINS AND ACUTE POISONINGS

It seems reasonable to assume that, if aflatoxins could exist in the Asian and African dietaries to an extent sufficient to induce hepatocellular carcinomas, occasionally the concentration of these toxins might be so great as to precipitate an acute response. Although in the liver cancer etiology studies, the consumption of aflatoxins was calculated on the amounts of toxin in cooked foods averaged over long periods, instances were observed where individual samples contained unusually high concentrations of aflatoxins, e.g., 1.3 ppm in peanuts in Mozambique,[106] 1.7 ppm in cassava in Uganda,[107] and 12.3 ppm in roasted peanuts in Thailand;[76] in each case, these were "food-grade" samples. There are a few reports from Asia and Africa which suggest that some outbreaks of putative acute poisonings could have resulted from ingestion of large amounts of aflatoxins over short periods of time; most of these outbreaks involved children.

A. Taiwan Outbreak

In 1967 a report was published presenting the results of a study of an outbreak of apparent poisoning of 26 persons in two Taiwan rural villages.[108] The victims came from households which had consumed moldy rice for up to 3 weeks; they developed edema of the legs and feet, abdominal pain, vomiting, and palpable livers, but no fever. The three fatal cases were children between the ages of 4 and 8 years; autopsies were not done, and the cause of death could not be established. In a retrospective analysis of the outbreak, a few rice samples from affected households were assayed for aflatoxins; two of the samples contained up to 200 µg aflatoxin B_1 per kilogram. This first report of a possible link between aflatoxin consumption and an acute toxic response in man alerted many researchers to consider both acute and chronic aspects of aflatoxin toxicity.

B. Ugandan Case

Aflatoxin B_1 was circumstantially associated with the death of a 15-year-old African boy in Uganda.[107] The youth, his younger brother, and his sister became ill at approximately the same time; the younger siblings survived, but the older boy died 6 days later with symptoms resembling the victims in the Taiwan outbreak discussed above. An autopsy revealed pulmonary edema, flabby heart, and diffuse necrosis of the liver. Histology demonstrated interstitial edema of the heart, centrilobular necrosis with a mild fatty liver, in addition to the edema and congestion in the lungs. Serck-Hanssen[107] stated that the main components of the diet of these children were cassava, beans, fish, and meat. A sample of the cassava contained 1.7 mg aflatoxins per kilogram (1.7 ppm) which Alpert and Serck-Hanssen[109] suggest may be lethal if such a diet were consumed over a few weeks; this estimate is based on studies of the acute toxicity of aflatoxin B_1 in the African monkey.[109]

C. Reye's Syndrome (Thailand)
1. Characteristics of the Disease

Considerably stronger evidence associating the aflatoxins with an acute toxic response in humans came from Thailand in 1971. While investigating in the northeast of Thailand what appeared to be a viral infection in children with clinical symptoms similar to viral encephalitis, a medical team from Udorn Provincial Hospital and SEATO Medical Research Laboratory in Bangkok discovered that the disease was identical to what Reye and co-workers[110] described in 1963 in Australia as "Encephalopathy and Fatty Degeneration of the Viscera" and is known more widely now as Reye's syndrome.[111] The disease affects young children, particularly between the ages of 3 and 8 years and is characterized by a short prodrome followed by a sudden onset of coma and convulsions; there is a rapid progression of the condition with deepening coma, irregular breathing, and a usually fatal outcome within 72 hr. Autopsy reveals epicardial petechiae, some myocardial fat deposits, interstitial hemorrhage and congestion of the lungs, enlarged, pale fatty liver with little necrosis, swollen pale fatty kidneys, and severe cerebral edema. Blood contains low levels of glucose and high concentrations of ammonia, and there is no evidence of infection or inflammation. After several years study, the researchers were unable to find any evidence to associate this disease with a virus.[112,113]

2. Implication of Aflatoxin

The disease occurs in Thailand with distinct seasonal and geographic distributions, with most cases seen during the rainy season in rural families living in the northeast of Thailand. These distributions are quite similar to those for aflatoxin contamination of Thai foodstuffs[76] and in part led the medical team to suggest a possible mycotoxin etiology for Reye's syndrome as it occurs in Thailand.

Small specimens of leftover foods eaten by two Thai children prior to the onset of Reye's syndrome were found to be heavily contaminated with aflatoxin and a variety of toxigenic molds. In one such case reported by Bourgeois and co-workers,[113] a 3-year-old Thia boy was brought to a provincial hospital after a 12-hr illness of fever, vomiting, coma, and convulsions. The child died 6 hr later, and an autopsy revealed marked cerebral edema with neuronal degeneration, severe fatty metamorphosis of the liver, kidneys, and heart, and lymphocytolysis in the spleen, thymus, and lymph nodes. Upon admission of the child to the hospital, a medical team traveled to the boy's home and obtained a small sample of steamed glutinous rice which had been cooked 2 days before the onset of the child's illness and reportedly had been the only food the family had eaten for the past 2 days. The small size of the sample precluded an accurate measurement of the amount of aflatoxins present but chemical assay indicated the amount was in the parts per million range. The rice also contained toxigenic strains of *A. flavus, A. clavatus, A. ochraceous,* and *A. niger*.[114] The *A. clavatus* strain was later shown to produce cytochalasin E[79,103,104] and two quinazolone tremorgens;[80] cytochalasin E appears to increase the permeability of cell membranes and produces severe cerebral edema.[104] This circumstantial association of Reye's syndrome with the ingestion of mycotoxins stimulated further studies in Thailand.

Although a few reports had been published on the acute toxicity of aflatoxins in monkeys, a detailed dose-response study was not done until this opportunity presented itself in Thailand. A total of 24 young macaque monkeys were given orally a single administration of 0.5 to 40.5 mg purified aflatoxin B_1 per kilogram body weight. Some animals receiving 4.5 mg/kg and all animals receiving 13.5 mg/kg aflatoxin showed a response strikingly similar to Reye's syndrome in children and died 67 to 148 hr after toxin administration.[113] Unchanged aflatoxin B_1 could be detected in tissues from animals dying as late as 6 days after receiving a single dose of the toxin;[114] this finding prompted an analysis for aflatoxins in tissues from Reye's syndrome victims.

Autopsy specimens from 23 Thai children who had died with Reye's syndrome and from 15 children and youths who had died from unrelated causes were assayed for aflatoxins B_1, B_2, G_1, G_2, and M_1.[115] The B_1 form was found in one or more specimens from 22 of the 23 Reye's syndrome cases, and in several instances, these aflatoxin concentrations were as high as those seen in specimens from the monkeys poisoned with the aflatoxins (see Table 13). Trace amounts of aflatoxin B_1 were found in tissues from 10 to 13 children aged 1 to 7 years, who died of causes unrelated to Reye's syndrome, and in one of two youths, who died in accidents. These 15 control cases lived and died in the same areas as the Reye's syndrome cases, areas in which aflatoxin contamination of human foodstuffs is widespread;[76] the trace amounts of aflatoxins in tissue specimens from control cases is thought to reflect chronic low-level ingestion of the mycotoxin in that area of Thailand.

3. Associations Outside Thailand

Others have also reported aflatoxin residues in autopsy specimens of children dying of Reye's syndrome. Becroft in New Zealand, who first suggested that contamination of foods by aflatoxin may have a role in the etiology of Reye's syndrome,[116] analyzed liver specimens from two children in Auckland, who died of Reye's syndrome.[117] Chloroform extracts from both liver specimens contained blue fluorescing material chromatographically similar to aflatoxins B_1 and G_1; the amount of aflatoxin B_1 present was estimated to be in the range of 5 to 50 µg/kg liver in each specimen. Analysis of ten control livers did not detect the presence of aflatoxins.

Dvoráčková et al.[118] reported finding aflatoxin B_1 in liver specimens from two infants, who died with liver damage and encephalopathy, and later[119] added four more cases of infants, who died similarly and also had detectable concentrations of aflatoxin B_1 in their livers. The infants ranged in age from 3 days to 12 months, but in each case the investigators had evidence to support the possibility of exposure to the mycotoxin.

Two reports in the U.S. have suggested an association between Reye's syndrome and aflatoxin exposure. Chaves-Carballo and co-workers[120] found fluorescing material chromatographically similar to aflatoxin G_2 in the fixed (formaldehyde) liver of a 15-year-old Reye's syndrome patient; similar material could not be found in seven other cases or in 12 controls. The fluorescent material had UV and IR spectral characteristics corresponding to aflatoxin B_1 and was present in the liver at a concentration of 22.5 µg/kg as aflatoxin B_1. Hogan and co-workers[121] have described an additional case.

4. Contributing Factors

Other agents have been proposed as responsible for Reye's syndrome. In the U.S., it has been generally assumed that the disease is a viral one, probably associated with outbreaks of chicken pox and influenza B.[122,123] A viral etiology would explain seasonal

Table 13
COMPARISON OF AFLATOXIN B₁ CONCENTRATIONS IN AUTOPSY SPECIMENS FROM REYE'S SYNDROME CASES AND MONKEYS POISONED WITH AFLATOXIN B₁[114,115]

	Aflatoxin B₁ concentrations (μg/kg specimen or mℓ fluid)	
Specimen	Human	Monkey (AFB₁ dose) (mg/kg)
Brain	1—4	30 (13.5)
		30 (40.5)
Liver	47	37 (13.5)
	93	163 (40.5)
Kidney	1—4	87 (40.5)
		162 (0.5)
Bile	8	125 (13.5)
		150 (13.5)
		163 (40.5)
Stool	13	
	42	
	108	
	123	
Stomach contents	116	
	127	
Intestinal contents	81	

and geographic variations and the occasional fever and upper respiratory tract infections seen; on the other hand, the lack of family involvement and lack of pathologic findings consistent with viral diseases are difficult to explain. Also, in areas where the disease occurs epidemically such as Thailand, vigorous attempts to associate a virus with Reye's syndrome have been consistently unproductive.[112]

Mycotoxins as the sole agents in the etiology of Reye's syndrome also fail to explain all aspects of the disease. Lack of sibling involvement, few reports of sublethal poisonings, and the improbability of ingesting sufficient toxin at one time to precipitate the disease are factors which prompt suggestion of a multicomponent etiology. Bradford and Latham[124] questioned whether Reye's syndrome is a single disease entity, and Bourgeois et al.[101] have proposed the pathogenesis may begin with a toxic, nutritional, or infectious insult to produce a clinically unmanifested liver injury such as a fatty metamorphosis; this would be followed by a second injury to the liver, perhaps acute aflatoxicosis or a viral infection, resulting in hypoglycemia, and accumulation of metabolites, such as ammonia and free fatty acids. A quick progression would follow: further liver injury, fatty degeneration of the liver, kidney, and heart, convulsions and cerebral edema because of severe hypoglycemia and hyperammonemia, coma, and death. This view has been expanded by Mullen,[125] who proposed that environmental exposure to various toxic compounds may be responsible for the initiation of the pathogenesis and alter the body's biochemical and immunological competence, rendering the body more susceptible to viral infection. Many of the complexities of the disease itself and suggested etiologic factors are reviewed in the published proceedings of a conference on Reye's syndrome held in 1974.[126]

D. Indian Childhood Cirrhosis

A childhood cirrhosis occurs at high frequency in certain areas in India; peak incidence occurs at 3 years of age. After symptoms of GI upset and anorexia, the children develop an enlarged liver with a characteristic leafy border; the disease may progress to jaundice, ascites, fibrosis, cirrhosis, and hepatic coma.[127,128] In 1967 Robinson[129] described a similar syndrome in Indian infants and analyzed foods, breast milk, and urine samples for the presence of aflatoxins. Some association was seen between cirrhosis cases and the presence in these samples of fluorescent materials similar to the aflatoxins. Two additional reports in 1970 suggested that aflatoxin consumption was associated with the children's disease,[127,130] but again the chemical evidence offered to support the suggestion was not strong.

Protein-caloric malnutrition is often treated in India with peanut protein flour supplements. In one episode,[128] children suffering from kwashiorkor were given peanut flour supplement for several days until it was discovered that the flour contained 300 μg aflatoxins per kilogram flour. The supplementation was stopped, and 20 children aged 1.5 to 5 years, who had eaten 1 to 2 oz peanut supplement daily for 5 days to 1 month, were studied for possible aflatoxin-induced liver damage. In 6 months to a year, some of the children developed lesions similar to Indian childhood cirrhosis, whereas this progression is not typical of kwashiorkor cases.

E. Indian Hepatitis Outbreak

In October 1974, unseasonal rains in western India resulted in extensive mold damage to standing corn crops. The people in these rural areas were poor and thus forced to eat the contaminated grain for lack of alternate foodstuffs. After a few weeks of consuming the moldy corn, many people in over 200 villages became ill with symptoms of liver injury.[131] The people became ill almost simultaneously, with vomiting, anorexia, and jaundice. Severe cases progressed to ascites and edema of lower extremities; in only a few cases was the liver enlarged and tender. Of the 397 patients studied, 106 died, in most instances with massive GI bleeding. No infants were involved, and most cases were between the ages of 6 and 30 years. Dogs who shared the food of affected households also developed ascites and jaundice and died a few weeks after onset; other domestic animals which did not share the family food were not affected.

Five specimens of corn were collected from affected households; they were contaminated with *A. flavus,* and chemical analysis revealed aflatoxin contents ranging from 6.25 to 15.6 mg/kg corn which is extremely heavy contamination. One liver specimen was assayed, but did not contain a detectable level of aflatoxin; also seven urine specimens were negative in the aflatoxin assay. Aflatoxin B_1 was detected in 2 of 7 serum samples collected from patients. Histopathologic examination of liver specimens revealed extensive bile duct proliferation, periportal fibrosis, and occasional multinucleated grain cells. The authors estimated that the patients had ingested 2 to 6 mg of aflatoxin each day for several weeks. Attempts to associate the disease with infectious agents were unsuccessful, and the circumstantial evidence suggests the hepatitis outbreak was caused by the large aflatoxin intakes.

F. Other Possible Cases

Two cases have been described in which chemical analyses of liver specimens from men dying of liver injury have suggested the presence of aflatoxins. Bosenberg[132] in Germany described the case of a 45-year-old man, who died a short time after an apparent gastric illness. He had eaten an unusually large amount of nuts, which were apparently quite moldy. The illness was diagnosed as acute yellow atrophy of the liver, but analysis of the liver revealed the presence of a blue fluorescing material which cochromatographed with aflatoxin B_1 on a thin layer chromatographic (TLC) plate. The author suggests the case may be one of acute aflatoxin poisoning.

In 1976 Phillips et al.[133] reported on the analysis of a liver specimen from a 56-year-old man with carcinoma of the rectum and liver. The liver contained 520 ng aflatoxin B_1 per gram liver; chemical identity was confirmed by mass spectroscopy. The source of the aflatoxin for this rural resident of Missouri could not be determined. The level of aflatoxin in this liver specimen is an order of magnitude greater than was found in the livers from Reye's syndrome victims discussed above.

It has generally been assumed that exposures to large amounts of aflatoxin and presumably most other mycotoxins occur only in areas of the world with a tropical climate and with agricultural practices not common to the technologically modern countries in the western hemisphere. With greater experience and at closer scrutiny, however, it appears this assumption may not be valid. Although most foodstuffs produced and sold in the advanced temperate countries appear to contain at most only trace amounts of aflatoxins, small local conditions may arise where the exposure could be considerable, e.g., the man in Germany, who apparently ate a lethal amount of moldy nuts. It is conceivable that silo operators when handling moldy grains could be exposed to high concentrations of airborne mycotoxins; similar conditions may also occur in grain processing plants and even food animal feedlots. A report from the Netherlands cautions on this problem;[134] in particular, employees in a nut handling plant were exposed to settled dusts containing up to 330 μg total aflatoxins per kilogram of dust, and in airborne dust, the levels were as high as 410 μg/kg. The authors estimated some employees could inhale as much as 2.5 μg aflatoxins per 45-hr work week, approximately 10% of the exposure seen in the dietary studies of Asia and Africa reviewed above. This level of exposure may not be inconsequential.

REFERENCES

1. **Barger, G.,** *Ergot and Ergotism,* Gurney and Jackson, Edinburgh, 1931.
2. **Bové, F. J.,** *The Story of Ergot,* S. Karger, Basel, 1970, 297.
3. **Groger, D.,** Ergot, in *Microbial Toxins,* Vol. 8, Kadis, S., Ciegler, A., and Ajl, S. J., Eds., Academic Press, New York, 1972, 321.
4. **Barger, G., Carr, F. H., and Dale, H. H.,** An active alkaloid from ergot, *Br. Med. J.,* 2, 1792, 1906.
5. **Kraft, F.,** Uber dass Mutterkorn, *Arch. Pharm., Berlin,* 244, 336, 1906.
6. **Stoll, A.,** Zur Kenntis der Mutterkornalkaloide, *Verh. Naturforsch. Ges. Basel,* 101, 190, 1920.
7. **Gabbai, Lisbonne, and Pourquier,** Ergot poisoning at Pont St. Esprit, *Br. Med. J.,* 2, 650, 1951.
8. **Mayer, C. F.,** Endemic panmyelotoxicosis in the Russian grain belt. I. The clinical aspects of alimentary toxic aleukia (ATA). A comprehensive review, *Mil. Surg.,* 113, 173, 1953.
9. **Mayer, C. F.,** Endemic panmyelotoxicosis in the Russian grain belt. II. The botany, phytopathology, and toxicology of the Russian cereal food, *Mil. Surg.,* 113, 295, 1953.
10. **Joffe, A. Z.,** Alimentary toxic aleukia, in *Microbial Toxins,* Vol. 7, Kadis, S., Ciegler, A., and Ajl, S. J., Eds., Academic Press, New York, 1971, 139.
11. **Mirocha, C. J. and Pathre, S.,** Identification of the toxic principle in a sample of poaefusarin, *Appl. Microbiol.,* 26, 719, 1973.
12. **Bamburg, J. R., Strong, F. M., and Smalley, E. B.,** Toxins from moldy cereals, *J. Agric. Food Chem.,* 17, 443, 1969.
13. **Wyatt, R. D., Weeks, B. A., Hamilton, P. B., and Burmeister, H. R.,** Severe oral lesions in chickens caused by ingestion of dietary fusariotoxin T-2, *Appl. Microbiol.,* 24, 251, 1972.
14. **Wyatt, R. D., Colwell, W. M., Hamilton, P. B., and Burmeister, H. P.,** Neural disturbances in chickens caused by dietary T-2 toxin, *Appl. Microbiol.,* 26, 757, 1973.

15. **Uraguchi, K.**, Neurotoxic mycotoxins of *Penicillium citreo-viride* Biourge, in *Pharmacology and Toxicology of Naturally Occurring Toxins*, Raskova, H., Ed., Academic Press, New York, 1970, 143.
16. **Uraguchi, K.**, Yellowed rice toxins: citreoviridin, in *Microbial Toxins*, Vol. 6, Ciegler, A., Kadis, S., and Ajl, S. J., Eds., Academic Press, New York, 1971, 367.
17. **Shank, R. C.**, Mycotoxins in the environment, in *Trace Substances and Health. A Handbook, Part I*, Newberne, P. M., Ed., Marcel Dekker, New York, 1976, 67.
18. **Moroeanu, S. B.**, Epidemiological observations on the endemic nephropathy in Rumania, in *The Balkan Nephropathy*, Wolstenholme, G. E. W. and Knight, J., Eds., Churchill, London, 1967, 4.
19. **Bulic, F.**, The possible role of genetic factors in the aetiology of the Balkan nephropathy, in *The Balkan Nephropathy*, Wolstenholme, G. E. W. and Knight, J., Eds., Churchill, London, 1967, 17.
20. **Danilović, V. and Stojimirović, B.**, Endemic nephropathy in Kolubaro, Serbia, in *The Balkan Nephropathy*, Wolstenholme, G. E. W. and Knight, J., Eds., Churchill, London, 1967, 44.
21. **Gaon, J. A.**, Endemic nephropathy in Bosnia, in *The Balkan Nephropathy*, Wolstenholme, G. E. W. and Knight, J., Eds., Churchill, London, 1967, 51.
22. **Barnes, J. M.**, Possible nephrotoxic agents, in *The Balkan Nephropathy*, Wolstenholme, G. E. W. and Knight, J., Eds., Churchill, London, 1967, 110.
23. **Krough, P., Hald, B., Pleština, R., and Čeovíc, S.**, Balkan (endemic) nephropathy and foodborne ochratoxin A: preliminary results of a survey of foodstuffs, *Acta. Pathol. Microbiol. Scand. Sect. B*, 85, 238, 1977.
24. **Puchlev, A.**, Endemic nephropathy in Bulgaria, in *The Balkan Nephropathy*, Wolstenholme, G. E. W. and Knight, J., Eds., Churchill, London, 1967, 28.
25. **Schofield, F. W.**, Damaged sweet clover: the cause of a new disease in cattle simulating hemorrhagic septicemia and blackleg, *J. Am. Vet. Med. Assoc.*, 64, 553, 1924.
26. **Link, K. P.**, The anticoagulant from spoiled sweet-clover hay, *Harvey Lect.*, 39, 162, 1943.
27. **Huebner, C. F. and Link, K. P.**, Studies on the hemorrhagic sweet clover disease. VI. The synthesis of the δ-diketone derived from the hemorrhagic agent through alkaline degradation, *J. Biol. Chem.*, 138, 529, 1941.
28. **Stahmann, M. A., Huebner, C. F., and Link, K. P.**, Studies on the hemorrhagic sweet clover disease. V. Identification and synthesis of the hemorrhagic agent, *J. Biol. Chem.*, 138, 513, 1941.
29. Proceedings of the Symposium on Psoralen and Radiant Energy, Brook Lodge, Michigan, March 27 to 28, 1958, *J. Invest. Dermatol.*, 32, 133, 1959.
30. **Becker, S. W., Jr.**, Use and abuse of psoralen, *JAMA*, 173, 1483, 1960.
31. **Elliott, J. A., Jr.**, Clinical experiences with methoxsalen in treatment of vitiligo, *J. Invest. Dermatol.*, 32, 311, 1959.
32. **Labby, D. H., Imbrie, J. D., and Fitzpatrick, T. B.**, Studies of liver function in subjects receiving methoxsalen, *J. Invest. Dermatol.*, 32, 273, 1959.
33. **Tucker, H. A.**, Clinical and laboratory tolerance studies in volunteers given oral methoxsalen, *J. Invest. Dermatol.*, 32, 277, 1959.
34. **Griffin, A. C., Hakin, R. E., and Knox, J.**, The wave length effect upon erythermal and carcinogenic response in psoralen treated mice, *J. Invest. Dermatol.*, 31, 289, 1958.
35. **Griffin, A. C.**, Methoxsalen in ultraviolet carcinogenesis in the mouse, *J. Invest. Dermatol.*, 32, 367, 1959.
36. **Urbach, F.**, Modification of ultraviolet carcinogenesis by photoactive agents, *J. Invest. Dermatol.*, 32, 373, 1959.
37. **Forbes, P. D. and Urbach, F.**, Experimental modification of photo-carcinogenesis. II. Fluorescent whitening agents and simulated solar UVR, *Food Cosmet. Toxicol.*, 13, 339, 1975.
38. **Birmingham, D. J., Key, M. M., Tubich, G. E., and Perone, V. B.**, Phototoxic bullae among celery harvesters, *Arch. Dermatol.*, 83, 73, 1961.
39. **Musajo, L., Caporale, G., and Rodighiero, G.**, Isolation of bergapten from celery and parsley, *Gazz. Chim. Ital.*, 84, 870, 1954; as cited in **Birmingham, D. J., Key, M. M., Tubich, G. E., and Perone, V. B.**, *Arch. Dermatol.*, 3, 73, 1961.
40. **Innocenti, G., Dall Acqua, F., and Caporale, G.**, Investigations of the content of furocoumarins in *Apium graveolens* and in *Petroselinum sativum*, *Planta Med.*, 29, 165, 1976.
41. **Scheel, L. D., Perone, V. B., Larkin, R. L., and Kupel, R. E.**, The isolation and characterization of two phototoxic furanocoumarins (psoralens) from diseased celery, *Biochemistry*, 2, 1127, 1963.
42. **Perone, V. B., Scheel, L. D., and Meitres, R. J.**, A bioassay for the quantitation of cutaneous reactions associated with pink rot celery, *J. Invest. Dermatol.*, 42, 267, 1964.
43. **Yu, H.**, Formation of phytotoxic substance in celery, *Kuo Li Tai-wan Ta Hsueh Chih Wu Ping Chung Hai Huseh Kan*, 4, 133, 1975; *Chem. Abstr.*, 84, 176825w, 1976.
44. **Cole, R. S.**, Light-induced cross-linking of DNA in the presence of a furocoumarin (psoralen). Studies with phage λ, *Escherichia coli*, and mouse leukemic cells, *Biochim. Biophys. Acta*, 217, 30, 1970.

45. **Doll, R., Muir, C., and Waterhouse, J., Eds.,** Cancer Incidence in Five Continents, Vol. 1, International Union Against Cancer, Geneva, 1966.
46. **Segi, M. and Kurihara, M.,** Cancer Mortality for Selected Sites in 21 Countries, 1966-1967, No. 6, Japan Cancer Society, Tokyo, 1972, 104.
47. **Higginson, J. and Oettlé, A. G.,** Cancer incidence in the Bantu and "Cape Colored" races of South Africa: report of a cancer survey in the Transvaal, 1953-55, *J. Natl. Cancer Inst.,* 24, 589, 1960.
48. **Oettlé, A. G.,** Cancer in Africa, especially in regions south of the Sahara, *J. Natl. Cancer Inst.,* 33, 383, 1964.
49. **Prates, M. D. and Torres, F. O.,** A cancer survey in Lourenco Marques, Portuguese-East Africa, *J. Natl. Cancer Inst.,* 35, 729, 1965.
50. **Berman, C.,** *Primary Carcinoma of the Liver,* Lewis, London, 1951.
51. **Stewart, H. L.,** Geographic distribution of hepatic cancer, in *Primary Hepatoma,* Burdette, W. S., Ed., University of Utah Press, Salt Lake City, 1965, 31.
52. **Yeh, S. and Cowdry, E. V.,** Incidence of malignant tumors in Chinese, especially in Formosa, *Cancer,* 7, 425, 1954.
53. **Shanmugaratnam, K.,** Primary carcinomas of the liver and biliary tract, *Br. J. Cancer,* 10, 232, 1956.
54. **Marsden, A. T. H.,** The geographical pathology of cancer in Malaya, *Br. J. Cancer,* 12, 161, 1958.
55. **Bhamarapravati, N. and Virranuvatti, V.,** Liver diseases in Thailand. An analysis of liver biopsies, *Am. J. Gastroenterol.,* 45, 267, 1966.
56. **Tuyns, A. J. and Obradovic, M.,** Unexpected high incidence of primary liver cancer in Geneva, Switzerland, *J. Natl. Cancer Inst.,* 54, 61, 1975.
57. **Oettlé, A. G.,** The aetiology of primary carcinoma of the liver in Africa: a critical appraisal of previous ideas with an outline of the mycotoxin hypothesis, *S. Afr. Med. J.,* 39, 817, 1965.
58. **Alpert, M. E., Hutt, M. S. R., and Davidson, C. S.,** Hepatoma in Uganda. A study in geographic pathology, *Lancet,* 1, 1265, 1968.
59. **Peers, F. G. and Linsell, C. A.,** Dietary aflatoxins and liver cancer — a population based study in Kenya, *Br. J. Cancer,* 27, 473, 1973.
60. **Simons, M. J., Yap, E. H., Yu, M., Seah, C. S., Chew, B. K., Fung, W. P., Tan, A. Y. O. and Shanmugaratnam, K.,** Australia antigen in Singapore Chinese patients with hepatocellular carcinoma, *Lancet,* 1, 1149, 1971.
61. **Keen, P. and Martin, P.,** Is aflatoxin carcinogenic in man? The evidence in Swaziland, *Trop. Geogr. Med.,* 23, 44, 1971.
62. **Peers, F. G., Gilman, G. A., and Linsell, C. A.,** Dietary aflatoxins and human liver cancer. A study in Swaziland, *Int. J. Cancer,* 17, 167, 1976.
63. **Van Rensburg, S. J., Van der Watt, J. J., Purchase, I. F. H., Coutinho, L. P., and Markham, R.,** Primary liver cancer rate and aflatoxin intake in a high cancer area, *S. Afr. Med. J.,* 48, 2508a, 1974.
64. **Viranuvatti, V. and Satapanakul, C.,** Primary carcinoma of liver: analysis of 90 cases, *Proc. 9th Pacific Sci. Cong., 1957,* 17, 416, 1962.
65. **Shank, R. C., Bhamarapravati, N., Gordon, J. E., and Wogan, G. N.,** Dietary aflatoxins and human liver cancer. IV. Incidence of primary liver cancer in two municipal populations of Thailand, *Food Cosmet. Toxicol.,* 10, 171, 1972.
66. **Shank, R. C.,** Epidemiology of aflatoxin carcinogenesis, *Adv. Mod. Toxicol.,* 3, 291, 1977.
67. **Campbell, T. C. and Stoloff, L.,** Implication of mycotoxins for human health, *J. Agric. Food Chem.,* 22, 1006, 1974.
68. **Shank, R. C.,** The role of aflatoxin in human disease, *Adv. Chem.,* 149, 51, 1976.
69. **Alpert, M. E., Hutt, M. S. R., Wogan, G. N., and Davidson, C. S.,** Association between aflatoxin content of food and hepatoma frequency in Uganda, *Cancer,* 28, 253, 1971.
70. **Eppley, R. M.,** A versatile procedure for the assay and preparatory separation of aflatoxins from peanut products, *J. Assoc. Off. Anal. Chem.,* 49, 1218, 1966.
71. **Andrellos, P. J. and Reid, G. R.,** Confirmatory tests for aflatoxin B_1, *J. Assoc. Off. Agric. Chem.,* 47, 801, 1964.
72. **Davies, J. N. P. and Owor, R.,** The diagnosis of primary carcinoma of the liver: an assessment of the reliability of clinical diagnosis in Uganda Africans, *East Afr. Med. J.,* 37, 249, 1960.
73. **Alpert, M. E., Hutt, M. S. R., and Davidson, C. S.,** Primary hepatoma in Uganda. A prospective clinical and epidemiologic study of forty-six patients, *Am. J. Med.,* 46, 794, 1969.
74. **Campbell, T. C. and Salamat, L.,** Aflatoxin ingestion and excretion by humans, in *Symposium on Mycotoxins in Human Health,* Purchase, I. F. H., Ed., Macmillan, London, 1971, 271.
75. **Shank, R. C., Wogan, G. N., and Gibson, J. B.,** Dietary aflatoxins and human liver cancer. I. Toxigenic molds in foods and foodstuffs in tropical South-east Asia, *Food Cosmet. Toxicol.,* 10, 51, 1972.

76. **Shank, R. C., Wogan, G. N., Gibson, J. B., and Nondasuta, A.**, Dietary aflatoxins and human liver cancer. II. Aflatoxins in market foods and foodstuffs of Thailand and Hong Kong, *Food Cosmet. Toxicol.*, 10, 61, 1972.
77. **Shank, R. C., Gordon, J. E., Wogan, G. N., Nondasuta, A., and Subhamani, B.**, Dietary aflatoxins and human liver cancer. III. Field survey of rural Thai families for ingested aflatoxins, *Food Cosmet. Toxicol.*, 10, 71, 1972.
78. **Shank, R. C. Siddhichai, P., Subhamani, B., Bhamarapravati, N., Gordon, J. E., and Wogan, G. N.**, Dietary aflatoxins and human liver cancer. V. Duration of primary liver cancer and prevalence of hepatomegaly in Thailand, *Food Cosmet. Toxicol.*, 10, 181, 1972.
79. **Büchi, G, Kitura, Y., Yuan, S. S., Wright, H. E., Clardy, J., Demain, A. L., Glinsukon, T., Hunt, N., and Wogan, G. N.**, The structure of cytochalasin E, a toxic metabolite of *Aspergillus clavatus*, *J. Am. Chem. Soc.*, 95, 5423, 1973.
80. **Clardy, J., Springer, J. P., Büchi, G., Matsuo, K., and Wightman R.**, Tryptoquivaline and tryptoquivalone, two tremorgenic metabolites of *Aspergillus clavatus*, *J. Am. Chem. Soc.*, 97, 663, 1975.
81. **Newberne, P. M., Harrington, D. H., and Wogan, G. N.**, Effects of cirrhosis and other liver insults on induction of liver tumors by aflatoxin in rats, *Lab. Invest.*, 15, 962, 1966.
82. **Newberne, P. M., Rogers, A. E., and Wogan, G. N.**, Hepatorenal lesions in rats fed a low lipotrope diet and exposed to aflatoxin, *J. Nutr.*, 94, 331, 1968.
83. **Madhaven, T. V. and Gopalan, C.**, Effect of dietary protein on aflatoxin liver injury in weanling rats, *Arch. Pathol.*, 80, 123, 1965.
84. **Madhaven, T. V. and Gopalan, C.**, The effect of dietary protein on carcinogenesis of aflatoxin, *Arch. Pathol.*, 85, 133, 1968.
85. **Rogers, A. E. and Newberne, P. M.**, Diet and aflatoxin B_1 toxicity in rats, *Toxicol. Appl. Pharmacol.*, 20, 113, 1971.
86. **Rogers, A. E. and Newberne, P. M.**, Nutrition and aflatoxin carcinogenesis, *Nature (London)*, 229, 62, 1971.
87. **Lee, D. J., Sinnhuber, R. O., Wales, J. H., and Putnam, G. B.**, Effect of dietary protein on the response of rainbow trout (*Salmo gairdneri*) to aflatoxin B_1, *J. Natl. Cancer Inst.*, 60, 317, 1978.
88. **Newberne, P. M. and Rogers, A. E.**, Rat colon carcinomas associated with aflatoxin and marginal vitamin A. *J. Natl. Cancer Inst.*, 50, 439, 1973.
89. **DeLuca, L., Little, E. P., and Wolf, G.**, Vitamin A and protein synthesis by rat intestinal mucosa, *J. Biol. Chem.*, 244, 701, 1969.
90. **DeLuca, L., Schumacher, M., and Wolf, G.**, Biosynthesis of a fucose-containing glypcopeptide from rat small intestine in normal and vitamin A deficient conditions, *J. Biol. Chem.*, 245, 4551, 1970.
91. **Bhamarapravati, N. and Thamavit, W.**, Animal studies on liver fluke infestation, dimethylnitrosamine, and bileduct carcinoma, *Lancet*, 1, 206, 1978.
92. **Prince, A. M., Leblanc, L., Kronn, K., Masseyeff, R., and Alpert, M. E.**, S. H. antigen and chronic liver disease, *Lancet*, 2, 717, 1970.
93. **Bagshawe, A. F., Parker, A. M., and Jindani, A.**, Hepatitis-associated antigen in liver disease in Kenya, *Br. Med. J.*, 1, 88, 1971.
94. **Tong, M. J., Sun, S. -C., Schaeffer, B. T., Chang, N. -K., Lo, K. -J., and Peters, R. L.**, Hepatitis-associated antigen and hepatocellular carcinoma in Taiwan, *Ann. Intern. Med.*, 75, 687, 1971.
95. **Vogel, C. L., Anthony, P. P., Sadikali, F., Barker, L. F., and Peterson, M. R.**, Hepatitis-associated antigen and antibody in hepatocellular carcinoma: results of a continuing study, *J. Natl. Cancer Inst.*, 48, 1583, 1972.
96. **Smith, J. B. and Blumberg, B. S.**, Viral hepatitis, postnecrotic cirrhosis, and hepatocellular carcinoma, *Lancet*, 2, 953, 1969.
97. **Wright, R., McCollum, R. W., and Klatskin, G.**, Australia antigen in acute and chronic liver disease, *Lancet*, 2, 117, 1969.
98. **Welsh, J. D., Brown, J. D., Arnold, K., Chandler, A. M., Mau, H. M., and Thuc, T. K.**, Hepatitis-associated antigen in hepatoma in South Vietnam, *Lancet*, 1, 592, 1972.
99. **Purchase, I. F. H. and Goncalves, T.**, Preliminary results from food analyses in the Inhambane area, in *Symposium on Mycotoxins in Human Health*, Purchase, I. F. H., Ed., Macmillan, London, 1971, 263.
100. **Ciegler, A.**, Bioproduction of ochratoxin A and penicillic acid by members of the *Aspergillus ochraceous* group, *Can. J. Microbiol.*, 18, 631, 1972.
101. **Bourgeois, C., Olson, L., Comer, D., Evans, H., Keschamras, N., Cotton, R., Grossman, R., and Smith, T.**, Encephalopathy and fatty degeneration of the viscera: a clinicopathologic analysis of 40 cases, *Am. J. Clin. Pathol.*, 56, 558, 1971.
102. **Glinsukon, T., Yuan, S. S., Wightman, R., Kitaura, Y., Büchi, G., Shank, R. C., Wogan, G. N., and Christensen, C. M.**, Isolation and Purification of Cytochalasin E and Two Tremorgens from *Aspergillus clavatus*, *Plant Foods Man*, 1, 113, 1974.

103. **Glinsukon, T., Shank, R. C., Wogan, G. N., and Newberne, P. M.,** Acute and subacute toxicity of cytochalasin E in the rat, *Toxicol. Appl. Pharmacol.,* 32, 135, 1975.
104. **Glinsukon, T., Shank, R. C., and Wogan, G. N.,** Effects of cytochalasin E on fluid balance in the rat, *Toxicol. Appl. Pharmacol.,* 32, 158, 1975.
105. **Lindenfelser, L. A., Lillenoj, E. B., and Burmeister, H. R.,** Aflatoxin and trichothecene toxins: skin tumor induction and synergistic acute toxicity in white mice, *J. Natl. Cancer Inst.,* 52, 113, 1974.
106. **Van Rensburg, S. J., Kirsipuu, A., Coutinho, L. P., and Van der Watt, J. J.,** Circumstance associated with the contamination of food by aflatoxin in a high primary liver cancer areas, *S. Afr. Med. J.,* 49, 877, 1975.
107. **Serck-Hanssen, A.,** Aflatoxin-induced fatal hepatitis? A case report from Uganda, *Arch. Environ. Health,* 20, 729, 1970.
108. **Ling, K. -H., Wang, J. -J., Wu, R., Tung, T. -C., Lin, C. -K., Lin, S. -S., and Lin, T. -M.,** Intoxication possibly caused by aflatoxin B_1 in the moldy rice in Shuang-Shih township, *J. Formosan Med. Assoc.,* 66, 517, 1967.
109. **Alpert, E. and Serck-Hanssen, A.,** Aflatoxin-induced hepatic injury in the African monkey, *Arch. Environ. Health,* 20, 723, 1970.
110. **Reye, R. D. K., Morgan, G., and Baral, J.,** Encephalopathy and fatty degeneration of the viscera. A disease entity in childhood, *Lancet,* 2, 749, 1963.
111. **Bourgeois, C. H., Keschamras, N., Comer, D. S., Harikul, S., Evans, H., Olson, L., Smith, T., and Beck, M. R.,** Udorn encephalopathy, fatal cerebral edema and fatty degeneration of the viscera in Thai children, *J. Med. Assoc. Thailand,* 52, 553, 1969.
112. **Olson, L. C., Bourgeois, C. H., Jr., Cotton, R. B., Harikul, S., Grossman, R. A., and Smith, T. J.,** Encephalopathy and fatty degeneration of the viscera in northeastern Thailand. Clinical syndrome and epidemiology, *Pediatrics,* 47, 707, 1971.
113. **Bourgeois, C. H., Shank, R. C., Grossman, R. A., Johnsen, D. O., Wooding, W. L., and Chandavimol, P.,** Acute aflatoxin B_1 toxicity in the macaque and its similarities to Reye's syndrome, *Lab. Invest.,* 24, 206, 1971.
114. **Shank, R. C., Johnsen, D. O., Tanticharoenyos, P., Wooding, W. L., and Bourgeois, C. H.,** Acute toxicity of aflatoxin B_1 in the Macaque monkey, *Toxicol. Appl. Pharmacol.,* 20, 227, 1971.
115. **Shank, R. C., Bourgeois, C. H., Keschamras, N., and Chandavimol, P.,** Aflatoxins in autopsy specimens from Thai children with an acute disease of unknown etiology, *Food Cosmet. Toxicol.,* 9, 501, 1971.
116. **Becroft, D. M. O.,** Syndrome of encephalopathy and fatty degeneration of viscera in New Zealand children, *Br. Med. J.,* 2, 135, 1966.
117. **Becroft, D. M. O. and Webster, D. R.,** Aflatoxins and Reye's disease, *Br. Med. J.,* 4, 117, 1972.
118. **Dvořáčková, I., Žilková, J., Brodský, F., and Cerman, J.,** Aflatoxin and liver damage with encephalopathy, *Sborník ředeckých praci Lékařské fakulty KU v Hradci Králové,* 15, 521, 1972.
119. **Dvořáčková, I.,** personal communication.
120. **Chaves-Carballo, E., Ellefson, R. D., and Gomez, M. R.** An aflatoxin in liver of a patient with Reye-Johnson syndrome, *Mayo Clin. Proc.,* 51, 48, 1976.
121. **Hogan, G. R., Ryan, N. J., and Hayes, A. W.,** Aflatoxin B_1 and Reye's syndrome, *Lancet,* 1, 561, 1978.
122. **Glick, T. H., Likosky, W. H., Levitt, L. P., Mellin, H., and Reynolds, D. W.,** Reye's syndrome: an epidemiologic approach, *Pediatrics,* 46, 371, 1970.
123. **Corey, L., Rubin, R. J., Hattwick, M. A. W., Noble, G. R., and Cassidy, E.,** A nationwide outbreak of Reye's syndrome. Its epidemiologic relationship to influenza B, *Am. J. Med.,* 61, 615, 1976.
124. **Bradford, W. D. and Latham, W. C.,** Acute encephalopathy and fatty hepatomegaly, *Am. J. Dis. Child.,* 114, 152, 1967.
125. **Mullen, P. W.,** Immunopharmacological considerations in Reye's syndrome: a possible xenobiotic initiated disorder, *Biochem. Pharmacol.,* 27, 145, 1978.
126. **Pollack, J. D., Ed.,** *Reye's Syndrome,* Grune & Stratton, New York., 1975, 470.
127. **Yadgiri, B., Reddy, V., Tulpule, P. G., Srikantia, S. G., and Gopalan, C.,** Aflatoxin and Indian childhood cirrhisis, *Am. J. Clin. Nutr.,* 23, 94, 1970.
128. **Amla, I., Kamala, C. S., Gopalakrishna, G. S., Jayaraj, A. P., Sreenivasamurthy, and Parpia, H. A. B.,** Cirrhosis in children from peanut meal contaminated by aflatoxin, *Am. J. Clin. Nutr.,* 24, 609, 1971.
129. **Robinson, P.,** Infantile cirrhosis of the liver in India with special reference to probable aflatoxin etiology, *Clin. Pediatr.,* 6, 57, 1967.
130. **Amal, I., Kumari, S., Sreenivasamurthy, V., Jayaraj, A. P., and Parpia, H. A. B.,** Role of aflatoxin in Indian childhood cirrhosis, *Indian Pediatr.,* 7, 262, 1970.
131. **Krishnamachari, K. A. V. R., Bhat, R. V., Nagarajan, V., and Tilak, T. B. G.,** Hepatitis due to aflatoxicosis. An outbreak in Western India, *Lancet,* 1, 1061, 1975.
132. **Bösenberg, H.,** Diagnostische Möglichkeiten zum Nachweis von Aflatoxin Vergiftungen, *Zentralbl. Bakteriol. Parasitenkd. Infektionskr. Hyg. Abt. 1 Orig. Reihe A,* 220, 252, 1972.

133. **Phillips, D. L., Yourtee, D. M., and Searles, S.,** Presence of aflatoxin B_1 in human liver in the United States, *Toxicol. Appl. Pharmacol.,* 36, 403, 1976.
134. **Van Nieuwenhuize, J. P., Herber, R. F. M., DeBruin, A., Meyer, Ir. P. B., and Duba, W. C.,** Epidemiologisch onderzoek naar carcinogeniteit bij langdurige "low level" expositie van een fabriekspopulatie. II. Eigen onderzoek, *T. Soc. Geneesk,* 51, 754, 1973.
135. **Alpert, M. E.,** personal communication.

Chapter 4

MYCOTOXINS: ASSESSMENT OF RISK

Ronald C. Shank

TABLE OF CONTENTS

I. Introduction ... 142

II. Carcinogenic Risk ... 142
   A. Estimate of Exposure ... 142
   B. Dose Response Relationships: Aflatoxins and the Rat/Trout ... 144
   C. Extrapolation from Animal Data to Human Risk ... 145
      1. Mathematical Models ... 145
      2. Toxicological Considerations for Low-Dose Extrapolation ... 149

III. Acute Poisoning ... 150

References ... 152

I. INTRODUCTION

Assessing the risk mycotoxins present to human health is limited here in large part to aflatoxin risks. Ergot alkaloids have long been recognized for their hazard and are still a potential problem, as the last outbreak of human ergot poisoning occurred as recently as 1951 in France. Epoxytrichothecenes may present another potential hazard, as suggested by outbreaks of alimentary toxic aleukia in Russia. Sterigmatocystin is about half as potent as a carcinogen as aflatoxin B_1, but so far has rarely been found in human foods. Patulin is a known contaminant in apple juice and other fruit products for human consumption and possibly may be one of the most serious mycotoxin threats to human health next to the aflatoxins, especially for children, who frequently drink fruit juices. The psoralens are recognized now as a minor nuisance, but their chronic toxicity is not well known. Dicumarol poisoning seems limited to only therapeutic accidents. Yellowed rice toxins certainly offer a potential hazard to human health, but not enough is known about the regular occurrence of these compounds in human foods to provide a basis for risk assessment. Recently ochratoxin A contamination of foodstuffs was found to occur more frequently in areas of Yugoslavia, where a fatal chronic nephropathy (Balkan endemic nephropathy) is prevalent, suggesting that the mycotoxin may play a role in the etiology of this human disease,[1] but again there is not enough evidence for a risk assessment.

There is sufficient information on the aflatoxins, however, upon which to base such an assessment. Quantitative carcinogenicity (dose-response) data are available for aflatoxin B_1 in two animal species; human exposures have been estimated several times from plate samples, and these exposures have been correlated with liver cancer rates; aflatoxin contamination is widespread, but usually at low concentrations, and acute poisonings should be rare. Aflatoxins are extremely potent carcinogens (the most potent known for the rat, see Table 1), and cancer is a real risk; this puts aflatoxin B_1 as the major mycotoxin for consideration in this risk assessment.

II. CARCINOGENIC RISK

A. Estimate of Exposure

Most emphasis can be placed on aflatoxin B_1 because it is the most frequently occurring and most potent carcinogen of all the aflatoxins found in the human food supply.

The consumption of aflatoxins has been measured for several populations in Africa and Southeast Asia. Daily intakes have ranged from less than 5 ng/kg body weight to over 200 ng/kg body weight. These measurements were made in relatively stable societies with little intranational migration; the diets were composed of primarily local unprocessed (commercially) foodstuffs.

The estimation of the aflatoxin consumption in European and North American populations is made more complex by the great variety of foodstuffs used in the dietary, the large influence of commercial food processing, the wide distribution of foodstuffs, and the mobility of the populations. Nevertheless, the importance of the problem of mycotoxins in human foodstuffs makes urgent some estimation as to how extensive this contamination might be.

A large variety of foods and foodstuffs have been assayed for aflatoxin contamination, and from these studies one concludes that the principal sources are peanuts and peanut products and some cereal grains, with occasional exposure from milk and dairy products. Assuming a "worst case" situation, a child may eat 1 oz (30 g) of peanut butter 5 days a week, an average of 21.4 g of peanut butter a day, and continue to eat peanut butter regularly to adulthood. Also assuming an average body weight from childhood to adult-

Table 1
POTENCIES OF VARIOUS CHEMICAL CARCINOGENS IN THE RAT

Compound	Route	Actual Dose[a]	μg/kg body weight/week	Target	Ref.
Aflatoxin B_1[b]	Oral	1 μg/kg diet	0.42	Liver	2
Bis(chloromethyl) ether	Inhalation	0.5 mg/m³, 6 hr/day 5 days/week	252	Lung, olfactory	3
Diethylnitrosamine	Oral	0.075 mg/kg body weight/day	525	Liver	4
Dimethylnitrosamine	Oral	2 mg/kg diet	840	Liver	5
Patulin[b]	s.c.	0.2 mg/rat, 2 times/week, 61 weeks	1,600	Local	6
Sterigmatocystin[b]	Oral	0.15 mg/rat/d	4,200	Liver	7
Lasiocarpine	i.p.	7.8 mg/kg body weight, 2 times/week for 4 weeks, 1 time/week for 5 weeks	7,800	Liver	8
Dimethylamino-azobenzene	Oral	1 mg/rat/day	28,000	Liver	9
2-Acetylamino-fluorene	Oral	0.01% in diet	42,000	Liver	10
Vinylchloride	Inhalation	130 mg/m³ 4 hr/day, 5 days/week	43,600	Liver, etc.	11
Benzidine	Oral	0.017% in diet	71,400	Liver	12

[a] Dose recalculated in units of μg/kg body weight/week assuming, where necessary, that rats weigh 250 g, eat a 15-g diet per day, and breathe 0.0042 m³/hr.
[b] Mycotoxin.

hood of 50 kg and an average aflatoxin B_1 content in the peanut butter of 10 ppb (10 ng/g) and assuming this represents the principal source of dietary aflatoxins, the average daily consumption of aflatoxin from childhood to adulthood would be

$$\frac{21.4 \text{ g peanut butter/day}}{50 \text{ kg BW}} \times \frac{10 \text{ ng aflatoxin } B_1}{\text{g peanut butter}} =$$

$$4.28 \text{ ng AFB}_1/\text{kg BW/day from the diet}$$

Concern has been expressed over the possible occurrence of aflatoxins in foods derived from animals fed aflatoxin-treated feeds.[13,14] It seems the general conclusion is that large concentrations of aflatoxins do not occur in foods of animal origin. Milk from dairy cows can contain aflatoxin M_1, a carcinogenic hydroxylated derivative of aflatoxin B_1, but it is unlikely that it could be detected in commercial pooled milk. Nonetheless, the threat to infants who depend upon milk remains to be defined.

Another source of aflatoxins and presumably other mycotoxins, and one that is usually overlooked, is air. This has been clearly demonstrated in a nut processing plant; inhalation of aflatoxins would also seem to be a problem where large amounts of mold-damaged grains are handled, such as in grain elevators and in feedlots.

Van Nieuwenhuize and co-workers[15] measured the exposure to aflatoxins for men working in an oil seed processing plant (see Table 2). Airborne peanut dust varied in particular size, with an average of 8.4% (by weight) of the total dust having a particle

Table 2
EXPOSURE OF WORKMEN TO AIRBORNE AFLATOXIN IN A PEANUT-PROCESSING PLANT[15]

Work area	Aflatoxin concentration in air (ng/m^3) Average	Range	Average exposure to aflatoxin (ng/man/45-hr work week)
	0.87	0.1—2.3	39.2
Press, extraction 1	0.52	0.4—0.7	—
Press, extraction 2	0.58	0.4—0.9	—
Bulk storage	1.2	0.05—6.5	54.0 (maximum = 292.5)
Bagging 1	40.8	19.8—61.5	1,836.0 3,262.5 (ave = 2500)

size less than 5 μm, and much of that dust had particle sizes less than 1.1 μm. Thus, the small particles in these inhaled dusts would reach the pulmonary alveoli, and the larger particles would be swept from the upper airways and swallowed or expectorated. The airborne dusts contained up to 410 μg aflatoxins per kilogram dust in the area where workmen collected peanut meal in bags (the peanut meals contained up to 495 μg aflatoxins per kilogram at the time the dust concentration was 410 μg/kg). The average exposure to airborne aflatoxins was calculated to range from 0.039 to 2.5 μg per worker per 45-hr work week, assuming an average respiration of 1 m^3 per work hour. The range reflects different air concentrations associated with different jobs in the same plant. On a kilogram body weight per day basis, these exposures range from 0.08 to 5.10 ng, or approaching the order of magnitude of the exposure estimated in the worst case analysis given above for dietary exposures.

B. Dose-Response Relationships: Aflatoxins and the Rat/Trout

Two careful dose-response studies have been done on the carcinogenicity of aflatoxin B_1. Wogan and co-workers[2] fed male Fischer rats agar gel-defined diet containing highly purified aflatoxin B_1. Control and test diets were prepared fresh weekly, and aflatoxin concentrations were verified by chemical assay; dietary concentrations were 0, 1, 5, 15, 50, and 100 μg aflatoxin per kilogram diet (ppb). Experimental treatment began when the animals weighed approximately 80 g and continued until clinical deterioration of the rats was observed; at the time, survivors of the treatment group were killed for histopathologic examination. The incidence of liver carcinoma that resulted from feeding these diets is given in Table 3.

The authors point out that the Fischer rat appears to be the most sensitive rodent to aflatoxin carcinogenicity and that this sensitivity may be increased by incorporating purified aflatoxin in a defined diet as opposed to the more often reported method of mixing contaminated peanut meals with commercial laboratory animal chows. (This increased sensitivity to carcinogenicity has also been observed in the same laboratory in a study on the carcinogenicity to Fischer rats of sodium nitrite incorporated in defined diets and in commercial rat chow.[16])

The Fischer rat is not unique in its sensitivity to aflatoxin; the trout has also demonstrated high sensitivity. Halver[17] fed young rainbow trout (more than 100 fish per group) a diet containing 0.5, 2, or 8 ppb crystalline aflatoxin B_1 for up to 20 months and obtained the dose-tumor response shown in Table 4.

Table 3
INCIDENCE OF LIVER CARCINOMAS IN RATS FED AFLATOXIN B$_1$[2]

Aflatoxin level (ppb)	Time of appearance of earliest tumor (weeks)	Duration of experiment (weeks)	Number of animals at risk[a]	Number of animals with hepatocellular carcinoma
0	—	74—109	18[b]	0
1	104	78—105	22	2
5	93	65—93	22	1
15	96	69—96	21	4
50	82	71—97	25	20
100	54	54—88	28	28

[a] Animals surviving longer than 50 weeks.
[b] Animals surviving for maximum period.

Table 4
INCIDENCE OF HEPATOMAS IN TROUT FED AFLATOXIN B$_1$[17]

Aflatoxin level (ppb)	Hepatoma incidence (%)
Reference diet	2[a]
0.5	44
2.0	60
8.0	65

[a] 1 out of 90 fish at risk after 12 months on study; concentration of aflatoxin B$_1$ in reference complete test diet not stated.

From these data, Halver estimated that less than 0.15 μg of aflatoxin B$_1$ ingested over a 20-month period (apparently assuming an average food intake of 0.5 g per fish per day for the 0.5 ppb diet) was sufficient to induce a high incidence of hepatoma after 16 to 20 months of exposure. Consequently he fed 200 to 300 trout 0.1 μg aflatoxin B$_1$ per fish over periods of up to 24 weeks and observed high incidences of hepatomas at 20 months (see Table 5).

The rainbow trout, then, seems to be more sensitive than the Fischer rat, 0.5 ppb in the diet producing a 44% tumor incidence in the fish in 20 months, whereas 1 ppb in the diet produced a 9% tumor incidence in rats in 23 months. Feeding fish a total of 0.1 μg aflatoxin B$_1$ over a 2 to 24 week period yielded a tumor incidence of 49 to 70% at 20 months; feeding rats 1 ng aflatoxin per kilogram diet (1 ppb) resulted in a 9% tumor incidence; these rats would have consumed approximately 11 μg toxin in 23 months.

On the other hand mice (C3HfB/HEN and C57Bl/6NB) fed 1 ppm aflatoxin B$_1$ in the diet for 20 months or Swiss mice fed 150 ppm aflatoxins B$_1$ + G$_1$ in the diet for 20 months failed to develop liver cancer.[18] Vesselinovich and co-workers[19] injected i.p. 6 μg/g body weight AFB$_1$ 3 times during the perinatal period of an F$_1$ hybrid strain (C57Bl × C3H) and in 80 weeks observed hepatocellular carcinomas in all 16 mice.

C. Extrapolation from Animal Data of Human Risk
1. Mathematical Models

In the last decade, several mathematical models have been proposed for use in extrapolating from laboratory animal carcinogenicity data in risk assessments for humans

Table 5
INCIDENCE OF HEPATOMAS IN TROUT FED 0.1 µg AFLATOXIN B₁ (TOTAL INTAKE) OVER VARYING PERIODS[17]

Diet treatment[a]	Trout with hepatoma at 20 months (%)
Reference Diet	0
0.1 µg B₁ per fish for 300 fish in 2 weeks	55
0.1 µg B₁ per fish for 300 fish in 4 weeks	62
0.1 µg B₁ per fish for 300 fish in 8 weeks	55
0.1 µg B₁ per fish for 200 fish in 12 weeks	59
0.1 µg B₁ per fish for 200 fish in 16 weeks	61
0.1 µg B₁ per fish for 200 fish in 20 weeks	70
0.1 µg B₁ per fish for 200 fish in 24 weeks	49

[a] Fish were fed diets containing aflatoxin 5 days/week.

exposed to the same environmental carcinogen. Extrapolations are usually necessary because of the lack of human epidemiological data, the impossibility of obtaining direct human data, and the problem toxicologists live with daily, the practical necessity to use small numbers of test animals and therefore high doses, to observe a significant cancer incidence, even though the test doses are usually several orders of magnitude higher than environmental concentrations to which human populations are exposed.

Most mathematical models assume a linear relationship between dose of carcinogen and incidence of a specific cancer in the population at risk; their principal difference is the method by which they derive the slope of the dose-response curve for the human population. These models will be discussed in detail in Chapter 8 assessing the risk of environmental N-nitroso compounds where little human data are available. For the aflatoxins, we are fortunate enough to have the results of four individual epidemiological studies on dietary aflatoxins and human liver cancer (see Chapter 3). The data from those studies provide an estimate of the slope of the dose-response curve which logically should be more accurate than any mathematical extrapolation from data obtained from animals exposed to high levels of aflatoxins.

Peers and Linsell[20] were the first to estimate the slope of the human dose-response curve, basing the estimate on their results correlating consumption of aflatoxins with liver cancer rates in defined geographical areas of Kenya. In the regression analysis of their data, the logarithm of the aflatoxin consumption was correlated with the arithmetical expression of liver cancer incidence in the number of annual cases occurring per 100,000 adults. This format was followed by Van Rensburg and colleagues[21] in their analysis of the results from their study in Mozambique combined with the Kenya results[20] and data from the study in Thailand.[22,23]

Such graphic techniques may not be justified when comparing responses over large ranges of doses because the regression lines may not be linear over such ranges. This is especially important if one wants to compare the human and animal data to determine whether the test animals were good predictors of human sensitivity.

It has long been the practice in pharmacology and toxicology to analyze dose-response relationships by correlating the log of the dose with an expression of the cumulative response, i.e., the percentage of the total test population responding to a given dose and all lower doses. Such dose-response curves (Figure 1) are sigmoid: linear, for all practical purposes, near the 50% response, and curvilinear at each extreme. In probability analysis,

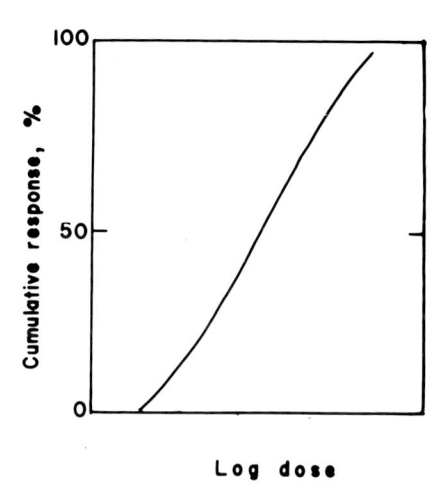

FIGURE 1. Typical dose-response curve for test compound given to a homogeneous population.

the cumulative response is expressed not arithmetically, but in probits, units of probability based on normal equivalent deviations in a Gaussian (bell-shaped) distribution of individual sensitivities to the test compound in the sample population. True Gaussian distributions seldom occur in biological populations, but they can be approximated in homogenous animal populations. The advantage of probability analysis is that it does provide a linear relationship for dose and response and reflects a diminishing probability of tumor induction with diminishing dose, i.e., lowering a dose by a given percentage near the 50% response will produce a much greater lowering in response than will the same percentage lowering of dose at either extreme of the curve. This relationship agrees well with most quantal dose response relationships; death is a quantal response, and since liver cancer is always fatal, liver cancer rates are in fact death rates and therefore quantal.

Figure 2 is a graphic representation of the carcinogenicity data obtained experimentally for the trout and the rat and epidemiologically for the human, presented as a log-probability relationship. The trout data yield a slope to the dose-response curve similar to that seen in the human data, but it would obviously be wrong to extrapolate directly from trout to human, in spite of the similar slopes of the dose-response curves, because the human response would be greatly overestimated, ranging from 45 to 70% of the population consuming 4 to 200 ng aflatoxin B_1 per kilogram body weight per day, rather than the observed incidence of 0.1 to 2%.

On the other hand, the dose-response curve for the rat appears almost to be a nonlinear extension of the human curve. Only 21 to 28 rats were treated per dose in the depicted study, and therefore the error associated with each point is large, but if the curve is assumed to be linear for all but the lowest dose, the slope of the rat curve is obviously much greater than for the human curve, and as a result, extrapolation from the rat curve to human aflatoxin consumption levels would underestimate the human cancer rate at the lower consumption levels.

A common comparator for species sensitivity to a chemical is the dose of the compound which will produce the quantal response in 50% of the population, in this case the dose which would produce a 50% liver cancer lifetime incidence, a tumor dose 50 (TD_{50}). From Figure 2, it can be seen that the TD_{50} for aflatoxin B_1 in rainbow trout is approximately 3.5 ng/kg body weight per day; for the male human, it can be estimated (by

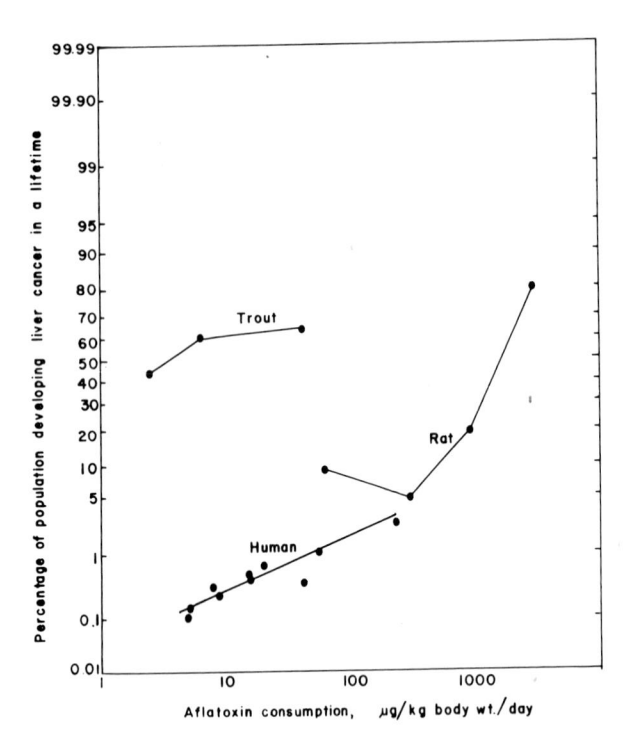

FIGURE 2. Dose-response relationships for aflatoxin B_1 and liver cancer in the trout, rat, and human. Incidences of hepatocellular carcinoma, expressed as cases per 100,000 population per life time and plotted on the probability coordinate, were taken from data provided by Halver[17] for trout and Wogan and co-workers[2] for rats; human data were taken from Shank and co-workers[22,23] (Thailand, open circles), Peers and Linsell[20] (Kenya, closed circles), Van Rensburg and co-workers[21] (Mozambique, open squares), and Peers and co-workers[24] (Swaziland, closed squares). Dose is expressed as the logarithm of the nanograms aflatoxin B_1 eaten per kilogram body weight per day.

linear extrapolation assuming no change in slope) to be approximately 100,000 ng/kg body weight per day. From such a comparison, one might conclude that man is two orders of magnitude more resistant to aflatoxin carcinogenicity than the rat and five orders of magnitude more resistant than the trout. This does not agree quantitatively with the conclusion drawn from the comparison of dose-response curve slopes for man and trout and does not agree even qualitatively with the conclusion similarly drawn from the comparison of slopes for man and rat. The one conclusion that can be drawn seems to be that neither the trout nor rat bioassay serves as an accurate predictor of the quantitative risk of aflatoxin consumption by man.

Three important assumptions are inherent in the above exercises in predicting the human response to dietary levels of aflatoxin B_1. The first is that the human population is near homogeneous and the sensitivities to aflatoxin carcinogenicity in the human population are normally distributed. The second assumption made is that all liver cancers occurring in the populations at risk are induced by aflatoxins; while this may be close to the truth in the test situations with the trout and rats, there is certainly no reason to believe this assumption is valid for the human situation; indeed, the raison d'etre for the second section of this book, consideration of risks presented by environmental N-nitroso compounds, is that those latter compounds may also play a role in the etiology of human cancer, including cancer of the liver. The third important assumption is that all tumors induced by dietary aflatoxins are cancers of the liver. In fact, aflatoxin B_1 is capable of producing neoplasms in the kidney,[25-27] glandular stomach,[28] and colon[29,30] in the rat and pulmonary tumors in A/He mice.[31]

2. Toxicological Considerations for Low-Dose Extrapolations

Several toxicological considerations pertinent to the assessment of risk presented by exposure to environmental carcinogens will be treated briefly here and in detail in Chapter 8 on the risk assessment for N-nitroso compounds. The availability of data on aflatoxin exposure and human liver cancer makes unnecessary the use of high-dose animal data for extrapolation to human response at low environmental concentrations. Nonetheless, for those who insist on this kind of extrapolation, it is important that they be aware of the defense mechanisms of the body and the mechanisms of chemical carcinogenesis in order to rationally evaluate the risk because of environmental carcinogens.

It is generally recognized that chemical carcinogens are highly reactive reagents and are generated at the site of tumor induction. Carcinogens exist in the environment in chemically stable forms (e.g., aflatoxin B_1 is relatively stable in foodstuffs) and are activated either by spontaneous decomposition in physiologic media or, as in the case of aflatoxin B_1, by metabolism (to the 2,3-oxide) in the target cell. As explained in Volume II, aflatoxin 2,3-oxide is unstable and rapidly breaks down to an electrophile which covalently binds to DNA, altering the genetic material of the cell which oxidized the parent aflatoxin B_1. Thus, in comparing the effect of aflatoxin B_1 and other parent carcinogens on various biological systems, such as laboratory animals and the human, it is important to know not only how much parent carcinogen is being administered, but also the extent to which the parent compound is being converted to the ultimate (active form) carcinogen.

For most compounds foreign to the body, several metabolic pathways exist; these pathways are followed at different rates and compete with each other for substrate (parent carcinogen). Several instances are now recognized where concentration of substrate governs in part the extent to which each pathway is followed. Examples can be given where, at low doses, detoxication pathways (e.g., glutathione conjugations) dominate and the compound is made more soluble in water and excreted in the urine and bile; at doses high enough to saturate and overload these pathways (e.g., deplete cellular stores of glutathione), other reactions occur, some of which can increase the toxicity (and presumably the carcinogenicity) of the substrate.

This dose-dependent aspect in the pharmacokinetics of toxin metabolism results in a nonlinear extrapolation from biochemical and toxicological responses at high doses to similar responses at low doses; in each case, the magnitude of the response observed at low doses should be less than would be predicted from a linear extrapolation from responses observed at high doses.

Advances in the studies of chemical carcinogenesis have been rapid, and much evidence now exists to support the hypothesis that most chemical carcinogens are electrophilic agents which covalently bind to DNA to create a non-fatal, heritable change in the genetic template. The altered template can be repaired enzymatically to restore its original fidelity. Recent evidence indicates that this process, DNA repair, like metabolic activation, is dose-dependent and that the rate of repair may be inversely proportional to the dose of carcinogen.[32-34] In such cases, permanent template alteration following exposure to low levels of carcinogen would be less than expected from linear extrapolations from responses at high levels.

The effectiveness of this repair system is seen clinically in patients with xeroderma pigmentosum, a genetic deficiency of the DNA repair enzymes in the skin; these patients are at a much higher risk for skin cancer induced by sunlight (which damages skin DNA) than are people with intact repair systems in skin.

Also important in these considerations is immunologic repression of cancerous cells. This is manifest clinically in the use of immunosuppression drugs to prevent rejection of tissue transplants, where patients so treated are at a relatively high risk for cancer having lost the immunological defense mechanism.[35]

The conclusion to be drawn from these arguments is that it is not consistent with biological mechanisms to assume a constant linearity (no change in slope) in dose-response curves over wide ranges of dose that include concentrations that inhibit, saturate, or overload natural defense mechanisms such as metabolic detoxication, DNA repair, and immunologic tumor repression. In essence, this is an argument, consistent with known biological mechanisms, that environmental exposure to a parent carcinogen at any concentration does not of necessity result in a frank tumor.

III. ACUTE POISONING

So little is known about human exposure to mycotoxins other than the aflatoxins that it is not possible at this time to predict the likelihood of acute poisonings in man due to ingestion of these compounds. Chapter 3 has reviewed the few instances of human mycotoxicoses which resulted from gross contamination of grain used to make bread: ergotism and alimentary toxic aleukia. The preceeding chapters have already documented the widespread occurrence of small quantities of aflatoxin in human food supply and associated this with human liver cancer. It is reasonable to assume that if small amounts of an environmental toxin can occur frequently enough to precipitate a chronic disease, perhaps on rare occasion the same toxin could occur in the environment in amounts large enough to bring about an acute response in highly sensitive individuals. There are a few reports of putative exposure to large amounts of aflatoxins associated with what appear to be fatal human poisonings.

Exposures to aflatoxin associated with human poisonings discussed in Chapter 3 are summarized here in Table 6. No direct evidence is available for determining the exposure to aflatoxins in the first two cases in Table 6.[37] Shank and co-workers[40] conducted an acute toxicity test on aflatoxin B_1 in the monkey and found 37 μg aflatoxin per kilogram liver in a monkey that died 73 hr after being poisoned with 13.5 mg aflatoxin per kilogram body weight; in another monkey poisoned with 40.5 mg/kg body weight and dying in 80 hr, the aflatoxin B_1 level in the kidney was 163 μg/kg. Thus, at least in the monkey, comparison of dose levels with hepatic residue levels suggests that 3 days following oral administration, 2.74 to 4.02 μg aflatoxin B_1 can be found in a kilogram of liver per milligram of aflatoxin B_1 administered per kilogram body weight. From this, if it is assumed that approximately 3 μg aflatoxin can be found in the livers of children for every milligram aflatoxin per kilogram body weight consumed, then the 2-year-old boy in Table 6 would have had to eat 93 μg aflatoxin per kilogram liver divided by 3 (micrograms aflatoxin found per milligram aflatoxin eaten) or 31 mg aflatoxin eaten per kilogram body weight; similarly, the 13-year-old boy would have had to consume 16 mg aflatoxin per kilogram body weight. Assuming the 2-year-old boy weighed 10 kg and the 13-year-old boy 40 kg, then the boys would have had to eat approximately 310 mg and 640 mg aflatoxin, respectively. It is most unlikely that such large amounts could have been consumed in a single day, for this would require that the food contain several hundred parts aflatoxin B_1 per million parts food which is about 100 times higher than

Table 6
SUMMARY OF REPORTED PUTATIVE AFLATOXIN POISONINGS IN HUMANS

Victim	Symptoms	Aflatoxins in liver	Estimated exposure to aflatoxins	Ref.
2-year-old boy	Reye's syndrome	93 μg B$_1$/kg	310 mg	36
13-year-old boy	Reye's syndrome	47 μg B$_1$/kg	640 mg	36
Children	Hepatitis	—	2—6 mg/day for 30 days	37
15-year-old girl	Reye's syndrome	22.5 μg?/kg	—	38
22-month-old boy	Reye's syndrome	5—50 μg B$_1$/kg	—	39
8-month-old girl	Reye's syndrome	5—50 μg B$_1$/kg	—	39

any food samples assayed in a 3-year study in Thailand.[22,41] If, instead of consuming all the toxin in one day, it was consumed over a month, then 10 to 20 mg aflatoxin would have had to be consumed each day; assuming all the toxin was ingested in 250 g of boiled peanuts and 500 g of boiled rice each day by the 2-year-old boy would require those foods to contain together 13 ppm aflatoxin, and for the 13-year-old boy, eating 250 g of peanuts and 1 kg of rice, 16 ppm aflatoxin would be required. These concentrations are 10 times higher than the highest seen in any plate samples assayed in Thailand.[22]

Such high levels of contamination apparently did occur in western India, however, and this has been associated with an outbreak of hepatitis in which 106 children and young adults (ages 5 to 30 years) died.[37] Maize samples collected from households of hepatitis victims contained aflatoxins in the range of 6.25 and 15.6 ppm. Assuming that the children ate 75 to 80 g of maize daily and adults ate 350 g daily, the ingestion of aflatoxin would have been 2 to 6 mg/day for several weeks. This is not too far from the 10 to 20 mg/day calculated for the Thai boys (310 to 640 mg eaten over 30 days rather than in a single day).

No estimates of exposure are given for the last three cases in Table 6. If one uses the same rationale as above for the two Thai cases, that 3 μg aflatoxin can remain in a kilogram of liver for every mg aflatoxin per kilogram body weight consumed, then the victims may have consumed as much as 1.7 to 16.7 mg of the toxin.

Lopez and Crawford[42] assayed 152 samples of peanuts sold for human consumption in Uganda. The samples were collected throughout all seasons, and 15% of the total contained aflatoxin B$_1$ in concentrations exceeding 1 ppm; two samples contained between 10 and 11 ppm aflatoxin B$_1$. The authors refer to a report by Rutishauser and co-workers[43] indicating that the average consumption of peanuts in five boy's schools in the Ugandan province of Buganda was 570 g per boy per week. From this consumption rate, Lopez and Crawford calculated that a meal containing 100 g of peanuts with 1 ppm aflatoxin B$_1$ would result in a dose of 0.1 mg or 2.5 μg/kg body weight; comparing this to toxic doses for ducklings, it would take 6 months of consuming such peanuts to constitute a toxic dose.

If the cases summarized in Table 6 do in fact represent acute aflatoxin poisonings, then one would conclude that such poisonings in children would require ingestion of at least several milligrams aflatoxin B$_1$ per day for several weeks. Considering that such doses far exceed the levels of toxin which could be obtained from usual human foodstuffs, acute aflatoxin poisoning in humans would be expected to occur only under the most unusual circumstances of gross contamination in the diet.

REFERENCES

1. **Krough, P., Hald, B., Pleština, R., and Ceović, S.,** Balkan (endemic) nephropathy and foodborne ochratoxin A: preliminary results of a survey of foodstuffs, *Acta. Pathol. Microbiol. Scand. Sect. B*, 85, 238, 1977.
2. **Wogan, G. N., Paglialunga, S., and Newberne, P. M.,** Carcinogenic effects of low dietary levels of aflatoxin B_1 in rats, *Food Cosmet. Toxicol.* 12, 681, 1974.
3. **Laskin, S., Kuschner, M., Drew, R. T., Cappiello, V. P., and Nelson, N.,** Tumors of the respiratory tract induced by inhalation of bis(chloromethyl) ether, *Arch. Environ. Health*, 23, 135, 1971.
4. **Druckrey, H., Preussmann, R., Ivankovic, S., and Schmahl, D.,** Organotrope carcinogene Wirkungen bei 65 verschiedenen N-nitroso-Verbindungen an BD-Ratten, *Z. Krebsforsch.*, 69, 103, 1967.
5. **Terracini, B., Magee, P. N., and Barnes, J. M.,** Hepatic pathology in rats on low dietary levels of dimethylnitrosamine, *Br. J. Cancer*, 21, 559, 1967.
6. **Dickens, F. and Jones, H. E. H.,** Carcinogenic activity of a series of reactive lactones and related substances, *Br. J. Cancer*, 15, 85, 1961.
7. **Purchase, I. F. H. and Van der Watt, J. J.,** Carcinogenicity of sterigmatocystin to rat skin, *Toxicol. Appl. Pharmacol.*, 26, 274, 1973.
8. **Svoboda, D. J. and Reddy, J. K.,** Malignant tumors in rats given lasiocarpine, *Cancer Res.*, 32, 908, 1972.
9. **Druckrey, H., Schmahl, D., Dischler, W., and Schildbach, A.,** Quantitative Analyse der experimentellen Krebserzeugung, *Naturwissenschaften*, 49, 217, 1962.
10. **Albert, R. E., Burns, F. J., Bilger, L., Gardner, D., and Troll, W.,** Cell loss and proliferation induced by N-2-fluorenylacetamide in the rat liver in relation to hepatoma induction, *Cancer Res.*, 32, 2172, 1972.
11. **Maltoni, C. and Lefemine, G.,** Carcinogenicity bioassays of vinyl chloride: current results, *Ann. N. Y. Acad. Sci.*, 246, 195, 1975.
12. **Boyland, E., Harris, J., and Horning, E. S.,** The induction of carcinoma of the bladder in rats with acetamidofluorene, *Br. J. Cancer*, 8, 647, 1954.
13. **Purchase, I. F. H.,** Aflatoxin residues in food of animal origin, *Food Cosmet. Toxicol.*, 10, 531, 1972.
14. **Schoental, R.,** Mycotoxicoses "by proxy," *Int. J. Environ. Stud.*, 8, 171, 1975.
15. **Van Nieuwenhuize, J. P., Herber, R. F. M., DeBruin, A., Meyer, I. P. B., and Duba, W. C.,** Epidemiologisch onderzoek naar carcinogeniteit bij langdurige "low level" expositie van een fabriekspopulatie. II. Eigen onderzoek, *T. Soc. Geneesk*, 51, 754, 1973.
16. **Newberne, P. M.,** personal communication. 1978.
17. **Halver, J. E.,** Aflatoxicosis and trout hepatoma, in *Aflatoxin: Scientific Background, Control, and Implications*, Goldblatt, L. A., Ed., Academic Press, New York, 1969, 265.
18. **Wogan, G. N.,** Aflatoxin carcinogenesis, in *Methods in Cancer Research*, Vol. 7, Busch, H., Ed., Academic Press, New York, 1973, 309.
19. **Vesselvinovitch, S. D., Mihailovich, N., Wogan, G. N., Lombard, L. S., and Rao, K. V. N.,** Aflatoxin B_1, a hepatocarcinogen in the infant mouse, *Cancer Res.*, 32, 2289, 1972.
20. **Peers, F. G. and Linsell, C. A.,** Dietary aflatoxins and liver cancer — a population based study in Kenya, *Br. J. Cancer*, 27, 473, 1973.
21. **Van Rensburg, S. J., Van der Watt, J. J., Purchase, I. F. H., Coutinho, L. P., and Markham, R.,** Primary liver cancer rate and aflatoxin intake in a high cancer area, *S. Afr. Med. J.*, 48, 2508a, 1974.
22. **Shank, R. C., Gordon, J. E., Wogan, G. N., Nondasuta, A., and Subhamani, B.,** Dietary aflatoxins and human liver cancer. III. Field survey of rural Thai families for ingested aflatoxins, *Food Cosmet. Toxicol.*, 10, 71, 1972.
23. **Shank, R. C., Bhamarapravati, N., Gordon, J. E., and Wogan, G. N.,** Dietary aflatoxins and human liver cancer. IV. Incidence of primary liver cancer, in two municipal populations of Thailand, *Food Cosmet. Toxicol.*, 10, 171, 1972.
24. **Peers, F. G., Gilman, G. A., and Linsell, C. A.,** Dietary aflatoxins and human liver cancer. A study in Swaziland, *Int. J. Cancer*, 17, 167, 1976.
25. **Epstein, S. M., Bartus, B., and Farber, E.,** Renal epithelial neoplasms induced in male Wistar rats by oral aflatoxin B_1, *Cancer Res.*, 29, 1045, 1969.
26. **Butler, W. H. and Barnes, J. M.,** Carcinogenic action of groundnut meal containing aflatoxin in rats, *Food Cosmet. Toxicol.*, 6, 135, 1968.
27. **Merkow, L. P., Epstein, S. M., Slifkin, M., Farber, E., and Pardo, M.,** Ultrastructure of renal neoplasms induced by aflatoxin B_1, *Proc. Am. Assoc. Cancer Res.*, 12, 44, 1971.
28. **Butler, W. H. and Barnes, J. M.,** Carcinoma of the glandular stomach in rats given diets containing aflatoxin, *Nature (London)*, 209, 90, 1966.
29. **Wogan, G. N. and Newberne, P. M.,** Dose-response characteristics of aflatoxin B_1 carcinogenesis in the rat, *Cancer Res.*, 27, 2370, 1967.
30. **Newberne, P. M. and Rogers, A. E.,** Rat colon carcinomas associated with aflatoxin and marginal vitamin A, *J. Natl. Cancer Inst.*, 50, 439, 1973.

31. **Wieder, R., Wogan, G. N., and Shimkin, M. B.**, Pulmonary tumors in strain A mice given injections of aflatoxin B_1, *J. Natl. Cancer Inst.*, 40, 1195, 1968.
32. **Pegg, A. E.**, Alkylation of rat liver DNA by dimethylnitrosamine: effect of dosage on O^6-methylguanine levels, *J. Natl. Cancer Inst.*, 58, 681, 1977.
33. **Kleihues, P. and Margison, G. P.**, Exhaustion and recovery of repair excision of O^6-methylguanine from rat liver DNA, *Nature (London)*, 259, 153, 1976.
34. **Maher, V. M., Dorney, D. J., Mendrala, A., Harvey, R. G., and McCormick, J. J.**, Comparing in normal diploid human fibroblasts the cytotoxicity and mutagenicity of the 7,8-dihydrodiol-9,10-oxide ("anti" isomer) and "K-region" epoxide of benzo(a)pyrene, *Proc. Am. Assoc. Cancer Res.*, 17, 190, 1977.
35. **Penn, I.**, Malignancies associated with immunosuppressive or cytotoxic therapy, *Surgery*, 83, 492, 1978.
36. **Shank, R. C., Bourgeois, C. H., Keschamras, N., and Chandavimol, P.**, Aflatoxins in autopsy specimens from Thai children with an acute disease of unknown etiology, *Food Cosmet. Toxicol.*, 9, 501, 1971.
37. **Krishnamachari, K. A. V. R., Bhat, R. B., Nagarajan, V., and Tilak, T. B. G.**, Hepatitis due to aflatoxicosis. An outbreak in Western India, *Lancet*, 1, 1061, 1975.
38. **Chaves-Carballo, E., Ellefson, R. D., and Gomez, M. R.**, An aflatoxin in liver of a patient with Reye-Johnson syndrome, *Mayo Clin. Proc.*, 51, 48, 1976.
39. **Becroft, D. M. O. and Webster, D. R.**, Aflatoxins and Reye's disease, *Br. Med. J.*, 4, 117, 1972.
40. **Shank, R. C., Johnsen, D. O., Tanticharoenyos, P., Wooding, W. L., and Bourgeois, C. H.**, Acute toxicity of aflatoxin B_1 in the Macaque monkey, *Toxicol. Appl. Pharmacol.*, 20, 227, 1971.
41. **Shank, R. C., Wogan, G. N., Gibson, J. B., and Nondasuta, A.**, Dietary aflatoxins and human liver cancer. II. Aflatoxins in market foods and foodstuffs of Thailand and Hong Kong, *Food Cosmet. Toxicol.*, 10, 61, 1972.
42. **Lopez, A. and Crawford, M. A.**, Aflatoxin content of groundnuts sold for human consumption in Uganda, *Lancet*, 2, 1351, 1967.
43. **Rutishauser, I. H. E., Dean, R. F. A., and Burgess, H. J. L.**, *East Afr. Med. J.*, 39, 478, 1962; as cited in **Lopez, A. and Crawford, M. A.**, *Lancet*, 2, 1351, 1967.

Chapter 5

OCCURRENCE OF N-NITROSO COMPOUNDS IN THE ENVIRONMENT

Ronald C. Shank

TABLE OF CONTENTS

I. Introduction 156

II. Methods of Chemical Analysis 156
    A. Lack of Specificity in Earlier Reports 156
    B. Quantitative Specific Analyses 157
        1. Thermionic Nitrogen Detector 157
        2. Thermal Energy Analyzer 157
        3. Gas Chromatograph—Mass Spectrometer 157
    C. Air Analyses 158
    D. Water Analyses 158
    E. Analyses of Biological Systems 159

III. Formation and Persistence in the Environment 159
    A. Formation: Kinetics 160
    B. Formation and Persistence in the Atmosphere 161
    C. Formation and Persistence in Water and Soil 161
    D. Formation in Foods and Gastric Juice 163
    E. Enhancement and Inhibition of Nitrosamine Formation 163

IV. Occurrence of N-Nitroso Compounds in Air 164
    A. Industrial Air 164
    B. Urban Air 165

V. Occurrence of N-Nitroso Compounds in Water 166

VI. Occurrence of N-Nitroso Compounds in Foods and Foodstuffs .. 166
    A. History 166
    B. Cured Meat Products 169
    C. Other Processed Foods 173
    D. Precursors in Foods 173
        1. Nitrates and Nitrites 174
        2. Secondary and Tertiary Amines 174

VII. Other Environmental Sources of N-Nitroso Compounds 175
    A. Drugs and Cosmetics 175
    B. Industrial and Agricultural Processes 176

References 178

I. INTRODUCTION

This chapter discusses where in the human environment hazardous N-nitroso compounds and their precursors occur, how they get there, and how their presence is reliably detected. The two chapters which follow discuss in detail exactly what kind of damage the N-nitroso compounds do to animals and humans and, as far as is known as of mid-1978, how the compounds effect their insult. The N-nitroso compounds have been studied more for their power to induce cancers than for any other biologic role; perhaps some 130 N-nitroso compounds have been tested for carcinogenicity, and at least 120 of them have given strikingly positive results. As a chemical class of compounds, this group is capable of causing cancer in every vital tissue (although no single compound within the group is carcinogenic to all vital tissues). Several of the N-nitroso compounds readily cause cancer after only a single exposure; several of these compounds are able to induce cancers across the placenta. Now it is known that these compounds are present in the human environment, and attempts must be made to determine the extent to which these compounds may be responsible for human cancers. The following chapters review what is currently known about this class of environmental carcinogens and discuss the risk they may present to human health.

II. METHODS OF ANALYSIS

A. Lack of Specificity in Earlier Reports

In the 1960s, a number of laboratories reported the presence of a variety of N-nitroso compounds in cured meat products, fish, cheese, mushrooms, alcoholic beverages, and cigarette tobacco, but in most cases, chemical analysis was not specific, and the possibility that the reported nitrosamines were artifacts cannot be dismissed. Several methods are now available which unequivocally determine at high sensitivity volatile nitrosamines, and one method can be used to measure nonvolatile nitrosamines in biological systems. A major review of the several methods for chemical analysis of N-nitroso compounds was published by the International Agency for Research on Cancer (IARC)[1] in 1972, and a more recent review has been published by Fiddler.[2]

Thin layer chromatography (TLC) has been one of the most frequently used analytical methods. This technique may be satisfactory if it can be shown that the sample itself after clean-up offers no interfering compounds and the sample is known to contain only one authentic N-nitroso compound such as in in vitro metabolism studies. The reagents often used to visualize N-nitroso compounds on TLC plates are palladium (II) chloride/diphenylamine[3] and sulfonilic acid/1-naphthylamine (Griess reagents). The palladium (II) chloride/diphenylamine reagent is not specific for N-nitroso compounds and forms interfering chromophores with other compounds, such as unsaturated hydrocarbons, dicarbonic acids, and compounds containing α-, β-unsaturated carbonyl groups;[3] propionic acid (CH_3CH_2COOH) often found in dairy products reacts with this reagent to interfere with diethylnitrosamine analysis.[4]

Several procedures rely on converting nitrosamines to secondary amines and nitrite, reducing them to hydrazines, or oxidizing them to nitramines; these conversions are difficult to do quantitatively and do not avoid the problem of interfering compounds in the sample. Wasserman[5] reviewed several of these methods and other techiques for nitrosamine analysis and pointed out the many drawbacks and pitfalls involved. Three techniques stand out as being the most quantitative, reliable, and sensitive and are discussed below in detail; these methods are:

1. Gas chromatography, (GC) using the thermionic nitrogen detector

2. The thermal energy analyzer (TEA), using either a gas or a high pressure liquid chromatograph
3. Gas chromatography, using a mass spectrometer (MS) as the detector

B. Quantitative Specific Analyses
1. Thermionic Nitrogen Detector

The thermionic nitrogen detector is a specialized alkali flame ionization detector for organically bound nitrogen. Samples of nitrosamines are pyrolyzed under conditions favoring the formation of cyan radicals, and such instruments are capable of detecting less than a picogram of organically bound nitrogen per injected sample. Although this rapid, relatively inexpensive technique is highly sensitive, it is not specific. As a screening technique, this instrument will permit detection of nitrosamines in food extracts or water at concentrations as low as 25 pg/mℓ,[6,7] yet the detector itself will not distinguish between nitrosamines, amines, or nitrites should they cochromatograph.

2. Thermal Energy Analyzer

In the last several years, a new instrument has been developed specifically for detecting and quantifying N-nitroso compounds.[8] The N-NO bond is the weakest bond in N-nitroso compounds; the TEA ruptures this bond selectively by rigid control of flash pyrolysis over a catalyst. The resulting nitrosyl radical ($\cdot$NO) is then oxidized in ozone to form an electronically excited nitrogen dioxide (NO_2). As the excited NO_2 rapidly decays to its ground state, it produces a characteristic emission of light in the near IR. The intensity of this emission is measured by a photomultiplier tube. The analyzer is quantitative, as the light emitted is proportional to the concentration of nitrosyl radical which in turn is proportional to the concentration of N-nitroso compound.

The TEA combines high sensitivity with high specificity. Fine and associates have reported several studies on the analysis of air,[9] water,[10] foods,[11] cosmetics,[12] lubricants,[13] etc., identifying specific nitrosamines at the parts-per-trillion level. Many interfering compounds are removed in a cold trap placed immediately after the pyrrolyzer, and transmission filters exclude luminescent contaminants produced in the oxidizer. The TEA has an added advantage in that it can be interfaced with either a gas chromatograph for volatile nitrosamines or a high pressure liquid chromatograph for nonvolatile nitrosamines.[8]

3. Gas Chromatograph — Mass Spectrometer

The IARC[14] has recommended that analyses for N-nitroso compounds include a MS confirmation of the presence of those compounds regardless of the original screening method used because few instruments offer the specificity of a MS, and there are many interfering compounds in biological specimens. This instrument fragments molecules and measures the mass to charge ratio for each fragment. The spectrum of ratios is essentially unique for each compound, giving the MS the ability to provide absolute identification providing the spectrum for an authentic standard is known. The spectrometer can be detected to monitor chromatographic peaks at specific mass to charge ratios characteristic of known compounds suspected of being in the sample; this adds high sensitivity (in the parts-per-billion range) to the merits of the instrument.

Until recently MS analysis was made difficult because of the great complexity and expense of the instrument, but now computerized bench models make operation relatively simple and affordable for many laboratories. The MS requires volatile samples and therefore is usually interfaced with a GC; this necessitates preparation of volatile derivatives of nonvolatile nitroso compounds prior to GC-MS analysis. Comparisons of the GC-MS and GC-TEA methods have recently been made.[15,16]

N-nitrosamides generally are of low volatility and/or are thermally unstable and are not readily analyzed by GC. Two methods have been developed for N-nitrosamide analysis by high pressure liquid chromatography; both methods utilize reverse phase columns. One method[17] uses detection by UV absorption with confirmation by high resolution mass spectroscopy. The second method[18] is less specific but simpler, detecting compounds which hydrolyze in dilute acid to give nitrite which is then measured colorimetrically. The initial reports on these methods did not describe the use in actual field samples of air, water or biological specimens.

C. Air Analyses

Analytical methods for determining N-nitroso compounds in air, which are present usually in only trace amounts, generally rely on concentration of the compounds in cold traps, on adsorption columns, or by dissolution in an alkaline aqueous medium. When carefully done, any of these methods will trap more than 90% of the N-nitroso compounds in the air. The alkaline traps are satisfactory only for N-nitrosamines, for the N-nitrosamides quickly decompose under such conditions.

A typical cold trap system for sampling air would consist of three glass vacuum traps in series, the first cold enough to remove water and the remaining two at $-80°C$. Air, at approximately 2 ℓ/min, is sampled for a few hours, and the combined condensates are extracted with a solvent such as dichloromethane.[9] Care must be taken to be sure nitrosation of trapped amines does not occur during collection; this reaction can be avoided by keeping the trap contents at high pH.

Adsorption columns have been used quite successfully in air sampling for nitrosamines. Tenax GC columns have been employed frequently by Environmental Protection Agency scientists, but carbon and silica gel adsorbants have also been used. Pellizarri and co-workers[19] have collected nitrosamines on Tenax GC 1.5 × 6 cm glass columns at a collection flow rate of approximately 1 ℓ of air per minute.

Impinger traps containing 1 N KOH have been used to remove nitrosamines from streams of air moving at up to 5 ℓ/min.[20] The nitrosamines are extracted from the alkaline aqueous solution into dichloromethane. Fine and co-workers[10] have combined this method with the use of cold traps with satisfactory results; the researchers pointed out that artifactual formation of dimethylnitrosamine could occur if ozone and dimethylamine coexist in the trapped air and that this formation was less in KOH traps than in the cryogenic system. In all the analyses described above, exceptional care must be taken to make certain reasonable recoveries are obtained without the catalysis of chemical reactions occurring and leading to the inadvertent formation of artifactual nitrosamines.

D. Water Analyses

The analysis of nitrosamines in water is straightforward. The water is extracted several times with dichloromethane which in turn is dried and concentrated prior to chromatography. Fine and co-workers[10] have extracted ten volumes of water three times with one volume fresh dichloromethane per extraction; the dichloromethane was dried over sodium sulfate and concentrated in a Kuderna-Danish evaporator-concentrator.

Water samples, such as sludge, sewage, and heavily polluted water which are rich in organics, may present dichloromethane extracts too complex for GC separation. Such samples may be steam distilled after the addition of sodium hydroxide to 3 M concentration. The distillate may be extracted with dichloromethane or passed through a sulfonic acid ion exchange column prior to extraction. Wasserman[5] offers an excellent review of several analytical procedures.

E. Analyses of Biological Specimens

A great many methods have been published for the quantitative determination of volatile nitrosamines in a variety of foodstuffs, and these methods are applicable to other biological specimens, including viscera, blood, urine, and feces. Several of these methods have been collected in a monograph by the IARC.[1] Most methods can be divided into two groups, based on the manner in which the nitrosamines are extracted. Sen and co-workers[21] reported a method which in several variations have become widely used. The sample is homogenized in 3 N KOH, and the homogenate is extracted with dichloromethane to isolate the nitrosamines. After filtration the dichloromethane extract is steam-distilled first to remove the organic solvent and then to evaporate and collect the nitrosamines. The acidified aqueous distillate is passed through a column (e.g., Rexyn 101/polyamide) to remove amines and pigments; the effluent is made alkaline once more and extracted with fresh dichloromethane. The organic extract is dried, concentrated into n-hexane, and passed through a basic alumina column; after washing the column with n-pentane, the purified nitrosamines are eluted with dichloromethane and concentrated in preparation for application to a GC. Goodhead and Gough[22] have simplified the cleanup procedure by modifying the above method as follows: steam distillation is done on a specimen-water homogenate; the distillate is acidified and extracted with dichloromethane; the extract is washed with sodium hydroxide, dried over sodium sulfate, and concentrated prior to GC analysis.

A second approach to the extraction of nitrosamines from biological samples was described by chemists at the U.S. Food and Drug Administration[23] and is widely used for food samples. The comminuted sample is digested in methanolic KOH; the filtered digest is steam distilled, and the distillate is extracted with dichloromethane; the extract is concentrated before analysis by GC. An important advantage of this method is that digestion of the sample avoids the uncertainty of nitrosamine recovery from minced or homogenized samples.

All the above methods have a sensitivity of approximately 1 μg nitrosamine per kilogram sample and require several hours to isolate and purify the nitrosamine extract prior to GC analysis. Fine and co-workers[11] have described a much simplified method for nitrosamine analysis in foods, with a sensitivity of 0.05—0.01 μg nitrosamine per kilogram sample. A mixture of minced food, sodium hydroxide, and crude mineral oil is heated under vacuum, and the distillate obtained is acidified and extracted with dichloromethane. The organic extract is concentrated in a Kuderna-Danish evaporator and analyzed by GC — TEA. The high selectivity of the thermal energy analyzer for nitrosamines reduces the need for extensive cleanup of the dichloromethane extracts.

III. FORMATION AND PERSISTENCE IN THE ENVIRONMENT

Until recently environmental exposure to N-nitroso compounds was thought to be limited to small groups of chemists working with these compounds, chemical plant employees preparing rocket fuel, and cancer researchers, who used the compounds experimentally as model carcinogens. In the mid-1960s, it became apparent that some of these nitroso compounds could exist in the human food supply, and in the mid-1970s, this observation was extended to include the human workplace and urban environment. Today, N-nitroso compounds are recognized as pollutants of specialized environments and occurring at low levels; precursors of the nitroso compounds occur far more extensively and at greater concentrations and can lead to the formation of nitrosamines and nitrosamides in the air, water, soil, food, and human GI tract.

A. Formation: Kinetics

The poisoning of sheep in Norway with nitrite-treated fish meal[24,25] focussed international attention on the problem of accidental formation of N-nitroso compounds in the environment. In 1961 and 1962 outbreaks of toxic hepatosis in ruminants in Norway were associated with feeding animals fish meal prepared from sodium nitrite-preserved herring.[24,25] TLC analysis demonstrated the presence of 30 to 100 ppm dimethylnitrosamine in six samples of herring meal known to have produced toxic hepatosis. The fish, especially the skin, contained di- and trimethylamine, compounds which impart the "fishy" smell when the product is not fresh. Dimethylnitrosamine was not found in fish not preserved with sodium nitrite; presumptive evidence was obtained for the presence of other nitroso compounds, but they were not identified.[25]

In retrospect this formation of dimethylnitrosamine seems obvious. The standard procedure for preparing dimethylnitrosamine in the laboratory has been to add acid and sodium nitrite to an aqueous solution of dimethylamine. As researchers from various disciplines examined the problem of nitrosamine formation, it became clear these same reactions were occurring in several aspects of the environment.

In the formation of N-nitrosamines, the reaction is a nitrosation of a secondary or a tertiary amine by nitrous anhydride (N_2O_3) and to lesser extents by nitrosyl halide (for example, NOCl), nitrous acidium ion ($H_2NO_2^+$) and nitrosyl thiocynate (ON-NCS).[26,27] In acid solution, nitrite is converted to nitrous acid which in turn forms nitrous anhydride as follows:

$$NO_2^- + H^+ \longrightarrow HNO_2$$

$$2HNO_2 \rightleftharpoons N_2O_3 + H_2O$$

The amine is nitrosated in the unionized form[28] at a rate proportional to the nitrous anhydride concentration and thus proportional to the square of nitrous acid concentration, hence the square of the nitrite concentration.

$$R_2NH + N_2O_3 \xrightarrow{k} R_2NNO + HNO$$

$$K \text{ is proportional to } \{N_2O_3\}, \{HNO_2\}^2, \text{ and } \{NO_2^-\}^2$$

Nitrosation occurs more readily with amines of weak basicity (e.g., morpholine) and more slowly with strongly basic amines (such as dimethylamine) because the nitrous anhydride attacks the unionized form of the amine and weakly basic amines ionize less readily than do strongly basic amines. The rate of nitrosation (k) is maximum near the hydrogen ion concentration at which nitrous acid is 50% ionized ($pK_a = 3.36$); at lower hydrogen ion concentrations (higher pH values), the inorganic nitrogen is present as nitrite rather than nitrous acid, and therefore the nitrosating agent is formed slowly. At higher hydrogen ion concentrations (lower pH values), much of the amine becomes ionized, and the unionized form becomes the limiting reagent.[26,27] Nitrosation of secondary and tertiary amines, then, would occur when the amine and nitrite are present in a solution where the pH is between 1 and 7, but maximum rates occur between pH 3.0 and 3.4 for most amines. In some cases, the maximum yield occurs at pH 4 to 5 because nitrous acid is unstable at low pH values, and if the amine nitrosates slowly, the nitrosating agent is lost before much of the amine is nitrosated.[27,29,30]

The formation of N-nitrosamides is slightly different as the principal nitrosating agent appears to be the nitrous acidium ion ($H_2NO_2^+$); this ion also forms in acid solutions of nitrite.[27]

$$NO_2^- + H^+ \longrightarrow HNO_2$$

$$HNO_2 + H^+ \rightleftharpoons H_2NO_2^+$$

$$RNH \cdot CO \cdot R' + H_2NO_2^+ \longrightarrow RN(NO) \cdot CO \cdot R' + H_2O + H^+$$

The nitrosation reaction rate does not show a pH maximum and is proportional to the hydrogen ion concentration and the nitrite concentration (not the square of the nitrite concentration as in the case of amine nitrosation).

Returning to the nitrite-treated herring instance, it is apparent that bacterial decomposition of the fish probably produced dimethylamine and acids, and addition of the sodium nitrite provided the source of nitrosating agent, in this case, nitrous anhydride. In spite of the strong basicity of dimethylamine, there was present in the fish sufficient acid and nitrite to produce amounts of dimethylnitrosamine acutely fatal to sheep. Since sodium nitrite is quantitatively a major food additive in the Western world, and nitrate, a precursor of nitrite, is widespread in the environment, a current concern is the extent to which these nitrosations take place in the human environment.

B. Formation and Persistence in the Atmosphere

The formation of nitrosamines in a gaseous environment has been demonstrated.[31,32] At concentrations of 50 to 100 ppm, diethylamine and NO_2 react rapidly to form measurable levels of diethylnitrosamine,[32] but the rates are strongly dependent on concentration of reactants, the presence of moisture and particles (or walls to facilitate surface rather than gaseous reactions), and the presence of light, especially UV light. Detailed studies have been carried out by Hanst and co-workers;[33] the formation of dimethylnitrosamine in air was measured by longpath IR spectroscopy. Dimethylamine (1 ppm), NO_2 (1 ppm), NO (4 ppm) in dry nitrogen gas reacted to nitrosate 1% of the dimethylamine per minute; when nitrous acid and water are also present (1 ppm dimethylamine, 0.5 ppm nitrous acid, 2 ppm nitrous oxide, 2 ppm NO_2, and 13000 ppm water in room air), 4% of the amine is nitrosated per minute. Dimethylnitrosamine undergoes photolysis and has a half-life of about 30 min in noontime sunlight on a clear day (half-life of about an hour on a cloudy day).[33] The authors point out that although nitrosation of dimethylamine in heavily polluted urban atmospheres might take place during the night, the reaction would cease in the morning, as the nitrous acid was destroyed by photolysis. The sunlight would also begin to decompose the nitrosamine formed during the night, so that by midday all the nitrosamine formed the previous night would have been destroyed. One would conclude from these studies then that formation of nitrosamines in urban atmospheres is not a serious problem and that most of the nitrosamines found in such atmospheres, especially in the afternoon, result from direct emissions.

C. Formation and Persistence in Water and Soil

Fertilizer, pesticides, and industrial and municipal wastes supply soils and waters with secondary and tertiary amines and nitrate which are precursors for nitro compounds. Controlled studies have shown that amines and nitrate or nitrite must be present in water or soil at relatively high concentrations to yield detectable amounts of nitrosamines.[34-37] Microorganisms appear to be required for the conversion of tertiary amines to secondary amines and for the reduction of nitrate to nitrite.[35,38] Nitrosamine formation was measured in soil at pH values from 3.8 to 6.5 and in water and sewage at pH values between 4 and 7.[38]

Only a few detailed reports have been published on the formation of N-nitroso compounds in water and soil, and these have been limited to laboratory studies in which samples were treated with nitrate or nitrite and various amines; a field survey of the occurrence of nitrosamines in water and soils has not yet been made. The studies discussed below offer presumptive evidence, however, that such carcinogens may occur in water and a variety of soils (only in one of these studies did the results rely on MS confirmation of identity of the nitrosamine).

Samples of municipal sewage at pH 4.0 to 7.0 were incubated at 30°C after the addition of approximately 200 or 2000 µg trimethylamine and 40 or 400 µg nitrate per milliliter.[35] Microorganisms were credited with the demethylation of trimethylamine to dimethylamine, as no secondary amine was detected in filter sterilized sewage. Nitrate was also readily reduced to nitrite, and as would be expected, small amounts of dimethylnitrosamine were formed. A similar study was done on lake water, demonstrating the conversion of trimethylamine to dimethylamine with subsequent nitrosation to form dimethylnitrosamine, if nitrite was available.[36] In this latter study, a fungicide, thiram [bis(dimethylthiocarbamoyl)disulfide] and nitrite were added to sewage at pH 4, and the formation of dimethylnitrosamine (identity confirmed by MS) was observed. The studies were continued to determine the persistence of dimethylnitrosamine, diethylnitrosamine, and di-*n*-propylnitrosamine in water.[39,40] There was no degradation of these nitrosamines added at concentrations of 20 to 30 ppm to Cayuga (New York) lake water (pH 8.2) over a 3.5-month incubation at 30°C in the dark. Under similar conditions, these carcinogens did slowly disappear from sewage at pH 6.0. Under field conditions, however, at the surfaces of waters (and soils) exposed to sunlight, photolysis would be expected to bring about rapid degradation of these compounds.

Quite similar studies have been done for soil samples. The addition of trimethylamine, nitrate, and nitrite leads to the formation of dimethylnitrosamine in spodosol, silty clay loam, and silt loam.[34] Apparently microbial action demethylates the tertiary amine to dimethylamine and reduces nitrate to nitrite. The pH of the various soils used in this study (pH 3.8, 5.8, 6.5, respectively) favored the nitrosation reaction. Maximum concentrations of dimethylnitrosamine were observed a few days after addition of approximately 1600 ppm dimethylamine and 330 ppm nitrite, but the nitrosamines did not persist.

As secondary amines are probably the limiting reactants in nitrosamine formation in most soils, Tate and Alexander[37] investigated whether various nitrogenous pesticides could be converted in soil to secondary amines known to be precursors of carcinogenic nitrosamines. Using Williamson silt loam (pH 6.4), pesticides were added to give a final concentration of 1000 ppm. After incubation at 30°C, the soils were analyzed for the presence of secondary amines and volatile nitrosamines. Dimethyldithiocarbamate appeared to slowly decompose to yield dimethylamine in the soil; approximately 0.25% of the pesticide was removed as dimethylamine after 2 weeks of incubation. Diethyldithiocarbamate also slowly decomposed to a secondary amine, presumably diethylamine. Both these amines would be expected to be nitrosated to powerful carcinogens if the acidic soils contained sufficient nitrate or nitrite.

In a Canadian study,[42] the herbicide, glyphosate (N-(phosphoromethyl)-glycine) was added to various soils with sodium nitrite. N-Nitrosoglyphosate was formed within 24 hr only when the additives were present in relatively high concentrations. It is not known whether this nitrosamine is carcinogenic. Seiler[43] treated glyphosate with acetic acid and sodium nitrite and isolated an unidentified ether-soluble compound which gave some evidence of weak mutagenicity when analyzed by the Ames test.

D. Formation in Foods and Gastric Juice

Nitrate, nitrite, and amines are natural constituents in foods, and given the proper conditions, nitrosamines can form. Certain vegetables, such as spinach, celery, and beets, can contain more than 600 ppm nitrate nitrogen (fresh vegetables generally contain little nitrite); lettuce and white potatoes are also large contributors of nitrate to the daily American diet, more so because of their frequent consumption rather than any unusual nitrate concentration.[44] Cured meats can also contain nitrate and nitrite. White[44] has estimated that vegetables contribute almost 90% of the daily ingested nitrate (approximately 86 mg nitrate per person per day) and about 2% of the nitrite; another 9% of the daily nitrate and 20% of the nitrite come from cured meats, and less than 1% of the ingested nitrate is from drinking water, unless the water source is private wells. Some wells can contain as much as 500 ppm nitrate nitrogen, supplying almost 95% of the daily ingested nitrate.[45] Table 1 summarizes White's data[44] for the various sources of ingested nitrate and nitrite.

In 1940 German investigators reported the presence of nitrate and nitrite as a normal component in human saliva.[47] The nitrate content is reported to depend directly upon the amount of nitrate ingested,[48,49] although recently it was suggested that some of the salivary nitrite and nitrate may originate from *de novo* synthesis in the intestine by heterotrophic nitrification of ammonia or organic nitrogen compounds.[50] The nitrate, be it from ingested foods or *de novo* synthesis, could be absorbed from the intestine and then secreted in the saliva; bacterial reduction in the oral cavity could convert nitrate to nitrite. White[44] estimated that up to 30 mg of nitrate and 9 mg of nitrite could appear in saliva over a 24-hr period, thus saliva may be a major source of ingested nitrite.

Sander and Burkle[51] demonstrated the in vivo nitrosation of secondary amines by ingested nitrite, a reaction favored by the acidic environment provided by gastric juice. Single boli of ingesta rich in nitrosatable amines and nitrite can be important in the rapid formation of carcinogenic nitrosamines in the stomach.[52] The importance of salivary nitrite in such reactions remains to be shown; although the amount of ingested nitrite contributed by saliva is large over a 24-hr period, the concentration of saliva-derived nitrite at any one time in the gastric juice is small, especially in comparison to nitrite concentrations resulting from ingestion of certain cured meats and vegetables; since nitrosation of amines is a function of the square of the nitrite concentration (see Section II.A above), it is the concentration of nitrite in the stomach at any one time, not the total amount of nitrite ingested over a day, that is critical to the amount of potentially harmful nitrosamine formed.

E. Enhancement and Inhibition of Nitrosamine Formation

Several chemicals can accelerate the rate of nitrosation of secondary amines.[53] Thiocyanate (NCS^-) is particularly effective in increasing the rate of nitrosation of morpholine, N-methylaniline, aminopyrine, and other amines.[54,55] The catalysis takes place by the formation of the nitrosating species $ON \cdot NCS$:

$$HNO_2 + H^+ + N\equiv C-S^- \rightleftharpoons O=N-N=C=S + H_2O$$

$$R_2NH + O=N-N=C=S \rightleftharpoons R_2N-N=O + H^+ + N\equiv C-S^-$$

This reaction[56] is proportional to the nitrite concentration, unlike the reaction relying on nitrous anhydride (N_2O_3) which is proportional to the square of the nitrite concentration;[57] thus the thiocyanate-catalyzed reaction is especially important under conditions of low nitrite concentration. The rate of nitrosation of sarcosine can be increased by a factor of up to 400 by the addition of thiocyanate under in vitro conditions.[54] This

Table 1
ESTIMATED AVERAGE DAILY INGESTION OF NITRATE AND NITRITE IN THE U.S.[a][44]

Source	Nitrite (mg)	Nitrate (mg)
Cured meats	2.38	9.4
Vegetables	0.20	86.1
Bread	0.02	2
Fruits, juices	0	1.4
Water[b]	0	0.7
Milk and products	0	0.2
Total	2.60	99.8

[a] Excluding what might be present in saliva.
[b] Assuming water is typical of most large public water supplies in the U.S.[46]

catalyst is normally a constituent of human saliva, especially that from cigarette smokers, of gastric juice, and of many foods.[53,58] Other similar enhancers of amine nitrosations include bromide, chloride, sulfate, and phosphate.[28,59]

Ascorbic acid,[60] sulfhydryl compounds such as glutathione,[61-64] and tannins[65] can inhibit the nitrosation of amines essentially by scavenging the nitrite before it can react with the amine. More precisely the ascorbic acid in an acidic environment (approximately pH 4) is thought to react with the nitrous anhydride derived from nitrate and form the nitrite ester of ascorbic acid; this may decompose to form the monodehydroascorbic acid and nitric oxide.[60] The reaction may continue as shown in Figure 1 to yield the (di)dehydroascorbic acid. A portion of the nitric oxide could revert to nitrite; hence a molar excess of ascorbic acid is required to completely block nitrosation of amines by nitrite in an acid environment. Ascorbic acid, in large molar excess compared to nitrite, has been shown to inhibit several in vivo nitrosation reactions;[61-64,66,67] however, rats fed a semipurified diet containing 50 ppm sodium nitrite (0.72 mmol/kg), 50 ppm morpholine, and 200 ppm ascorbic acid (1.14 mmol/kg) still developed liver cancers identical to those produced by nitrosomorpholine itself.[68]

IV. OCCURRENCES OF N-NITROSO COMPOUNDS IN AIR

A. Industrial Air

Few reports have been published on the detection of a nitrosamine in the ambient air of an industrial facility. Air inside the plant of a German factory producing dimethylamine was reported to contain 0.001 to 0.43 ppb (3.03 to 1303 ng/m^3) dimethylnitrosamine, although the air-sampling technique did not rule out the possibility of nitrosation of dimethylamine in the trapping system.[31] This possibility was minimized in a later study reported by Fine and co-workers.[69] A chemical factory in Baltimore, Md. which reduced dimethylnitrosamine to unsymmetrical dimethylhydrazine (a rocket propellant) was targeted for dimethylnitrosamine contamination of the environment. Ambient air at various locations on the site of the chemical factory (it is not clear if any of the measurements were made inside the factory where workmen could be exposed) contained from 1900 ng dimethylnitrosamine per cubic meter air (0.627 ppb) to as much as 36,000 ng/m^3 (11.88 ppb). The analysis involved GC with a thermal energy analyzer as the detector. The presence of dipropylnitrosamine in one air sample was suggested in one measurement, but was not confirmed; its relationship to processes in the chemical plant was not explained.

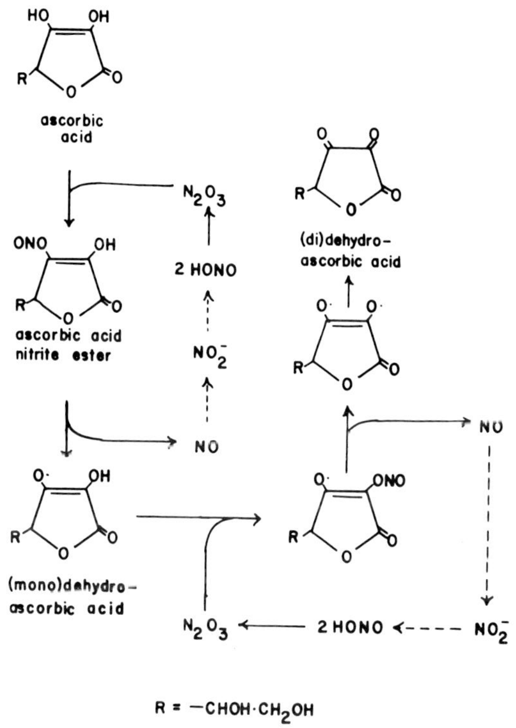

FIGURE 1. Scavenging of nitrite by ascorbic acid in an acidic environment.[58,60]

Dimethylnitrosamine was used industrially as both a precursor to the propellant dimethylhydrazine (unsymmetrical) and as a solvent in the rubber industry. A few cases of workman exposures to dimethylnitrosamine have appeared in unpublished reports, but no data on nitrosamine concentrations in the ambient air of these workplaces seem to exist. In addition, at the time these exposures took place, before 1960, the chemical procedures for nitrosamine analysis were not of sufficient specificity to have been useful in assessing exposure.

B. Urban Air

Several reports have been published on the occurrence of N-nitrosamines as air pollutants in the air supplies of industrial cities. Aside from the study in Germany,[31] the urban sites have been limited to Baltimore, Md., Belle, W. Va., New York City, Wilmington, Del., and Philadelphia, Pa. An extensive nationwide study of nitrosamines as air pollutants has recently been undertaken by the U.S. Environmental Protection Agency.

Baltimore has been studied more thoroughly because of the presence of the factory which used dimethylnitrosamine as a chemical intermediate in the production of rocket fuel, as described above. The levels of dimethylnitrosamine in Baltimore air over the fall and winter of 1975 were as high as 300 ng/m³ (0.990 ppb); these measurements were made approximately 800 m from the dimethylhydrazine plant.[9,69,70] The 24-hr average for dimethylnitrosamine concentrations in the air in this area was 1070 ng/m³ (0.353 ppb). Within population centers of Baltimore and Arundel County far removed from the factory, the average dimethylnitrosamine air concentration was 100 ng/m³ (0.033 ppb), with a high of 760 ng/m³ (0.251 ppb).

In Belle, W. Va., near a chemical plant which manufactured dimethylamine, all six air samples analyzed contained dimethylnitrosamine, based on GC and TEA; the highest concentration reported was 0.051 ppb (155 ng/m^3) dimethylnitrosamine. The average of the concentrations for all six samples was 0.020 ppb (61 ng/m^3) dimethylnitrosamine.[69] Of 6 air samples taken in Philadelphia, 1 contained dimethylnitrosamine (0.025 ppb; 76 ng/m^3). A single sample of air taken in New York City contained 160 ng/m^3 (0.053 ppb) dimethylnitrosamine.[71]

Trace amounts of what may have been other N-nitroso compounds have been reported, but the suspect compounds were not identified. It has never been explained why dimethylnitrosamine and only this nitroso compound has been detected repeatedly in urban air not in the vicinity of a point source of contamination, such as a chemical factory handling dimethylamine or dimethylnitrosamine. The results of these urban air analyses are summarized in Table 2.

V. OCCURRENCE OF N-NITROSO COMPOUNDS IN WATER

Only a few conclusive studies on the occurrence of N-nitroso compounds in waterways and water supplies have been reported. The 1976 Scientific and Technical Assessment Report (STAR) on nitrosamines assembled by the U.S. Environmental Protection Agency refers to a report on the finding of nitrosamines in a Russian reservoir, but no details were available.[71]

In their study on contamination of urban air in Baltimore by nitrosamines, Fine and co-workers[69] also examined water sources. Drinking water did not contain detectable nitrosamines, but effluent from a sewage treatment plant and the salt water in the estuary in Baltimore were contaminated. Again, only dimethylnitrosamine was found: 2700 ng/ℓ in the sewage effluent and between 90 and 940 ng/ℓ in the estuary. The higher estuary levels were observed in the vicinity of the chemical plant that used dimethylnitrosamine in the manufacturing process. In a drainage ditch and in mud puddles adjacent to this plant, as much as 5900 ng dimethylnitrosamine per liter of water and 6,000,000 ng per kilogram mud (wet weight basis) were found. The source(s) of the nitrosamine in the water samples could not be determined, but it is not unlikely that the high concentrations found in close proximity to the chemical plant could be attributed to the plant itself as a point source of contamination.

Two reports have demonstrated the presence of nitrosamines in deionized water.[72,73] The results, summarized in Table 3, are based on only those analyses confirmed by MS. Data were presented which suggests the nitrosamine contamination stems from the anion exchange resins, or more exactly, the nitrosamines may originate in the technical grade aliphatic amines used to prepare the resins. These findings, from two laboratories widely respected for their excellence in analytical chemistry, are important in two aspects:

1. The results should caution analytical chemists to contamination of samples being analyzed for nitrosamines (necessitating the use of control samples).
2. Municipal water treatment facilities which use ion exhange techniques in water purification must be alerted to the possibility of contaminating public water supplies and even commercial bottled water.

VI. OCCURRENCE OF N-NITROSO COMPOUNDS IN FOODS AND FOODSTUFFS

A. History

The discovery of dimethylnitrosamine in sheep fodder in the early 1960s (discussed in Section III.A) created a great deal of concern around the world as to whether car-

Table 2
DIMETHYLNITROSAMINE IN SAMPLES OF URBAN AIR

				Dimethylnitrosamine concentrations				
				Mean[b]		Range		
Site	Date	Number of samples	Number of samples containing DMN[a]	ng/m³	ppb	ng/m³	ppb	Ref.
Baltimore, Md.	August 1975	5	3	666	0.219	100—2918	0.033—0.960	9, 69
	October 1975	8	5	618	0.204	130—1800	0.043—0.594	69
	November 1975	11	11	819	0.270	26—3000	0.009—0.990	69
	December 1975	22	22	63	0.021	16—212	0.005—0.070	69
Philadelphia, Pa.	August 1975	6	1	—	—	76	0.025	9
Wilmington, Del.	August 1975	6	0	—	—	—	—	9
Belle, W. Va	August 1975	6	6	61	0.020	t—155[c]	t—0.051	9
Waltham, Mass.	July 1975	2	0	—	—	—	—	9

[a] DMN = dimethylnitrosamine.
[b] Arithmetic average of all samples.
[c] t = trace, approximately 1/10¹², 3 ng/m³.

Table 3
NITROSAMINES IN DEIONIZED WATER[72,73]

Type of water	Number of samples	Number of samples contaminated	Concentration of nitrosamines (μg/kg) DMN	DEN	Ref.
Untreated tap	Several	0	ND	ND	72
	20	2	0.01	0.05	73
Distilled	1	0	ND	ND	72
Deionized	6	2	0.01	ND	72
	25	13	0.08—0.34	0.33—0.83	73
Deionized, distilled	5	2	0.02—0.06	ND	72

cinogenic N-nitroso compounds could occur in human food products processed with sodium nitrite. The potential problem was not of small scale, for in many countries such as the U.S., sodium nitrite has been one of the most abundantly used food additives, based on annual tonnage of the salt produced for the food industry.

Many investigators began analyzing various commodities for the presence of nitrosamines. In Germany[74] wheat grain and flour and Tilsit cheese were reported to contain diethylnitrosamine, a potent carcinogen, but these findings could not be confirmed by English investigators.[75] Both groups used TLC to isolate the suspect compound and UV light to degrade the compound to "nitrite spots"; in some instances, the presence of diethylnitrosamine was "confirmed" by GC. It is now known, however, that such techniques are not sufficiently specific to reliably demonstrate the presence of N-nitroso compounds (see Section II). Similar problems were also true of early reports of nitrosamines in food products, for instance, the report from Norway[76] of up to 40 μg dimethylnitrosamine per kilogram in smoked fish, 6.5 μg/kg in smoked meats, and 30 μg/kg in certain mushrooms; the Dutch report[77] of diethylnitrosamine in spinach, albeit only after unusual storage conditions of artifically high concentrations of nitrite and hydrogen ion. Herdboys in the Transkei of South Africa prepare milk curds, one of their major daily foods, using the juice of the fruit of a solanaceous bush; using TLC, GC, nuclear magnetic resonance, and IR spectrophotometry, researchers found what appeared to be dimethylnitrosamine,[78] however, this finding has not been confirmed by MS. In Zambia eight samples of a distilled alcoholic beverage prepared by fermenting corn husks and sugar were reported to contain 1 to 3 ppm dimethylnitrosamine; the methodology included polarography and TLC, but not MS.[79] The reports of the occurrence of N-nitrosamines in food samples discussed in the following section are limited to those in which identification of the compounds relied on MS and/or TEA.

A most important result of these early reports of food-borne nitrosamines was the serious examination of the necessity of using sodium nitrite as a food additive. Nitrate-containing salts had been used to cure meats for thousands of years. In the early 1900s, it was determined that nitrite was responsible for the desirable color and flavor of cured meats. Not until 1941 was the bacteriostatic property of the salt appreciated, and even those studies did not focus on the inhibition of botulinum toxin production.[80] In 1925 the U.S. Department of Agriculture established a maximum allowable residual of nitrite (200 ppm as sodium nitrite). It was not until 1973, however, that experimental evidence was obtained regarding the amount of nitrite actually needed to prevent the production of botulinum toxin in vacuum-packed foods. In one study, ground ham containing 2.5% NaCl, 0.5% dextrose, and 0.02% sodium isoascorbate was inoculated with *Clostridium botulinum* spores, vacuum packed in cans, heated to an internal temperature of 68.5°C (standard industry practice), chilled, and then stored at 27°C for up to 24 weeks.[81] The results, summarized in Table 4, demonstrated that 150 ppm added sodium nitrite, but not 50 ppm, was sufficient to prevent toxin production in the canned meat when the level of spore contamination was low (90 spores per gram of meat); when spore contamination was high (5000 spores per gram of meat), more than 400 ppm added sodium nitrite was required to prevent toxin formation in the canned meat stored at the elevated temperature. In a second study,[82] wieners inoculated with *C. botulinum* spores and stored at 27°C did not become toxic when the sausages contained 100 ppm added sodium nitrite, but did contain botulinum toxin when only 50 ppm sodium nitrite were added (see Table 5). These reports apparently were the first scientific attempts to determine the actual levels of residual nitrite needed to protect consumers from botulinum poisoning associated with eating vacuum packed meats.

Because botulism toxin is the most potent poison known to man[83] and can occur in a variety of processed foods when the packing technique is not adequate, the use of nitrite as an additive to low-acid vacuum-packed foods seems justified. On the other hand, this benefit must be weighed with the evidence that the formation of carcinogenic nitrosamines in foods are almost totally related to the use of nitrite as a food additive (see Chapter 10).

B. Cured Meat Products

Several reports have documented the presence of the carcinogenic nitrosamine, N-nitrosopyrrolidine, in cooked bacon, but the manner in which it is formed during the cooking process remains unclear.[85-88] Uncooked bacon seldom contains N-nitrosopyrrolidine, but with frying at temperatures approaching 400°F, the carcinogen readily forms; concentrations in the bacon drippings may exceed those found in the lean,[86,89] and the carcinogen is also present in the fumes produced during the frying.[90] Dimethylnitrosamine is also often found in cooked bacon, drippings, and frying fumes, but at lower concentrations than those for N-nitrosopyrrolidine; approximately 70% of the total dimethylnitrosamine and 50% of the total nitrosopyrrolidine produced during cooking is volatilized into the fumes.[90] The amounts of dimethylnitrosamine, diethylnitrosamine, nitrosopyrrolidine, and nitrosopiperidine found in various bacon samples are summarized in Table 6.

The precursors for nitrosopyrrolidine formation in bacon during cooking have not yet been conclusively demonstrated. The amino acid, proline, a component of collagen, can be nitrosated to nitrosoproline which, upon heating can be decarboxylated to nitrosopyrrolidine.[101-103] Pyrrolidine can also be formed by ring closure of the diamine, putrescine, upon heating; similarly, spermidine could also be a precursor for nitrosopyrrolidine.[101] These reactions are summarized in Figure 2. Hwang and Rosen[104] prepared bacon with added nitrite, proline, putrescine, spermidine, and nitrosoproline, and measured the formation of nitrosopyrrolidine during controlled frying; only proline and nitrosoproline gave detectable yields of nitrosopyrrolidine. Hansen and co-workers[105] found that uncooked bacon contained about 0.5 μmol nitrosoproline per kilogram, and after frying, the bacon contained about 0.4 μmol nitrosopyrrolidine; if nitrosoproline was the sole precursor of the nitrosamine, approximately 80% of the nitrosoproline would have to be decarboxylated during frying. Thus, current evidence suggests that both free proline and nitrosoproline, formed during the curing process, are the most likely precursors of nitrosopyrrolidine in fried bacon.

As shown in Table 6, a wide variety of cured meats and sausages contain nitrosamines. Nitrosopyrrolidine, which occurs at the highest concentrations, induces cancers in the liver,[106] nasal cavities, and testis.[107,108] Dimethylnitrosamine, which occurs most frequently, but usually at low concentration, is a potent carcinogen for the liver and kid-

Table 4
EFFECT OF SODIUM NITRITE ON THE DEVELOPMENT OF TOXICITY IN GROUND HAM INOCULATED WITH *CLOSTRIDIUM BOTULINUM* SPORES[81]

Added nitrite (μ/g meat)	No. toxic samples[a]
Inoculum: 90 spores/g meat	
50	21 in 3 weeks storage
150	0 in 24 weeks storage
300	0 in 24 weeks storage
500	0 in 24 weeks storage
Inoculum: 5000 spores/g meat	
0	34 in 2 weeks storage
100	17 in 9 weeks storage
200	4 in 14 weeks storage
400	1 in 3 weeks storage

[a] A total of 40 samples at each nitrite concentration were stored at 27°C for up to 24 weeks.

Table 5
EFFECT OF SODIUM NITRITE ON THE DEVELOPMENT OF TOXICITY IN WIENERS INOCULATED WITH *CLOSTRIDIUM BOTULINUM* SPORES[82]

Added nitrite (μg/g wiener)	No. toxic samples[a]
0	5 in 2 weeks storage
50	1 in 8 weeks storage
100	0 in 8 weeks storage
150	0 in 8 weeks storage
200	0 in 8 weeks storage
300	0 in 8 weeks storage

[a] A total of five wieners at each nitrite concentration were inoculated with 620 spores and stored at 27°C for up to 8 weeks.

ney.[109] Diethylnitrosamine which is found occasionally in cured meats and sausages is carcinogenic to the liver and esophagus,[106] kidney,[110] and lung.[111] N-Nitroso-di-*n*-butylamine and nitrosopiperidine which have been found infrequently produce cancers in the urinary bladder, liver, and esophagus and the nasal cavities, esophagus, larynx, trachea, and liver, respectively.[106] Recently, Sen and co-workers[112] have demonstrated marked reduction in the formation of nitrosopyrrolidine during cooking when raw bacon has been treated with nitrite scavengers, such as propyl gallate, piperazine, ascorbyl palmitate, or sodium ascorbate; hopefully such additives might prove to be useful in reducing the formation of all nitrosamines in human foods where nitrite is present.

Table 6
CARCINOGENIC N-NITROSO COMPOUNDS IN HUMAN FOODS AND FOODSTUFFS

Nitrosamine concentration (μg/kg)[a]

Food product	Dimethylnitrosamine	Nitrosopyrrolidine	Diethylnitrosamine	Nitrosopiperidine	Ref.
Bacon					
Uncooked	(1/8) 1—4 30	Up to 40 0	1		85 87
Fried	(6/8) 2—5	(7/8) 10—108 4—25			86 87
Fried			1.5		83
Fried	(50/52) up to 5	(50/50) 1—200		(50/50) 0.08—0.25	91, 92
Ham	5				86
Frankfurter	(3/34) 11—84				93
Salami	20—80				94
Dry sausage	10—20				94
Souse	(6/7) 3—63	(1/7) 19			95
Blood and tongue	(2/2) 7, 45				95
Various cured meats[b]	(29/80) 2—35	(17/80) 13—105 <1	(9/80) 2—25		96 91, 92
Salami, Hungarian	1—4				85
Mettwurst sausage		105		20—60	97
Luncheon meat	1—4				85
Pork, Danish	1—4				85
Meat curing mix	850	2,500—6,000		7,000—25,000	97
Liver, pig raw			1.5		83
Fish, fried	Up to 1	(1/70) 0.01			91, 92
Haddock, fried	1—9				85
Hake, fresh or fried	1—9				85
Codfish, fresh, salted, fried	4				85
Cod		6			83
White herring	40—100				98
Yellow croaker	10—60				98
Anchovies	20				98
Croaker	20				98

Table 6 (continued)
CARCINOGENIC N-NITROSO COMPOUNDS IN HUMAN FOODS AND FOODSTUFFS

Nitrosamine concentration (μg/kg)[a]

Food product	Dimethylnitrosamine	Nitrosopyrrolidine	Diethylnitrosamine	Nitrosopiperidine	Ref.
Sable					
Raw	(4/4) 4				99
Smoked	(3/3) 4—9				99
Smoked, NO_3^- treated	(4/4) 12—14				99
Smoked, NO_2^- treated	(2/2) 8, 9				99
Smoked, NO_3^--NO_2^-	(2/2) 20, 26				99
Salmon					
Smoked	(1/2) 5				99
Smoked, NO_3^- treated	(2/2) 16, 17				99
Smoked, NO_2^- treated	(2/2) 4, 6				99
Shad, smoked, NO_3^- treated	(2/2) 10, 12				99
Cheese, Cheshire			1.5		83
Cheese, Grinland, Danish blue, Gouda, St. Paulin, Norwegian Tilsit, and Norwegian goat's milk	1—4				85
Fruit, canned	(5/12) <0.1	(1/12) 0.09			91, 92
Soup[c]	(3/30) <0.09	(1/30) 0.06		(1/30) 0.06	91, 92
Complete meals	(5/36) <0.2	(1/36) 0.08	(3/36) <0.03		91, 92
	(4/36) <0.5		(4/36) 0.2		
Sandwich and beer	1.27				100

[a] Number of positive samples/number of samples analyzed in brackets.
[b] Only five samples confirmed by MS.
[c] Also contained di(n)butylnitrosamine, 2/30 samples: 0.2 and 0.5 μg/kg.

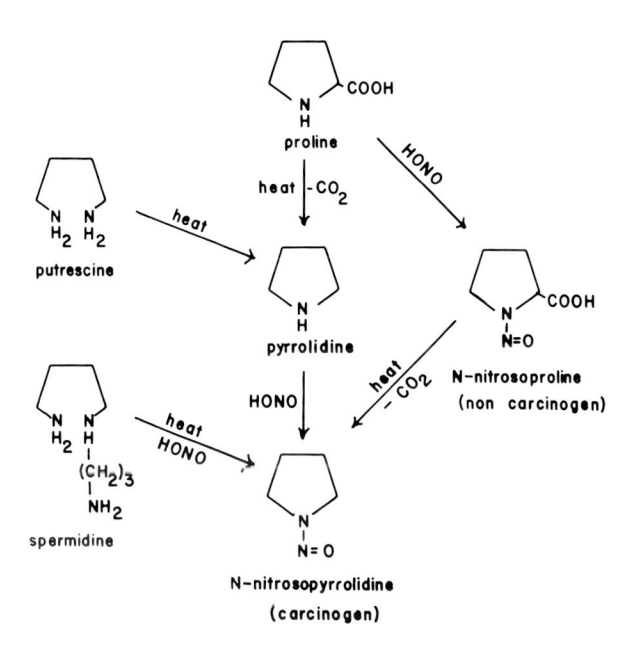

FIGURE 2. Proposed mechanisms for the formation of N-nitrosopyrrolidine during the curing and frying of bacon.

C. Other Processed Foods

There are several other varieties of processed foods, in addition to cured meats, that have been shown to contain carcinogenic nitrosamines; among the most frequently contaminated is processed fish. Table 6 summarizes results from five reports of dimethylnitrosamine, diethylnitrosamine, and nitrosopyrrolidine in fried, salted, and smoked fish products. Presumably, the relatively high frequency at which dimethylnitrosamine occurs is explained by the presence in fish of the free amines, dimethylamine and trimethylamine, compounds which contribute to the "fishy" smell of the product; that the concentrations of dimethylnitrosamine are usually low can probably be explained by the relatively strong basicity of dimethylamine, making it a low activity substrate for nitrosation by nitrous anhydride (see Section III.A). Other foods reported to have contained nitrosamines have been various cheeses, soup, canned fruit, raw pig liver, and a lunch consisting of beer and bacon, spinach, and tomato sandwich (see Table 6). In almost all cases, the amounts of nitrosamine, usually dimethylnitrosamine, have been quite low, a few micrograms per kilogram of food. It must be kept in mind, however, that the analytical techniques used in the reports summarized here were capable of detecting only the nitrosamines of relatively high volatility, those that could be separated by GC. The higher molecular weight nitrosamines and the nitrosamides have not been the subject of such intensive searches in human foods as have been the volatile compounds. Data on nonvolatile compounds should be forthcoming soon as the reliability of the TEA becomes more established, as that instrument can interface with a liquid chromatograph which readily separates nonvolatile nitrosamines.[113] A nitrosamide-specific detector for use with liquid chromatographs has recently been described;[18] this should facilitate environmental analysis for carcinogens, such as methylnitrosourea, N-methyl-N'-nitro-N-nitroso-guanidine, and nitrosocarbaryl.

D. Precursors in Foods

If one were to carry out an epidemiological study on the relation between dietary N-nitroso compounds and human cancer, it would not be sufficient, as was done in the aflatoxin dietary studies described in Chapter 3, to measure only the nitroso compounds present in plate samples of foods just before consumption. As discussed in Section III of this chapter, the acidity of the gastric secretions in the stomach provide an environment which strongly favors the nitrosation of secondary and tertiary amines found in foods; even if the N-nitroso compounds are not present in the ingested food, if the precursors for these compounds are present in the food and/or gastric secretions and saliva, the nitrosation reaction can take place in the stomach, and the carcinogens are formed where they escape detection by the most ardent investigator. The precursors for these carcinogens are simply nitrite ion and a variety of secondary and tertiary amines.

1. Nitrates and Nitrites

Recently, the U.S. National Academy of Science published a monograph on nitrates and nitrites in the environment; the report concluded that the principal route of human exposure to these ions is ingestion, and, except for individuals drinking water from polluted wells (where nitrate concentrations can be as high as 500 mg nitrate nitrogen or 2200 mg nitrate per liter), that vegetables are the major course of ingested nitrate, whereas most of the ingested nitrite is from cured meats and saliva.[44] Using U.S. Department of Agriculture data on per capita food consumption and chemical analyses of foodstuffs, White[44] estimated that the average American ingests 99.1 mg of nitrate per day from vegetables, fruit and juices, dairy products, bread, and cured meats and 2.6 mg nitrite per day from cured meats, vegetables, and bread, and another 8.62 mg nitrite from saliva. The presence of nitrite in saliva has been explained by bacterial reduction in the mouth of nitrate in ductal saliva and that the salivary nitrate is a direct reflection of the amount of nitrate ingested.[48,49] In spite of the relatively large amount of nitrite derived from saliva over a 24-hr period, cured meats are more likely to be of greater importance in the nitrosation of amines in the stomach. The nitrite in meats and certain vegetables is present in the stomach at one time, whereas the amount of nitrite from the saliva entering the stomach at any one time is quite small;[45] the time course for nitrite concentrations in the stomach is important because the nitrosation of amines is a function of the square of the concentration of nitrite,[26,27] not the total amount of nitrite entering the stomach over a period of many hours.

Concentrations of nitrate and nitrite in foods vary a great deal, depending upon species, climate, soil, and handling. Spinach, celery, lettuce, radishes, and beets can contain more than 2640 mg nitrate (600 mg nitrate N) per kilogram, while other vegetables, such as tomatoes, cucumbers, and asparagus, usually contain only a few milligrams of nitrate per kilogram.[114,115] Few vegetables contain more than 1 mg nitrite per kilogram; however, beets usually contain approximately 6 mg nitrite per kilogram; spinach, 2.7 mg/kg; and corn 2 mg/kg.[44] Improper handling of vegetables can promote bacterial reduction of nitrate to nitrite, e.g., spinach containing 30 mg nitrite when fresh contained 3550 mg nitrite per kilogram after 4 days of storage.[116] Several cured meats, fish, and cheese products contain sodium nitrite added as a preservative and color and flavor enhancer. The U.S. Department of Agriculture and the Food and Drug Administration announced in August 1978 the intention to eliminate nitrate as a food additive in cured meats and to greatly limit the use of nitrite as a food additive; this was in response to the large body of evidence that carcinogenic nitrosamines may form in foods or after the ingestion of foods to which nitrite has been added and by the observation that ingestion of nitrite itself appears to lead to cancers of the lymphatic system.[68]

2. Secondary and Tertiary Amines

Data on the occurrence of nitrosatable secondary and tertiary amines and amides are scant. There are more than 120 known carcinogenic N-nitroso compounds, yet little is known about where and how often the precursors for these compounds occur in the human food supply. Singer and Lijinsky[117] have reported on the natural occurrence of a few nitrosatable secondary amines found in cured meats, beverages, and fish products. The results of that report are summarized in Table 7.

Dimethylnitrosamine was found in all food samples reported and at relatively high concentrations in fish products. Perhaps more surprising was the frequent occurrence of morpholine; this amine has been used as an additive to boiler water to retard the corrosion of steam pipes in food processing plants, and its presence in the large variety of food products analyzed may reflect contamination of the foods and their packages by morpholine entrapped in steam.

VII. OTHER ENVIRONMENTAL SOURCES OF N-NITROSO COMPOUNDS

A. Drugs and Cosmetics

A large number of drugs are secondary or tertiary amines, and some concern has been expressed over the possibility of carcinogenic nitrosamines forming in the stomach following ingestion of nitrite-containing foods and a medication containing a nitrosatable amine. In 1972 a report[118] indicated that five drugs (tertiary amines) did nitrosate under laboratory conditions. Table 8 lists these drugs and the nitrosamines which formed during an incubation of 250 to 4000 mg of drug and 500 to 3500 mg of sodium nitrite at pH 2.0 to 4.7, 37° for 2 to 4 hr. Oxytetracycline and aminopyrine gave relatively high yields of dimethylnitrosamine, whereas the yields of diethylnitrosamine from disulfiram and nikethamide were small. In an extension of this work on nitrosation of drugs in laboratory model systems, Lijinsky[119] reported on the formation of dimethylnitrosamine from nitrite and the tranquilizers chlorpromazine and dextropropoxyphene, the antihistaminics chlorpheniramine and methapyrilene, and the narcotic methadone; diethylnitrosamine was formed from the antimalarial quinacrine and the antischistosomal lucanthone; cyclizine, the antihistaminic used to treat motion sickness, yielded dinitrosopiperazine. Wogan and coworkers[120] nitrosated ephedrine (a sympathomimetic) with sodium nitrite in hydrochloric acid and injected newborn mice i.p. on the 1st, 4th, and 7th day after birth with 200 mg nitrosoephedrine per kilogram body weight. The earliest liver cell carcinoma developed in males 50 weeks after dosing; the final incidence of liver cancers was 100% (tumors in all 15 animals treated) in male mice and 87% (13 of 15) in females.

Because ascorbate ion effectively competes with many nitrogen compounds for the nitrite in acid medium, Mirvish and co-workers[62] proposed that nitrosatable drugs should be formulated with sufficient amounts of ascorbic acid to block the nitrosation reaction. In chemical studies with aminopyrine, an analgesic used in Europe, 6.2 mM solution of ascorbic acid completely blocked dimethylnitrosamine formation from 20 mM aminopyrine and 25 mM nitrite.[121] In 1975 Mirvish[122] reviewed the evidence for the effect of ascorbic acid on in vitro and in vivo nitrosation of drugs and other amines and concluded that ascorbic acid should be a useful adjunct to drug formulations where the formation of carcinogenic nitrosamines is possible.

Triethanolamine can readily be nitrosated to form N-nitrosodiethanolamine, a known hepatocarcinogen.[106,123] Fan and co-workers[12] reported the presence of N-nitrosodiethanolamine in six samples of cosmetics and beauty aids, five samples of skin-care lotions, and five samples of shampoos; two of the cosmetic products contained 25 and 49 ppm of the carcinogen. The source of nitrosamine was presumed to be the nitrosation of triethanolamine which may have been added to these products as a wetting agent and emulsifier.

Table 7
NATURALLY OCCURRING NITROSATABLE SECONDARY AMINES IN FOODSTUFFS[117]

Food product	Dimethylamine	Pyrrolidine	Morpholine	Piperidine	Others
Baked ham	2	Trace	0.5	0.2	Trace: di-*n*-propylamine
Frankfurters	1			0.4	
Evaporated milk	3	0.7	0.2	0.3	0.2 methyl, *n*-butylamine
Whole milk	0.2			0.11	
Coffee	2	6	1	1	0.5 methyl, ethylamine
Tea	0.7		Trace		
Canned beer	0.6		0.4		
Bottled beer	0.7	Trace	0.2		
Wine	0.07	0.06	0.7		
Canned tuna	23		0.6		
Frozen ocean perch	180		9		
Frozen cod	740		Trace		
Spotted trout[c]	7		6		0.4 di-*n*-propylamine
Small mouth bass	110		0.7		0.2 di-*n*-propylamine
Salmon	82		1.0		

[a] Average values of several determinations each; number of individual food samples not reported.
[b] Trace is less than 0.1 ppm in most cases.
[c] Only one sample analyzed.

B. Industrial and Agricultural Processes

Shortly before finding N-nitrosodiethanolamine in cosmetics, Fan and co-workers[13] reported finding this carcinogen in synthetic cutting fluids used commercially. These fluids contained up to 45% triethanolamine and 18% sodium nitrite; the pH values ranged from 9 to 11. Eight different brands were selected from many available and specifically analyzed for the nitrosamine by high pressure liquid chromatography and high resolution MS. Three samples contained between 200 and 600 ppm N-nitrosodiethanolamine; another three contained between 1800 and 4200 ppm, and the two highest contained 10,400 and 29,900 ppm (2.99%). Even though these oils are diluted 10 to 100 times before use, the levels of contamination are high enough to make the potential for exposure to industrial workers a concern in occupational health.

Several agricultural chemicals have been shown to make adequate precursors for the formation of carcinogenic nitroso compounds.[56] Elespuru and Lijinsky[124] demonstrated that such compounds as dimethylphenylurea and tetramethylthiuram disulfide nitrosate to form dimethylnitrosamine, and the pesticide, carbaryl (N-methylnaphthylcarbamate), forms N-nitrosocarbaryl when mixed with nitrite in an acidic medium. This nitrosated insecticide has been shown to be mutagenic[124,125] and carcinogenic.[125-127] Dithiocarbamic acid derivative fungicides, such as bis (dimethyldithiocarbamato) zinc (ziram; $(CH_3)_2N$-CS-C$_2$ Zn) are nitrosated in the stomach of rats when ingested with sodium nitrite;[128] depending upon which fungicide is used, the nitrosation product can be dimethylnitrosamine or N-nitrosopyrrolidine.

Atrazine is a widely used pesticide in the U.S.; it is moderately persistent in the environment. In many cases, it is applied agriculturally with nitrogen fertilizers. When atrazine is mixed with sodium nitrite under conditions which simulate the environment of the human stomach, the compound becomes nitrosated (2-chloro-4(N-nitroso-N-ethyl amino)-6-isopropylaminotriazine).[41] The carcinogenicity of this compound which has the potential for wide exposure to man, as is the case with nitrosocarbaryl, is being studied.

Table 8
NITROSATION OF VARIOUS DRUGS WITH NITRITE[118]

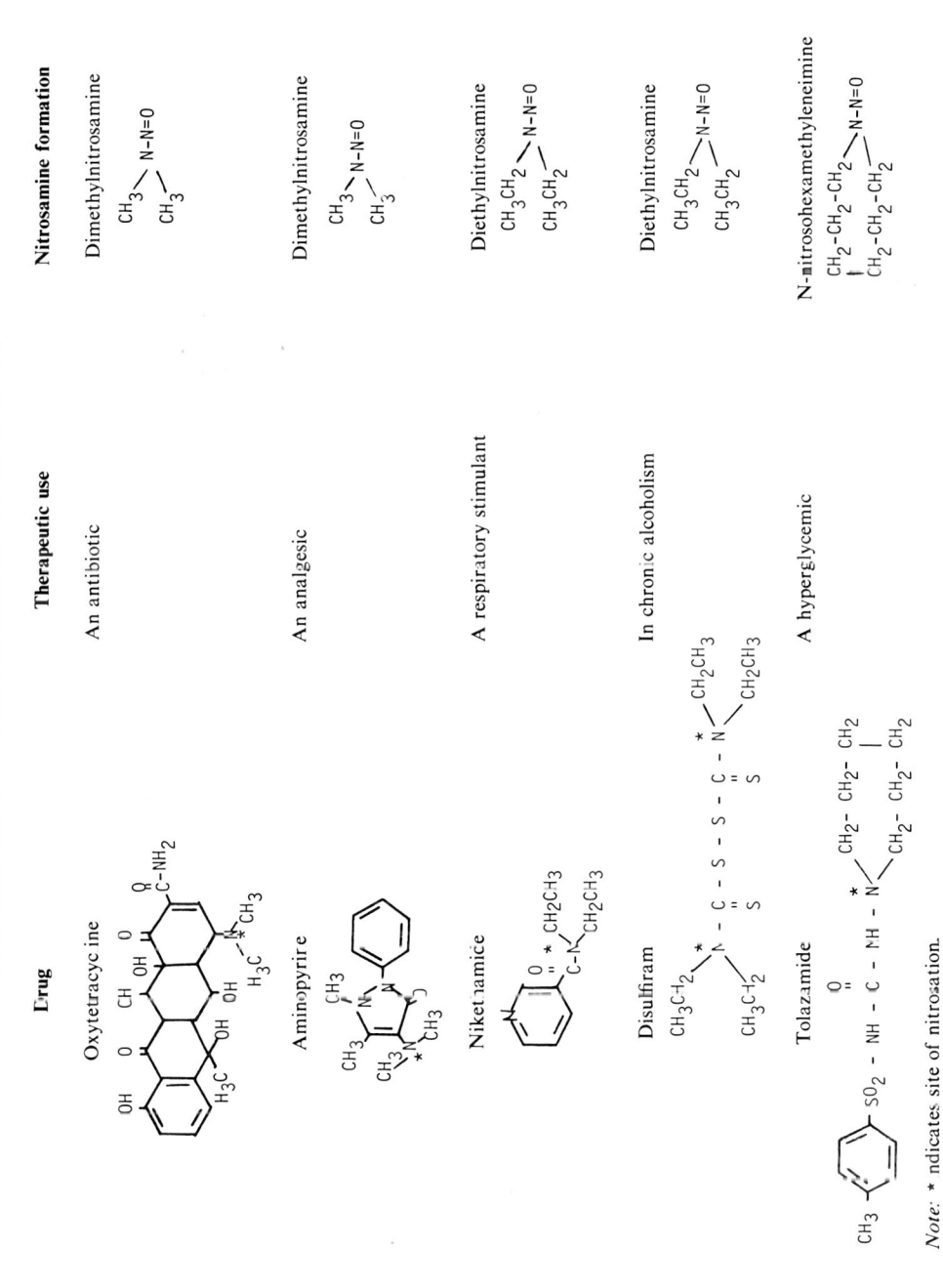

Drug	Therapeutic use	Nitrosamine formation
Oxytetracycline	An antibiotic	Dimethylnitrosamine $\begin{array}{c}CH_3\\CH_3\end{array}\!\!>\!\!N\text{-}N\text{=}O$
Aminopyrine	An analgesic	Dimethylnitrosamine $\begin{array}{c}CH_3\\CH_3\end{array}\!\!>\!\!N\text{-}N\text{=}O$
Nikethamide	A respiratory stimulant	Diethylnitrosamine $\begin{array}{c}CH_3CH_2\\CH_3CH_2\end{array}\!\!>\!\!N\text{-}N\text{=}O$
Disulfiram	In chronic alcoholism	Diethylnitrosamine $\begin{array}{c}CH_3CH_2\\CH_3CH_2\end{array}\!\!>\!\!N\text{-}N\text{=}O$
Tolazamide	A hyperglycemic	N-nitrosohexamethyleneimine

Note: * indicates site of nitrosation.

Studies on the nitrosation of pesticides have not been limited to observations on incubating amines with nitrite in acidic solutions. Ross and co-workers[129] analyzed six commercial herbicides for nitrosamines using GC or high pressure liquid chromatography coupled with a TEA. Three of the samples were popular weed killer preparations sold as home lawn care products in which the active ingredients were 2,4-dichlorophenoxyacetic acid, 2,4,5-trichlorophenoxypropionic acid, 3,6-dichloro-o-anisic acid, and/or 2-(2-methyl-4-chlorophenoxy) propionic acid, all as the dimethylamine salts; a small amount of dimethylnitrosamine (0.30 mg/ℓ) was found in one of these commercial products. Two industrial herbicides containing higher concentrations of 2,3,6-trichlorobenzoic acid and related polychlorobenzoic acids, all as dimethylamine salts, contained approximately 200 and 600 mg dimethylnitrosamine per liter, while a major agricultural herbicide containing α,α,α-trifluoro-2,6-dinitro-N,N-dipropyl-p-toluidine contained 154 mg dipropylnitrosamine per liter. As these concentrations could present exposures to carcinogenic nitrosamines greater than associated with foods and tobacco smoke, it would seem reasonable to make efforts to control such exposures. The authors[129] suggested that nitrosation may have resulted from a reaction between nitrite (in the active form) and the dimethylamine portion of the salt (sodium nitrite apparently is added to these products as a rust inhibitor for the cans used in distribution); if this is the case, it would seem justifiable to urge that these herbicide formulations be altered, so that the active ingredients are present not as the dimethylamine salts and that no sodium nitrite is added to the package. Many of the environmental sources of carcinogenic N-nitroso compounds discussed in this chapter are of such a nature as to be difficult to control, but simple changes in pesticide (and drug) formulations may present a relatively uncomplicated opportunity to significantly reduce the public's exposure to such potentially hazardous compounds.

REFERENCES

1. International Agency for Research on Cancer, *N-Nitroso Compounds: Analysis and Formation*, Scientific Publication No. 3, International Agency for Research on Cancer, Lyon, 1972.
2. **Fiddler, W.,** The occurrence and determination of N-nitroso compounds, *Toxicol. Appl. Pharmacol.*, 31, 352, 1975.
3. **Preussmann, R., Neurath, B., Wulf-Lorentzen, G., Daiber, D., and Hengy, H.,** Anfarbenmethoden und Dunnschichtchromatographie von organischen N-Nitrosoverbin dungen, *Z. Anal. Chem.*, 202, 187, 1964.
4. **Van Ginkel, J. C.,** Report of the Commission for Chemical Analysis, in 53rd Annual Meeting of the International Dairy Federation, 1968; as cited in Wasserman, A. E., *N-Nitroso Compounds: Analysis and Formation*, Scientific Publication No. 3, International Agency for Research on Cancer, Lyon, 1972, 14.
5. **Wasserman, A. E.,** A survey of analytical procedures for nitrosamines, in *N-Nitroso Compounds: Analysis and Formation*, Scientific Publication No. 3, International Agency for Research on Cancer, Lyon, 1972, 10.
6. **Dure, G., Weil, L., and Quentin, K. E.,** Determination of nitrosamines in natural water and waste water, *Z. Wasser Abwassen. Forsch.*, 8, 20, 1975.
7. **Hedler, L., Kaunitz, H., Marquadt, P., Fales, H., and Johnson, R. E.,** Detection of N-nitroso compounds by gas chromatography (nitrogen detector) in soybean oil extract, in *N-Nitroso Compounds: Analysis and Formation*, Scientific Publication No. 3, International Agency for Research on Cancer, Lyon, 1972, 71.
8. **Fine, D. H., Lieb, D., and Rufeh, F.,** Principle of operation of the thermal energy analyzer for the trace analysis of volatile and non-volatile N-nitroso compounds, *J. Chromatogr.*, 107, 351, 1975.
9. **Fine, D. H., Rounbehler, D. P., Belcher, N. M. and Epstein, S. S.,** N-nitroso compounds: detection in ambient air, *Science*, 192, 1328, 1976.

10. **Fine, D. H., Rounbehler, D. P., Sawicki, E., and Krost, K.**, Determination of dimethylnitrosamine in air and water by thermal energy analysis: validation of analytical procedures, *Environ. Sci. Technol.*, 11, 577, 1977.
11. **Fine, D. H., Rounbehler, D. P., and Oettinger, P. E.**, A rapid method for the determination of sub-part per billion amounts of N-nitroso compounds in foodstuffs, *Anal. Chim. Acta*, 78, 383, 1975.
12. **Fan, T. Y., Goff, U., Song, L. D. H., Arsenault, G. P., and Biemann, K.**, N-nitrosodiethanolamine in cosmetics, lotions and shampoos, *Food Cosmet. Toxicol.*, 15, 423, 1977.
13. **Fan, T. Y., Morrison, J., Rounbehler, D. P., Ross, R., Fine, D. H., Miles, W., and Sen, N. P.**, N-nitrosodiethanolamine in synthetic cutting fluids: a parts-per-hundred impurity, *Science*, 196, 70, 1977.
14. International Agency for Research on Cancer, Recommendations, in *N-Nitrosamines in the Environment*, Scientific Publication No. 9, Proceedings of a Working Conference held in Lyon, 1973, Bogovski, P. and Walker, E. A., Eds, International Agency for Research on Cancer, Lyon, 1975, 242.
15. **Fine, D. H., Rounbehler, D. P., and Sen, N. P.**, A comparison of some chromatographic detectors for the analysis of volatile N-nitrosamines, *J. Agric. Food Chem.*, 24, 980, 1976.
16. **Gough, T. A., Webb, K. S., Pringuer, M. A., and Wood, B. J.**, A comparison of various mass spectrometric and a chemiluminescent method for the estimation of volatile nitrosamines, *J. Agric. Food Chem.*, 25, 663, 1977.
17. **Heyns, K. and Röper, H.**, Analytik von N-Nitroso-Verbindungen. II. Trennung und quantitative Bestimmung von homologen N-Nitroso-N-Alkylharnstoffen und N-Nitroso-N-Alkylurethanen durch schnelle Hockdruckflussigheits-chromatographie, *J. Chromatogr.*, 93, 429, 1974.
18. **Singer, G. M., Singer, S. S., and Schmidt, D. G.**, A nitrosamide-specific detector for use with high pressure liquid chromatography, *J. Chromatogr.*, 133, 59, 1977.
19. **Pellizzari, E. D.**, Identification and Estimation of N-nitrosodimethylamine and Other Pollutants in the Baltimore, Maryland and Kanawha Valley Areas, Progress Report prepared by Research Triangle Institute under Contract No. 68-02-1220, U.S. Environmental Protection Agency, Research Triangle Park, N.C., January 1975.
20. U.S. Environmental Protection Agency, Summary Report on Nitrosamines, U.S. Environmental Protection Agency, Health Effects Research Laboratory, Environmental Research Center, Research Triangle Park, N.C., January 1976, 227.
21. **Sen, N. P., Smith, D. C., Schwinghamer, L., and Marleau, J. J.**, Diethylnitrosamine and other N'-nitrosamines in foods, *J. Assoc. Off. Anal. Chem.*, 52, 47, 1969.
22. **Goodhead, K. and Gough, T. A.**, The reliability of a procedure for the determination of nitrosamines in food, *Food Cosmet. Toxicol.*, 13, 307, 1975.
23. **Howard, J. W., Fazio, T., and Watts, J. O.**, Extraction and gas chromatographic determination of N-nitrosodimethylamine in smoked fish: application to smoked nitrite-treated chub, *J. Assoc. Off. Anal. Chem.*, 53, 269, 1970.
24. **Ender, G., Havre, G., Helgebostad, A., Koppang, N., Madsen, R., and Ceh, L.**, Isolation and identification of a hepatotoxic factor in herring meal produced from sodium nitrite preserved herring, *Naturwissenschaften*, 51, 637, 1964.
25. **Sakshaug, J., Sognen, E., Hansen, M. A., and Koppang, N.**, Dimethylnitrosamine; its hepatoxic effect in sheep and its occurrence in toxic batches of herring meal, *Nature (London)*, 206, 1261, 1965.
26. **Mirvish, S. S.**, Kinetics of dimethylamine nitrosation in relation to nitrosamine carcinogenesis, *J. Natl. Cancer Inst.*, 44, 633, 1970.
27. **Mirvish, S. S.**, Formation of N-nitroso compounds — chemistry, kinetics, and *in vivo* occurrence, *Toxicol. Appl. Pharmacol.*, 31, 325, 1975.
28. **Ridd, J. H.**, Nitrosation, diazotisation, and deamination, *Q. Rev. Chem. Soc.*, 15, 418, 1961.
29. **Turney, T. A. and Wright, G. A.**, Nitrous acid and nitrosation, *Chem. Rev.*, 59, 497, 1959.
30. **Mirvish, S. S.**, N-nitroso compounds: their chemical and *in vivo* formation and possible importance as environmental carcinogens, *J. Toxicol. Environ. Health*, 2, 1267, 1977.
31. **Bretschneider, K. and Matz, J.**, Nitrosamine (NA) in der atmospharischen und in der Luft am Arbeitplatz, *Arch. Geschwulstforsch.*, 43, 36, 1974.
32. **Neurath, G., Pirmann, B., Luttich, H., and Wirchern, H.**, Zur Frage der N-Nitrosoverbindungen in Tabakrauch. II. *Beitr. Tabakforsch.*, 3, 251, 1965; as cited in U.S. Environmental Protection Agency, Scientific and Technical Assessment Report on Nitrosamines, EPA 660/6-77 001, U.S. Environmental Protection Agency, Government Printing Office, Washington, D.C., 1976.
33. **Hanst, P. L., Spence, J. W., and Miller, M.**, Atmospheric chemistry of N-nitrosodimethylamine, *Environ. Sci. Technol.*, 11, 403, 1977.
34. **Ayanaba, A., Verstraete, W., and Alexander, M.**, Formation of dimethylnitrosamine, a carcinogen and mutagen, in soils treated with nitrogen compounds, *Soil Sci. Soc. Am. Proc.*, 37, 565, 1973.
35. **Ayanaba, A., Verstraete, W., and Alexander, M.**, Possible microbial contribution to nitrosamine formation in sewage and soil, *J. Natl. Cancer Inst.*, 50, 811, 1973.

36. **Ayanaba, A. and Alexander, M.,** Transformation of methylamines and formation of hazardous produce, dimethylnitrosamine, in samples of treated sewage and lake water, *J. Environ. Qual.,* 3, 83, 1974.
37. **Tate, R. L. and Alexander, M.,** Formation of dimethylamine and diethylamine in soil treated with pesticides, *Soil Sci.,* 118, 317, 1974.
38. **Mills, A. L. and Alexander, M.,** Factors affecting dimethylnitrosamine formation in samples of soil and water, *J. Environ. Qual.,* 5, 437, 1976.
39. **Tate, R. L. and Alexander, M.,** Stability of nitrosamines in samples of lake water, soil and sewage, *J. Natl. Cancer Inst.,* 54, 327, 1975.
40. **Tate, R. L. and Alexander, M.,** Resistance of nitrosamines to microbial attack, *J. Environ. Qual.,* 5, 131, 1976.
41. **Wolfe, M. L., Zepp, R. G., Gordon, J. A., and Fincher, R. C.,** N-nitrosamine formation from atrozine, *Bull. Environ. Contam. Toxicol.,* 15, 342, 1976.
42. **Khan, S. U. and Young, J. C.,** N-nitrosamine formation in soil from the herbicide glyphosate, *J. Agric. Food Chem.,* 25, 1430, 1977.
43. **Seiler, J. P.,** Nitrosation *in vitro* and *in vivo* by sodium nitrite and mutagenicity of nitrogenous pesticide, *Mutat. Res.,* 48, 225, 1977.
44. **White, J. W., Jr.,** Relative significance of dietary sources of nitrate and nitrite, *J. Agric. Food Chem.,* 23, 886, 1975; corrections *J. Agric. Food Chem.,* 24, 202, 1976.
45. National Research Council, Nitrates: An Environmental Assessment, Panel on Nitrates of the Coordinating Committee for Scientific and Technical Assessments of Environmental Pollutants, Commission on Natural Resources, Washington, D.C., National Academy of Sciences, 1978, 723.
46. **Durfor, C. N. and Becker, E.,** Public Water Supplies of the 100 Largest Cities in the United States, Water Supply Paper 1812, U.S. Geological Survey, Washington, D.C.
47. **Varady, J. and Szanto, G.,** Untersuchungen uber den Nitritgehalt des Speichels, des Magensaffes und des Marnes, *Klin. Wochenschr.,* 19, 200, 1940.
48. **Spiegelhalder, B., Eisenbrand, G., and Preussmann, R.,** Influence of dietary nitrate on nitrite content of human saliva: possible relevance to *in vivo* formation of N-nitroso compounds, *Food Cosmet. Toxicol.,* 14, 545, 1976.
49. **Tannenbaum, S. R., Weisman, M., and Fett, D.,** The effect of nitrate intake on nitrite formation in human saliva, *Food Cosmet. Toxicol.,* 14, 549, 1976.
50. **Tannenbaum, S. R., Fett, D., Young, V. R., Land, P. D., and Bruce, W. R.,** Nitrite and nitrate are formed by endogenous synthesis in the human intestine, *Science,* 200, 1487, 1978.
51. **Sander, J. and Burkle, G.,** Induktion maligner Tumoren be: Ratten durch gleichzeitige Verfutterung von Nitrit und sekundaren Aminen, *Z. Krebsforsch.,* 73, 54, 1969.
52. **Mysliwy, T. S., Wick, E. L., Archer, M. C., Shank, R. C., Newberne, P. M.,** Formation of N-nitrosopyrrolidine in a dog's stomach, *Br. J. Cancer,* 30, 279, 1974.
53. **Boyland, R.,** The catalysis of nitrosation by thiocyanate from saliva, *Food Cosmet. Toxicol.,* 9, 639, 1971.
54. **Mirvish, S. S., Sams, J., Fan, T. Y., and Tannenbaum, S. R.,** Kinetics of nitrosation of the amino acids proline, hydroxyproline and sarcosine, *J. Natl. Cancer Inst.,* 51, 1833, 1973.
55. **Boyland, E. and Walker, S. A.,** Catalysis of the reaction of aminopyrine and nitrite by thiocyanate, *Arzneim.-Forsch.,* 24, 1181, 1974.
56. **Feuer, H., Ed.,** *The Chemistry of the Nitro and Nitroso Groups,* Interscience, New York, 1969.
57. **Fan, T. Y. and Tannenbaum, S. R.,** Factors influencing the rate of formation of nitrosomorpholine from morpholine and nitrite: acceleration by thiocyanate and other anions, *J. Agric. Food Chem.,* 21, 237, 1973.
58. **Lane, R. P. and Bailey, M.E.,** Effect of pH on dimethylnitrosamine formation in human gastric juice, *Food Cosmet. Toxicol.,* 11, 851, 1973.
59. **Mirvish, S. S.,** Kinetics of nitrosamide formation from alkylureas, N-alkylurethane, and alkylguanidines: possible implications for the etiology of human gastric cancer, *J. Natl. Cancer Inst.,* 46, 1183, 1971.
60. **Dahn, H., Loewe, L., and Bunton, C. A.,** Uber die oxydation von Ascorbinsaure durch salpetrige Saure. VI. Ubersicht und Diskussion der Ergebnisse, *Helv. Chem. Acta,* 43, 320, 1960.
61. **Sen, N. P. and Donaldson, B.,** The effect of ascorbic acid and glutathione on the formation of nitrosopiperazines from piperazine adipate and nitrite, in *N-Nitroso Compounds: Analysis and Formation,* IARC Scientific Publication No. 3, Bogovski, P., Preussman, R., and Walker, E. A., Eds., International Agancy for Research on Cancer, Lyon, 1972, 103.
62. **Mirvish, S. S., Wallcave, L., Eagen, M., and Shubik, P.,** Ascorbate-nitrite reaction: possible means of blocking the formation of carcinogenic N-nitroso-compounds, *Science,* 177, 65, 1972.
63. **Fiddler, W., Pensabene, J. W., Kushnir, I., and Piotrowski, E. G.,** Effect of frankfurter cure ingredients on N-nitrosodimethylamine formation in a model system, *J. Food Sci.,* 38, 714, 1973.

64. **Fiddler, W., Pensabene, J. W., Piotroski, E. G., Doerr, R. C., and Wasserman, A. E.,** Use of sodium ascorbate or erythrobate to inhibit formation on N-nitrosodimethylamine in frankfurters, *J. Food Sci.*, 38, 1084, 1973.
65. **Bogovski, P., Castergnare, M., Pignatelli, B., and Walker, E. A.,** The inhibiting effect of tannins on the formation of nitrosamines, in *N-Nitroso Compounds: Analysis and Formation*, IARC Scientific Publication No. 3, Bogovski, P., Preussmann, R., and Walker, E. A., Eds., International Agency for Research on Cancer, Lyon, 1972, 127.
66. **Ivankovic, S., Preussmann, R., Schmahl, D., and Zeller, J.,** Prevention by ascorbic acid of *in vivo* formation of N-Nitroso compounds, in *N-Nitroso Compounds in the Environment*, IARC Scientific Publication No. 9, Bogovski, P., Walker, E. A., and Davis, W., Eds., International Agency for Research on Cancer, Lyon, 1974, 101.
67. **Ivankovic, S. and Druckrey, H.,** Transplacentare Erzeugung malinger Tumoren bei den Nachkommen nach einmaliger injektion von Athylnitrosoharnstoff an Schwangere Ratten, *Z. Krebsforsch.*, 71, 320, 1968.
68. **Shank, R. C. and Newberne, P. M.,** Dose-response study of the carcinogenicity of dietary sodium nitrite and morpholine in rats and hamsters, *Food Cosmet. Toxicol.*, 14, 1, 1976.
69. **Fine, D. H., Rounbehler, D. P., Rounbehler, A., Silvergleid, A., Sawicki, E., Krost, K., and DeMarrois, G. A.,** Determination of dimethylnitrosamine in air, water and soil by thermal energy analysis: measurements in Baltimore, Md., *Environ. Sci. Technol.*, 11, 581, 1977.
70. **Fine, D. H., Rounbehler, D. P., Pellizari, E. D., Bunch, J. E., Berkley, R. W., McCrae, J., Bursey, J. T., Sawicki, E., Krost, K., and DeMarrais, G. A.,** N-Nitrosodimethylamine in air, *Bull. Environ. Contam. Toxicol.*, 15, 739, 1976.
71. **U.S. Environmental Protection Agency,** Scientific and Technical Assessment Report on Nitrosamines, EPA-600/6-77-001, U.S. Environmental Protection Agency, U.S. Government Printing Office, Washington, D.C., 1977, 209.
72. **Gough, T. A., Webb, K. S., and McPhail, M. F.,** Volatile nitrosamines from ion-exchange resins, *Food Cosmet. Toxicol.*, 15, 437, 1977.
73. **Fiddler, W., Pensabene, J. W., Doerr, R. C., and Dooley, C. J.,** The presence of dimethyl and diethylnitrosamines in deionized water, *Food Cosmet. Toxicol.*, 15, 441, 1977.
74. **Hedler, L. and Marquardt, P.,** Occurrence of diethylnitrosamine in some samples of food, *Food Cosmet. Toxicol.*, 6, 341, 1968.
75. **Thewlis, B. H.,** Testing of wheat flour for the presence of nitrite and nitrosamines, *Food Cosmet. Toxicol.*, 5, 333, 1967.
76. **Ender, F. and Ceh, L.,** Occurrence of nitrosamines in foodstuffs for human and animal consumption, *Food Cosmet. Toxicol.*, 6, 569, 1968.
77. **Keybets, M. J. H., Groot, E. H, and Keller, G. H. M.,** An investigation into the possible presence of nitrosamines in nitrite-bearing spinach, *Food Cosmet. Toxicol.*, 8, 167, 1970.
78. **DuPleissis, L. S., Nunn, J. R., and Roach, W. A.,** Carcinogen in a Transkeian Bantu food additive, *Nature (London)*, 222, 1198, 1969.
79. **McGlashan, N. D., Walters, C. L., and McLean, A. E. M.,** Nitrosamines in African alcoholic spirits and oesophageal cancer, *Lancet*, 2, 1017, 1968.
80. **Tarr, H. L. A.,** Bacteriostatic action of nitrates, *Nature (London)*, 147, 417, 1941.
81. **Christiansen, L. N., Johnston, R. W., Kautter, D. A., Howard, J. W., and Aunan, W. J.,** Effect of nitrite and nitrate on toxin production by *Clostridium botulinum* and on nitrosamine formation in perishable canned comminuted cured meat, *Appl. Microbiol.*, 25, 357, 1973.
82. **Hustad, G. O., Cerveny, J. G., Trenk, H., Diebel, R. H., Kautter, D. A., Fazio, T., Johnston, R. W., and Kolari, O. E.,** Effect of sodium nitrite and sodium nitrate on botulinal toxin production and nitrosamine formation in wieners, *Appl. Microbiol.*, 26, 22, 1973.
83. **Lamanna, C.,** The most poisonous poison, *Science*, 130, 763, 1959.
84. **Alliston, T. G., Cox, G. B., and Kirk, R. S.,** Determination of steam-volatile N-nitrosamines in foods by formation of electron-capturing derivation from electrochemically derived amines, *Analyst*, 97, 915, 1972.
85. **Crosby, N. T., Forman, J. K., Palframan, J. F., and Sawyer, R.,** Estimation of steam-volatile N-nitrosamines in foods at the 1 μg/kg level, *Nature (London)*, 238, 342, 1972.
86. **Fazio, T., White, R. H., Dusold, L. R., and Howard, J. W.,** Nitrosopyrrolidine in cooked bacon, *J. Assoc. Off. Anal. Chem.*, 56, 919, 1973.
87. **Sen, N. P., Donaldson, B. A., Iyengar, J. R., and Panalaks, T.,** Nitrosopyrrolidine and dimethylnitrosamine in bacon, *Nature (London)*, 241, 473, 1973.
88. **Pensabene, J. W., Fiddler, W., Gates, R. A., Fagan, J. C., and Wasserman, A. E.,** Effect of frying and other cooking conditions on nitrosopyrrolidine formation in bacon, *J. Food Sci.*, 39, 314, 1974.

89. **Fiddler, W., Pensabene, J. W., Fagan, J. C., Thorne, E. J., Piotrowski, E. G., and Wasserman, A. E.,** A research note. The role of lean and adipose tissue on the formation of nitrosopyrrolidine in fried bacon, *J. Food. Sci.,* 39, 1070, 1974.
90. **Sen, N. P., Seaman, S., and Miles, W. F.,** Dimethylnitrosamine and nitrosopyrrolidine in fumes produced during the frying of bacon, *Food Cosmet. Toxicol.,* 14, 167, 1976.
91. **Gough, T. A., McPhail, M. F., and Webb, K. S., Wood, B. J., and Coleman, R. F.,** An examination of some foodstuffs for the presence of volatile nitrosamines, *J. Sci. Food Agric.,* 28, 345, 1977.
92. **Gough, T. A., Webb, K. S., Coleman, R. F.,** Estimate of the volatile nitrosamine content of UK food, *Nature (London),* 272, 161, 1978.
93. **Wasserman, A. E., Fiddler, W., Doerr, R. C., Osman, S. F., and Dooley, D. J.,** Dimethylnitrosamine in frankfurters, *Food Cosmet. Toxicol.,* 10, 681, 1972.
94. **Sen, N. P.,** The evidence for the presence of dimethylnitrosamine in meat products, *Food Cosmet. Toxicol.,* 10, 219, 1972.
95. **Fiddler, W., Feinberg, J. I., Pensabene, J. W., Williams, A. C., and Dooley, C. J.,** Dimethylnitrosamine in souse and similar jellied cured-meat products, *Food Cosmet. Toxicol.,* 13, 654, 1975.
96. **Panalaks, T., Iyengar, J. R., and Sen, N. P.,** Nitrate, nitrite, and dimethylnitrosamine in cured meat products, *J. Assoc. Off. Anal. Chem.,* 56, 621, 1973.
97. **Sen, N. P., Miles, W. F., Donaldson, B., Panalaks, T., and Iyengar, J. R.,** Formation of nitrosamines in a meat curing mixture, *Nature (London),* 245, 104, 1973.
98. **Fong, Y. Y. and Chan, W. C.,** Dimethylnitrosamine in Chinese marine salt fish, *Food Cosmet. Toxicol.,* 11, 841, 1973.
99. **Fazio, T., Damico, J. N., Howard, J. W., White, R. H, and Watts, J. O.,** Gas chromatographic determination and mass spectrometric confirmation of N-nitrosodimethylamine in smoke-processed marine fish, *J. Agric. Food Chem.,* 19, 250, 1971.
100. **Fine, D. H., Ross, R., Rounbehler, D. P., Silvergleid, A., and Song, L.,** Formation *in vivo* of volatile N-nitrosamines in man after ingestion of cooked bacon and spinach, *Nature (London),* 265, 753, 1977.
101. **Bills, D. D., Hildrum, K. I., Scanlan, R. A., and Libbey, L. M.,** Potential precursors of N-nitrosopyrrolidine in bacon and other fried foods, *J. Agric. Food Chem.,* 21, 876, 1973.
102. **Huxel, E. T., Scanlan, R. A., and Libbey, L. M.,** Formation of N-nitropyrrolidine from pyrrolidine-ring-containing compounds at elevated temperatures, *J. Agric. Food Chem.,* 22, 698, 1974.
103. **Kushnir, I., Feinberg, J. I., Pensabene, J. W., Piotrowski, E. G., Fiddler, W., and Wasserman, A. E.,** Isolation and identification of nitrosoproline in uncooked bacon, *J. Food Sci.,* 40, 427, 1975.
104. **Hwang, L. S. and Rosen, J. D.,** Nitrosopyrrolidine formation in fried bacon, *J. Agric. Food Chem.,* 24, 1152, 1976.
105. **Hansen, T., Iwaoka, W., Green, L., and Tannenbaun, S. R.,** Analysis of N-nitrosoproline in raw bacon. Further evidence that nitrosoproline is not a major precursor of nitrosopyrrolidine, *J. Agric. Food Chem.,* 25, 1423, 1977.
106. **Druckrey, H., Preussmann, R., Ivankovic, S., and Schmahl, D.,** Organotrope carcinogene Wirkungen bei 65 verschiedenen N-Nitroso-Verbindungen an BD-Ratten, *Z. Krebsforsch.,* 69, 103, 1967.
107. **Greenblatt, M. and Lijinsky, W.,** Nitrosamine studies: neoplasms of liver and genital mesothelium in nitrosopyrrolidine-treated MRC rats, *J. Natl. Cancer Inst.,* 48, 1687, 1972.
108. **Garcia, H. and Lijinsky, W.,** Tumorigenicity of five cyclic nitrosamines in MRC rats, *Z. Krebsforsch.,* 77, 257, 1972.
109. **Magee, P. N. and Barnes, J. M.,** The production of malignant primary hepatic tumors in the rat by feeding dimethylnitrosamine, *Br. J. Cancer,* 10, 114, 1956.
110. **Druckrey, H., Steinhoff, D., Preussmann, R., and Ivankovic, S.,** Carcinogenesis in rats by a single administration of methylnitrosourea and various dialkylnitrosamines, *Z. Krebsforsch.,* 66, 1, 1964; *Chem. Abstr.,* 61, 1075h.
111. **Dontenwill, W. and Mohr, U.,** Carinome des Respirationstractus nach Behandlung von Goldhamstern mit Diathylnitrosamin, *Z. Krebsforsch.,* 64, 305, 1961.
112. **Sen, N. P., Donaldson, B., Seaman, J., Iyengar, J. R., and Miles, W. F.,** Inhibition of nitrosamine formation in fried bacon by propyl gallate and L-ascorbyl palmitate, *J. Agric. Food Chem.,* 24, 397, 1976.
113. **Fine, D. H., Ross, R., Rounbehler, D. P., Silvergleid, A., and Song, L.,** Analysis of nonionic nonvolatile N-nitroso compounds in foodstuffs, *J. Agric. Food Chem.,* 24, 1069, 1976.
114. **Jackson, W. A., Steel, J. S., and Boswell, V. R.,** Nitrates in edible vegetables and vegetable products, *Proc. Am. Soc. Hortic. Sci.,* 90, 349, 1967.
115. **Brown, J. R. and Smith, G. E.,** Nitrate Accumulation in Vegetable Crops as Influenced by Soil Fertility Practices, Research Bulletin 920, University of Missouri Agriculture Experiment Station, Columbia, Mo., 1967.
116. **Schuphan, W.,** Der Nitratgehalt von Spinat (*Spinacia oleracea,* L.) in Beziehung zur Methamoglobinamie der Sauglinge, *Z. Ernaehrungswiss.,* 5, 207, 1965.

117. **Singer, G. M. and Lijinsky, W.,** Naturally occurring nitrosatable compounds. I. Secondary amines in foodstuffs, *J. Agric. Food Chem.,* 24, 550, 1976.
118. **Lijinsky, W., Conrad, E., and Van de Bogart, R.,** Carcinogenic nitrosamines formed by drug/nitrite interactions, *Nature (London),* 239,165, 1972.
119. **Lijinsky, W.,** Reaction of drugs with nitrous acid as a source of carcinogenic nitrosamines, *Cancer Res.,* 34, 255, 1974.
120. **Wogan, G. N., Paglialunga, S., Archer, M. C., and Tannenbaum, S. R.,** Carcinogenicity of nitrosation products of ephedrine, sarcosine, folic acid, and creatinine, *Cancer Res.,* 35, 1981, 1975.
121. **Mirvish, S. S., Gold, B., Eagen, M., and Arnold, S.,** Kinetics of the nitrosation of aminopyrine to give dimethylnitrosamine, *Z. Krebsforsch.,* 82, 259, 1974.
122. **Mirvish, S. S.,** Blocking the formation of N-nitroso compounds with ascorbic acid *in vitro* and *in vivo, Ann. N.Y. Acad. Sci.,* 258, 175, 1975.
123. **Lijinsky, W., Keefer, L., Conrad, E., and Van de Bogart, R.,** Nitrosation of tertiary amines and some biologic implications, *J. Natl. Cancer Inst.,* 49, 1239, 1972.
124. **Elespuru, R. K. and Lijinsky, W.,** The formation of carcinogenic nitroso compounds from nitrite and some types of agricultural chemicals, *Food Cosmet. Toxicol.,* 11, 807, 1973.
125. **Eisenbrand, G., Ungerer, O., and Preussmann, R.,** The reaction of nitrite with pesticides. II. Formation, chemical properties and carcinogenic activity of the N-nitroso derivative of N-methyl-1-naphthyl carbamate (carbaryl), *Food Cosmet. Toxicol.,* 13, 365, 1975.
126. **Lijinsky, W. and Taylor, H. W.,** Carcinogenesis in Sprague-Dawley rats by N-nitroso-N-alkylcarbamate esters, *Cancer Lett.,* 1, 275, 1976.
127. **Eisenbrand, G., Schmahl, D., and Preussmann, R.,** Carcinogenicity in rats of high oral doses of N-nitrosocarbaryl, a nitrosated pesticide, *Cancer Lett.,* 1, 281, 1976.
128. **Eisenbrand, G., Ungerer, O., and Preussmann, R.,** Rapid formation of carcinogenic N-nitrosamines by interaction of nitrite with fungicides derived from dithiocarbamic acid *in vitro* under stimulated gastric conditions and *in vivo* in the rat stomach, *Food Cosmet. Toxicol.,* 12, 229, 1974.
129. **Ross, R. D., Morrison, J., Rounbehler, D. P., Fan, S., and Fine, D. H.,** N-nitroso compound impurities in herbicide formulations, *J. Agric. Food Chem.,* 25, 1416, 1977.

Chapter 6

TOXICITY AND CARCINOGENICITY OF N-NITROSO COMPOUNDS

Ronald C. Shank and Peter N. Magee

TABLE OF CONTENTS

I.	Introduction	186
II.	Acute Toxicity	186
III.	Carcinogenicity	186
IV.	Mutagenicity	207
V.	Teratogenicity	207
VI.	Metabolism	208
VII.	Mechanism of Action	209
References		209

I. INTRODUCTION

The toxicology of N-nitroso compounds has been the subject of numerous reviews in the last several years,[1-6] and although research on these compounds has intensified since the earlier of these reviews, little that is new has surfaced regarding the toxicity of these agents, with the one exception of some fresh approaches to the metabolic activation of dimethylnitrosamine and some cyclic nitroso compounds. More nitrosamines and nitrosamides have been bioassayed since publication of the above reviews; most were shown to be strong carcinogens, and some have a real potential for environmental exposure. This chapter will summarize the most important generalities pertaining to the toxicology of N-nitroso compounds, especially in those areas that have relevance to their possible role as environmental carcinogens for man.

II. ACUTE TOXICITY

As discussed in Chapter 7, the acute toxicity of the simplest nitrosamine, dimethylnitrosamine, was described in man well before it was demonstrated in laboratory animals. The first report of experimental nitrosamine poisoning was made by Barnes and Magee in 1954;[7] a single fatal dose of dimethylnitrosamine to rats, mice, rabbits, and dogs produced severe liver damage in all species, and death occurred 1 to several days after administration of the poison. The hemorrhagic centrilobular necrosis of the liver described for dimethylnitrosamine later proved to be typical of most dialkylnitrosamines[8] and can be explained largely by the dependency upon the cytochrome P-450 monooxygenase for the metabolic activation of nitrosamines to form the ultimate toxin.[9-12] Organs other than the liver are also affected by nitrosamines; kidney,[13] lungs,[8] and testis[14] are included.

A second class of N-nitroso compounds, the nitrosamides, does not require metabolic activation to an ultimately toxic form, but rather, these compounds decompose rapidly at physiological pH[2] and damage cells at the site of administration. The nitrosamides are also quite toxic to tissues with a high rate of cell proliferation and turnover, e.g., bone marrow, lymphoid tissues, intestinal crypts.

Little work has been done on structure-activity relationships for the acute toxicity of N-nitroso compounds. For the limited data available, however, it appears that acute toxicity decreases with chain length of dialkylnitrosamines (see Table 1). Methylbenzylnitrosamine and dimethylnitrosamine appear to be among the most acutely toxic nitroso compounds, diethylnitrosamine is an order of magnitude less toxic, and ethyl-2-hydroxyethylnitrosamine is relatively nontoxic. Detailed studies on the acute toxicity of the nitroso compounds are lacking, probably because most investigators have focused on the striking carcinogenic properties of these compounds which have served almost 25 years as models in the field of chemical carcinogenesis and played a significant role in developing concepts of mechanism in cancer.

III. CARCINOGENICITY

More than 130 N-nitroso compounds have been tested for carcinogenicity since Magee and Barnes[15] first demonstrated the production of hepatocellular carcinomas in rats fed dimethylnitrosamine; at least 120 of these have been shown to be strong carcinogens (see Table 2). These agents have some remarkable properties for tumor production. As a chemical class, they are capable of producing tumors in virtually every vital tissue, although no one agent can induce cancer in all tissues. Table 3 provides a list of more than 30 tissues and representative nitrosamines and nitrosamides for which they are

Table 1
ACUTE TOXICITY OF SELECTED N-NITROSO COMPOUNDS

Compound	LD$_{50}$	Ref.
Dimethylnitrosamine	27—41	8
Diethylnitrosamine	216	8
Di-*n*-propylnitrosamine	480	2
Di-*n*-butylnitrosamine	1200	62
Di-*n*-amylnitrosamine	1750	41
Methyl-*n*-butylnitrosamine	130	8
Methyl-*t*-butylnitrosamine	700	8
Ethyl-*n*-butylnitrosamine	380	63
Ethyl-*t*-butylnitrosamine	1600	63
Ethyl-2-hydroxyethylnitrosamine	>7500	63
Di-2 hydroxyethylnitrosamine	>5000	64
Nitrososarcosine	5000	2
Nitrososarcosine, ethyl ester	4000	2
Methylphenylnitrosamine	200	8
Methylbenzylnitrosamine	18	63
Nitrosoazetidine	1600	65
Nitrosopyrrolidine	900	2
Nitrosomorpholine	282	41
Nitrosothiomorpholine	800	66
Nitrosopiperazine	2260	66
1-Methyl-4-nitrosopiperazine	1000	?
Nitrosopiperidine	200	2
2-Methyl-4-nitrosopiperidine	600	67
Methylnitrosourea	180	41
Methylnitrosourethane	240	68
Nitrosohexamethyleneimine	336	69
Nitrosoheptamethyleneimine	283	29
Nitrosooctamethyleneimine	566	29

Note: LD$_{50}$ units: mg/kg body weight, single oral dose, adult rats.

Table 2
CARCINOGENIC N-NITROSO COMPOUNDS

Compounds	Species	Target organs	Ref.
A. Methylnitrosamines			
1. Dimethylnitrosamine CH$_3$N (NO) CH$_3$	Rat	Liver	15
		Kidney	16
		Lung	70
		Nasal sinus	17, 71, 72
	Mouse	Liver	73
		Kidney, lung	74
	Hamster	Liver	75, 76
		Kidney	76
		Glandular stomach	77
	Mink	Liver	78
	Trout	Liver	79
2. Methoxymethylnitrosamine CH$_3$N (NO) CH$_2$—O—CH$_3$	Rat	Lung	80
3. 1-(Methoxyethyl)— methylnitrosamine CH$_3$N (NO) CH (CH$_3$) O—CH$_3$	Rat	Esophagus, lung	80

Table 2 (continued)
CARCINOGENIC N-NITROSO COMPOUNDS

Compounds	Species	Target organs	Ref.

A. Methylnitrosamines

Compounds	Species	Target organs	Ref.
4. Nitrososarcosine CH_3N (NO) CH_2-COOH	Rat Mouse	Esophagus Liver	63 81
5. Nitrososarcosine ethyl ester CH_3N (NO) $CH_2-COO-C_2H_5$	Rat	Forestomach, tongue Esophagus	63 2
6. Acetoxymethyl-methyl-nitrosamine CH_3N (NO) $O-CO-CH_3$	Rat	Intestine	82, 83
7. Ethylmethylnitrosamine CH_3N (NO) CH_2CH_3	Rat	Liver	2
8. 2-Chloroethyl-methyl-nitrosamine CH_3N (NO) C_2H_4-Cl	Rat	Liver	2
9. Dinitrosodimethyl-ethylene diamine CH_3N (NO) C_2H_4-N (CH_3) NO	Rat ·	Esophagus	63
10. Acetonitrilemethyl-nitrosamine CH_3N (NO) CH_2-CN	Rat	Liver	2
11. Vinylmethylnitrosamine CH_3N (NO) $CH=CH_2$	Rat	Nose Esophagus, pharynx, tongue	71, 72 2
12. n-Propylmethylnitrosamine CH_3N (NO) CH_2 CH_2 CH_3	Rat Hamster	Nasal cavities, liver, esophagus Nasal cavities, trachea, lung, liver	84 30
13. 4-(N-methyl-N-nitrosamino)-1-(3-pyridyl)-1-butanone CH_3N (NO) CH_2 CH_2 CH_2 | C=O (pyridyl)	Rat Mouse	Nasal cavities, liver, lung Lung	85 85
14. Nitrosoephedrine CH_3N (NO) CH (CH_3) CH (OH) (phenyl)	Rat Mouse	Liver, lung, forestomach Liver	86 81
15. Allylmethylnitrosamine CH_3N (NO) CH_2 $CH=CH_2$	Rat	Nose, kidney Esophagus	2, 72 2
16. n-Butylmethylnitrosamine CH_3N (NO) CH_2 $(CH_2)_2$ CH_3	Rat	Liver	8
17. 4-Hydroxybutyl-methyl-nitrosamine CH_3N (NO) CH_2 $(CH_2)_2$ CH_2 OH	Rat	Bladder	87
18. n-Pentylmethylnitrosamine CH_3H (NO) CH_2 $(CH_2)_3$ CH_3	Rat	Esophagus	2
19. Undecylmethylnitrosamine CH_3N (NO) CH_2 $(CH_2)_9$ CH_3	Rat	Liver, lung	88

Table 2 (continued)
CARCINOGENIC N-NITROSO COMPOUNDS

Compounds	Species	Target organs	Ref.

A. Methylnitrosamines

20. Dodecylmethylnitrosamine
    $CH_3N(NO)CH_2(CH_2)_{10}CH_3$

	Rat	Bladder	89
	Hamster	Bladder (Urinary)	90
	Guinea pig	Liver	88

21. Cyclohexylmethylnitrosamine

    $CH_3N(NO)$–⬡

    Rat — Esophagus — 2

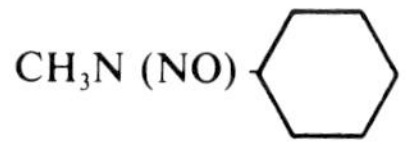

22. Phenylmethylnitrosamine
    (also methylnitrosoaniline)

    $CH_3N(NO)$–⬡

    Rat — Esophagus — 91

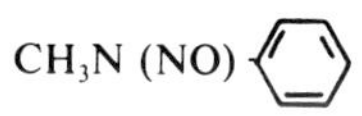

23. Benzylmethylnitrosamine

    $CH_3N(NO)CH_2$–⬡

Rat	Esophagus	63, 92
	Pharynx	92

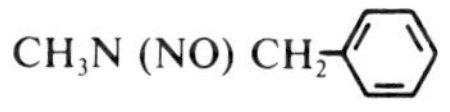

24. 2-Methylbenzyl-methyl nitrosamine

    $CH_3N(NO)CH_2$–⬡–CH_3

    Rat — Esophagus, pharynx — 92

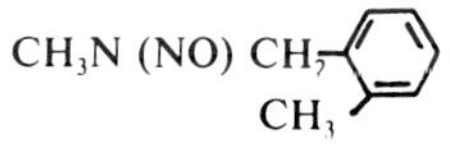

25. 3-Methylbenzyl-methyl nitrosamine

    $CH_3N(NO)CH_2$–⬡–CH_3

    Rat — Esophagus, pharynx — 92

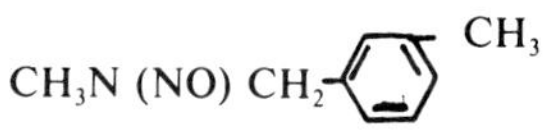

26. 4-Methylbenzyl-methyl nitrosamine

    $CH_3N(NO)CH_2$–⬡–CH_3

    Rat — Esophagus, pharynx — 92

27. 2-Phenylethyl-methyl nitrosamine

    $CH_3N(NO)CH_2CH_2$–⬡

    Rat — Esophagus — 2

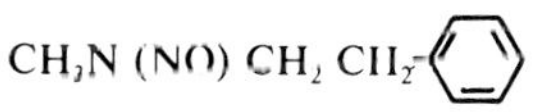

28. 3-(N-Nitrosomethyl-amine)-sulfolane

    $CH_3N(NO)$–⬠–SO_2

    Rat — Esophagus — 2

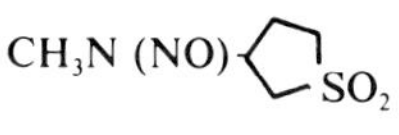

29. N^6-Methylnitrosoadenosine

    $CH_3N(NO)$–(adenosine ring with N-ribose)

    Mouse — Thymus, lung — 93

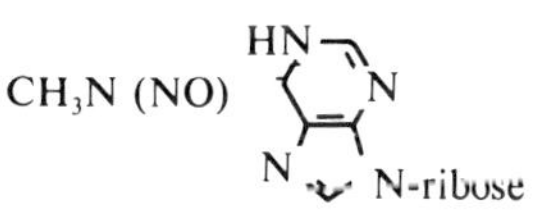

B. Ethylnitrosamines

30. Methoxymethyl-ethylnitrosamine
    $C_2H_5N(NO)CH_2-O-CH_3$

    Rat — Liver — 80

Table 2 (continued)
CARCINOGENIC N-NITROSO COMPOUNDS

Compounds	Species	Target organs	Ref.
B. Ethylnitrosamines			
31. 1-(Methoxy)ethyl-ethyl-nitrosamine C_2H_5N (NO) CH (CH$_3$) OCH$_3$	Rat	Lung, liver	80
32. Diethylnitrosamine C_2H_5N (NO) C_2H_5	Rat	Liver	2, 94
		Kidney	17, 71
		Esophagus	2, 64, 95
		Nasal sinus, pharynx	2
	Mouse	Liver	96
		Esophagus, stomach	97
		Nose	98
	Hamster	Liver, trachea, bronchus	99
		Lung, bronchus	100
		Nose	101
		Gland, stomach	77
	Guinea pig	Liver	102, 103
		Lung	103
	Rabbit	Liver	104, 105
	Hedgehog	Liver	106
	Dog	Liver	107
	Pig	Liver	108
	Monkey	Liver	109, 110
	Grass parakeet	Liver	108
	Trout	Liver	111
	Zebra fish	Liver	112
33. 2-Hydroxyethyl-ethylnitrosamine C$_2$	Rat	Liver, esophagus	2, 63
		Kidney	17, 113
34. Dihydroxyethylnitrosamine HOCH$_2$CH$_2$N (NO) CH$_2$CH$_2$OH	Rat	Liver	2
35. Vinylethylnitrosamine	Rat	Esophagus	2, 63
		Nose	2
C_2H_5N (NO) CH=CH$_2$	Hamster	Respiratory system, forestomach, pancreas	114
36. Bis (acetoxyethyl)-nitrosamine CH$_3$COO−C$_2$H$_5$N (NO) CH$_2$ | CH$_2$ | O | CH$_3$−CO	Rat	Liver	2
37. i-Propylethylnitrosamine C_2H_5N (NO) CH (CH$_3$) CH$_3$	Rat	Liver, esophagus	63
38. n-Butylethylnitrosamine C_2H_5N (NO) CH$_2$ (CH$_2$)$_2$ CH$_3$	Rat	Esophagus, liver	2, 63
		Kidney	17
	Mouse	Forestomach	115
39. 4-Hydroxybutyl-ethylnitrosamine C_2H_5N (NO) CH$_2$ (CH$_2$)$_2$ CH$_2$OH	Rat	Bladder	87

Table 2 (continued)
CARCINOGENIC N-NITROSO COMPOUNDS

Compounds	Species	Target organs	Ref.

B. Ethylnitrosamines

40. 3-Carboxypropyl-ethyl-nitrosamine
    C_2H_5N (NO) $CH_2CH_2CH_2COOH$ — Rat — Bladder — 87
41. 4-Picolyl-ethylnitrosamine — Rat — Esophagus, nose, lung — 2

   C_2H_5N (NO) CH_2 —⟨pyridine ring⟩

C. Propylnitrosamines

42. Di-n-propylnitrosamine — Rat — Liver, esophagus, tongue — 2, 41
    C_3H_7N (NO) C_3H_7 — Hamster — Nasal cavities, trachea — 116
43. Di-i-propylnitrosamine — Rat — Liver — 2
    $CH_3CH(CH_3)N$ (NO) $CH(CH_3)CH_3$
44. 2-Hydroxypropylpropyl-nitrosamine — Rat — Nasal cavities, lung, esophagus, liver — 84
    C_3H_7N (NO) $CH_2CH(OH)CH_3$ — Hamster — Nasal cavities, trachea, lung, liver — 30
45. 2-Oxo-n-propyl-propyl-nitrosamine — Hamster — Nasal cavities, respiratory tract, liver, kidney — 117
    C_3H_7N (NO) CH_2COCH_3
46. Di (2-hydroxypropyl)-nitrosamine — Rat — Nasal cavities, lungs, thyroid, esophagus, liver, kidneys — 118, 119

    $CH_3CH(OH)CH_2N$ (NO)
    |
    $CH_3CH(OH)CH_2$ — Hamster — Pancreas, liver, kidney, respiratory system — 120
    — Guinea pig — Liver — 121
47. Di (2-oxopropyl)nitrosamine — Hamster — Pancreas — 122, 123
    CH_3COCH_2N (NO) CH_2COCH_3
48. Butylpropylnitrosamine — Rat — Liver, esophagus — 87
    C_3H_7N (NO) C_4H_9
49. Di (2-acetoxypropyl)-nitrosamine — Hamster — Pancreas, liver, respiratory tract, kidney, vagina, gall bladder — 124

    $CH_3CH(CH_3COO)CH_2N$ (NO)
    |
    $CH_3CH(CH_3COO)CH_2$
50. Dinitroso-N,N'-bis-(1-hydroxymethyl-propyl) ethylenediamine — Mouse — Lung — 125

    $[HOCH_2-CH(CH_2CH_3)N(NO)CH_2]_2$

51. Diallynitrosamine — Hamster — Respiratory tract — 126

    $CH_2=CH-CH_2N$ (NO)
    |
    $CH_2=CH-CH_2$

Table 2 (continued)
CARCINOGENIC N-NITROSO COMPOUNDS

Compounds	Species	Target organs	Ref.

D. Butylnitrosamines

52. Di-*n*-butylnitrosamine	Rat	Liver	41, 62
		Bladder, esophagus	62, 127
C_4H_9N (NO) C_4H_9	Mouse	Liver, lung	128
	Hamster	Nasal cavities	129
		Lung	130
53. 4-Hydroxybutyl-butylnitrosamine	Rat	Bladder	62
	Mouse	Liver, lung	128
C_4H_9N (NO) CH_2 $(CH_2)_2$ CH_2OH			
54. 2-Oxobutyl-butylnitrosamine	Rat	Liver	87
C_4H_9N (NO) $CH_2COCH_2CH_3$			
55. 3-Oxobutyl-butylnitrosamine	Rat	Liver	87
C_4H_9N (NO) $CH_2CH_2COCH_3$			
56. Pentylbutylnitrosamine	Rat	Liver	2
C_4H_9N (NO) CH_2 $(CH_2)_3$ CH_3			

E. Other Aliphatic Nitrosamines

57. Di-*n*-pentylnitrosamine	Rat	Liver	41
$C_5H_{11}N$ (NO) C_5H_{11}		Lung	131
58. Diacetonitrilenitrosamine	Rat	Liver	2
$NC-CH_2N$ (NO) CH_2CN			
59. Nitrosotrimethylhydrazine	Rat	Liver, kidney	2
$(CH_3)_2 N-N$ (NO) CH_3			

F. Cyclic Nitrosamines

60. Nitrosoazetidine	Rat	Liver, lung	65
61. Nitrosopyrrolidine	Rat	Liver	2
		Testis	132
	Mouse	Lung	133
	Hamster	Trachea, lung	134
62. Nitroso-2,5-dimethylpyrrolidine	Rat	Liver	135
63. Nitroso-3,4-dichloropyrrolidine	Rat	Esophagus, olfactory	135

Table 2 (continued)
CARCINOGENIC N-NITROSO COMPOUNDS

Compounds	Species	Target organs	Ref.

F. Cyclic Nitrosamines (continued)

Compounds	Species	Target organs	Ref.
64. Nitrosonornicotine	Rat	Nasal cavities, esophagus	136
	Mouse	Lung	85
	Hamster	Trachea	85
65. Nitrosopyrroline	Rat	Liver	137
66. Nitrosooxazolidine	Rat	Liver	80
67. Nitrosoimidazolidone	Rat	Kidney	138
68. 1-Nitrosohydantoin	Rat	Tongue, pharynx, forestomach, kidney, lymphomas	139
69. Nitrosopiperidine	Rat	Nose, esophagus, liver, upper GI tract	2, 41, 72, 140
	Mouse	Forestomach, liver, esophagus, lung	141, 142
	Hamster	Trachea, lung, larynx	143, 144
	Monkey	Liver	145
70. Nitroso-2-methylpiperidine	Rat	Nasal turbinates, liver, upper GI tract, peripheral nervous system	140, 146
71. Nitroso-3-methylpiperidine	Rat	Nasal turbinates, upper GI tract	140

Table 2 (continued)
CARCINOGENIC N-NITROSO COMPOUNDS

Compounds	Species	Target organs	Ref.

F. Cyclic Nitrosamines (continued)

72. Nitroso-4-methylpiperidine — Rat — Nasal turbinates, upper GI tract — 140

73. Nitroso-2,6-dimethylpiperdine — Rat — Nasal turbinates, liver, upper GI tract — 140

74. Nitroso-3-piperidinol — Rat — Nasal cavities, tongue, pharynx, esophagus, liver — 140

75. Nitroso-4-piperidinol — Rat — Nasal cavities, tongue, pharynx, esophagus, liver — 140

76. Nitroso-4-piperidinone — Rat — Nasal cavities, tongue, pharynx, esophagus, liver — 140

77. Nitrosoanabasine — Rat — Esophagus — 147

78. Nitrosotetrahydropyridine — Rat — Liver — 140

79. Nitrosomorpholine — Rat — Liver — 2, 41, 72, 148, 149

		Kidney	72
		Ovary	113
		Blood vessels	148
		Nose	2, 72
	Mouse	Liver	149
	Hamster	Liver, blood vessels	148
		Lung	143

Table 2 (continued)
CARCINOGENIC N-NITROSO COMPOUNDS

Compounds	Species	Target organs	Ref.

F. Cyclic Nitrosamines (continued)

80. 2,6-Dimethylnitrosomorpholine | Hamster | Pancreas | 150, 151

81. Nitrosothiomorpholine | Rat | Esophagus | 66, 137
| | | Tongue, nasal cavities | 66

82. Nitrosopiperazine | Rat | Olfactory, nasal cavities, liver | 152

83. N-Nitroso-N'-methylpiperazine | Rat | Nose | 2

84. Carbethoxynitrosopiperzine | Rat | Liver, nose | 2

85. Dinitrosopiperazine | Rat | Esophagus, liver, nose | 2, 72, 152
| | | Upper GI tract | 153
| | Mouse | Liver, lung | 154

86. 2-Methyldinitrosopiperazine | Rat | Nasal turbinates, upper GI tract | 150

87. 2,5-Dimethyldinitrosopiperazine | Rat | Nasal turbinates, upper GI tract | 150

88. 2,6-Dimethyldinitroso-piperazine | Rat | Nasal turbinates, upper GI tract | 150

Table 2 (continued)
CARCINOGENIC N-NITROSO COMPOUNDS

Compounds	Species	Target organs	Ref.

F. Cyclic Nitrosamines (continued)

Compounds	Species	Target organs	Ref.
89. 1-Nitroso-5,6-dihydrouracil	Rat	Liver	155
90. Nitrosohexamethyleneimine	Rat	Liver, tongue, esophagus, nasal cavities	156, 157
	Mouse	Lung	158
	Hamster	Trachea	158
91. Dinitrosohomopiperazine	Rat	Nasal turbinates, upper GI tract	150
92. Nitrosoheptamethyleneimine	Rat	Lung, esophagus	159
		Tongue, trachea	159
	Hamster	Esophagus, pharynx, forestomach, trachea, nasal cavities	160
93. Nitrosooctamethyleneimine	Rat	Lung, esophagus, tongue, trachea	159

G. Nitrosamides

Compounds	Species	Target organs	Ref.
94. Methylnitrosoacetamide CH$_3$N (NO) COCH$_3$	Rat	Forestomach	2
95. Methylnitrosourea CH$_3$N (NO) CONH$_2$	Rat	Forestomach, intestine, kidney	19, 161
		Brain, spinal cord, skin, jaw, bladder, uterus, ovary	2, 162
	Mouse	Lung, forestomach, hematopoietic system, kidney, skin, brain, liver (newborn)	163—172
	Hamster	Intestine, pharynx, esophagus, trachea, bronchi, oral cavity, skin	173—175
	Guinea pig	Stomach, pancreas, ear ducts	176
	Rabbit	Brain, intestine, skin	177—180
	Dog	Brain, peripheral nerves	181—182

Table 2 (continued)
CARCINOGENIC N-NITROSO COMPOUNDS

Compounds	Species	Target organs	Ref.
G. Nitrosamines (continued)			
96. 1,3-Dimethylnitrosourea $CH_3N\,(NO)\,CO-NHCH_3$	Rat Mouse	Brain, peripheral nerves, spinal cord, kidney Hematopoietic system	2 30
97. Methylnitrosobiuret $CH_3N\,(NO)\,CO-NH-CONH_2$	Rat	Stomach, brain, peripheral nerves, kidney	183
98. Trimethylnitrosourea $CH_3N\,(NO)\,CO-N\,(CH_3)_2$	Rat	Brain, peripheral nerves Spinal cord, kidney, skin	2, 184 2
99. 1-Methyl-3-acetyl-1 nitrosourea $CH_3N\,(NO)\,CO-NH-COCH_3$	Rat	Stomach, nervous system	185
100. Hydrazodicarbonylbis-(Methylnitrosamide) $[CH_3N\,(NO)\,CO-NH-]_2$	Rat	Injection site	2
101. Methylnitrosourethane $CH_3N\,(NO)\,CO-O-CH_2CH_3$	Rat Mouse Guinea pig	Stomach, esophagus Lung Kidney Ovary Stomach, lung Stomach, pancreas	161, 186, 187 2, 188 2, 113 2 189 176
102. Ethylnitrosourea $C_2H_5N\,(NO)\,CONH_2$	Rat Mouse	Central and peripheral nervous system, kidney, hematopoietic system, skin, intestine, ovary, uterus Hematopoietic system, lung, central and peripheral nervous system, kidney	2, 190—193 194—199
103. 2-Hydroxyethylnitrosourea $HO-CH_2CH_2N\,(NO)\,CONH_2$	Rat	Bone	200
104. Ethylnitrosourethane $C_2H_5N\,(NO)\,CO-O-C_2H_5$	Rat	Intestine Stomach	201 2, 189
105. n-Propylnitrosourea $C_3H_7N\,(NO)\,CONH_2$	Rat	Mammary, hematopoietic system, ovary, thyroid, adrenal	202
106. n-Propylnitrosourethane $C_3H_7N\,(NO)\,CO-O-C_2H_5$	Rat	Upper Gi tract	203
107. n-Butylnitrosourea $C_4H_9N\,(NO)\,CONH_2$	Rat	Hematopoietic system Kidney Mammary, stomach, ovary, intestine, ear duct	204 205 206

Table 2 (continued)
CARCINOGENIC N-NITROSO COMPOUNDS

Compounds	Species	Target organs	Ref.

G. Nitrosamines (continued)

Compounds	Species	Target organs	Ref.
108. Butyl-dimethylnitrosourea $C_4H_9N\,(NO)\,CO-N\,(CH_3)_2$	Rat	Hematopoietic system, vagina	207
109. n-Butylnitrosourethane $C_4H_9N\,(NO)\,CO-O-C_2H_5$	Rat	Oral cavity, pharynx, esophagus, forestomach	208
	Mouse	Esophagus, forestomach	209
110. N-Methyl-N-nitroso-N'-nitroguanidine $CH_3N\,(NO)\,C\,(NH)\,NH-NO_2$	Rat	Forestomach	210
		Glandular stomach	211
		Small intestine	212
		Large intestine	213
		Skin	214
		Injection site	2
	Dog	Stomach	213
111. N-Ethyl-N-nitroso-N'-nitroguanidine $C_2H_5N\,(NO)\,C\,(NH)\,NH-NO_2$	Dog	Esophagus, stomach	215—217
112. N-n-Propyl-N-nitroso-N'-nitroguanidine $C_3H_7N\,(NO)\,C\,(NH)\,NH-NO_2$	Rat	Glandular stomach	30
	Hamster	Glandular stomach	30
113. N-n-Butyl-N-nitroso-N'-nitroguanidine $C_4H_9N\,(NO)\,C\,(NH)\,NH-NO_2$	Rat	Forestomach	30
114. N-i-Butyl-N-nitroso-N'-nitroguanidine $CH_3CH\,(CH_3)\,CH_2N\,(NO)\,C=NH$ $\quad\vert$ $\quad NH$ $\quad\vert$ $\quad NO_2$	Rat	Forestomach	30
115. N-n-Pentyl-N-nitroso-N'-nitroguanidine $C_5H_{11}N\,(NO)\,C\,(NH)\,NH-NO_2$	Rat	Forestomach	30
116. N-n-Hexyl-N-nitroso-N'-nitroguanidine $C_6H_{13}N\,(NO)\,C\,(NH)\,NH-NO_2$	Rat	Forestomach	30
117. Ethylnitrosocyanamide $C_2H_5N\,(NO)\,C\equiv N$	Rat	Esophagus, nose	139
118. Streptozotocin (N-Nitroso-N'-D-glucosyl-2-methylurea	Rat	Kidney	218
	Hamster	Liver	219

Table 2 (continued)
CARCINOGENIC N-NITROSO COMPOUNDS

Compounds	Species	Target organs	Ref.

G. Nitrosamines (continued)

119. Nitrosobenzthiazuron

    CH₃N (NO) CO—NH—[benzothiazole]

    Rat — Forestomach, kidney — 220

120. Nitrosocarbaryl
    (N-Methyl-1-naphthyl-
    N-nitrosocarbamate)

    CH₃N (NO) CO—O—[naphthyl]

    Rat — Forestomach — 221
    Injection site — 222

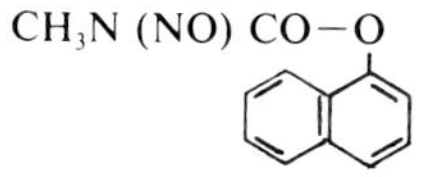

Table 3
TISSUE SUSCEPTIBILITY TO CARCINOGENICITY OF 120 N-NITROSO COMPOUNDS[a]

Liver: 53 Compounds

A. Aliphatic Nitrosamines (33)

Dimethylnitrosamine
Nitrososarcosine
Ethylmethylnitrosamine
2-Chloroethyl-methylnitrosamine
Acetonitrile-methylnitrosamine
n-Propylmethylnitrosamine
4-N-methyl-N-nitrosamino)-1-
 (3-pyridyl)-1-butanone
Nitrosoephedrine
n-Butylmethylnitrosamine
Undecylmethylnitrosamine
Methoxymethyl-ethylnitrosamine
1 (Methoxy)ethyl-ethylnitrosamine
Diethylnitrosamine
2-Hydroxyethyl-ethylnitrosamine
Dihydroxyethylnitrosamine
Bis-(acetoxyethyl)nitrosamine
i-Propylethylnitrosamine

n-Butylethylnitrosamine
Di-n-propylnitrosamine
Di-i-propylnitrosamine
2-Hydroxypropyl-propylnitrosamine
2-Oxo-n-propyl-propylnitrosamine
Di-(2-hydroxypropyl)propylnitrosamine
Butylpropylnitrosamine
Di-(2-acetoxypropyl)nitrosamine
Di-n-butylnitrosamine
4-Hydroxybutyl-butylnitrosamine
2-Oxobutyl-butylnitrosamine
3-Oxobutyl-butylnitrosamine
Pentylbutylnitrosamine
Dipentylnitrosamine
Diacetonitrilenitrosamine
Nitrosotrimethylhydrazine

B. Cyclic Nitrosamines (18)

Nitrosoazetidine
Nitrosopyrrolidine
Nitroso-2,5-dimethylpyrrolidine
Nitrosopyrroline
Nitrosooxyazolidine
Nitrosopiperidine
Nitroso-2-methylpiperidine
Nitroso-2,6-dimethylpiperidine
Nitroso-3-piperidinol

Nitroso-4-piperidinol
Nitroso-4-piperidinone
Nitrosotetrahydropyridine
Nitrosomorpholine
Nitrosopiperazine
Carbethoxynitrosopiperazine
Dinitrosopiperazine
1-Nitroso-5,6-dihydrouracil
Nitrosohexamethyleneimine

Table 3 (continued)
TISSUE SUSCEPTIBILITY TO CARCINOGENICITY OF 120 N-NITROSO COMPOUNDS[a]

C. Nitrosamides (2)

Methylnitrosourea (newborn)   Streptozotocin

Esophagus: 43 Compounds

A. Aliphatic Nitrosamines (26)

1-(Methoxyethyl)methylnitrosamine
Nitrososarcosine
Nitrososarcosine ethyl ester
Dinitrosodimethylenediamine
Vinylmethylnitrosamine
n-Propylmethylnitrosamine
Allylmethylnitrosamine
n-Pentylmethylnitrosamine
Cyclohexylmethylnitrosamine
Phenylmethylnitrosamine
  (also methylnitrosoaniline)
Benzylmethylnitrosamine
2-Methylbenzyl-methylnitrosamine
3-Methylbenzyl-methylnitrosamine
4-Methylbenzyl-methylnitrosamine

2-Phenylethyl-methylnitrosamine
3-(N-Nitrosomethylamine)sulfolane
Diethylnitrosamine
2-Hydroxyethyl-ethylnitrosamine
Vinylethylnitrosamine
i-Propylethylnitrosamine
n-Butylethylnitrosamine
4-Picolylethylnitrosamine
Di-n-propylnitrosamine
2-Hydroxypropyl-propylnitrosamine
Di-(2-hydroxypropyl)nitrosamine
Butylpropylnitrosamine

B. Cyclic Nitrosamines (12)

Nitroso-2,5-dimethylpyrrolidine
Nitrosonornicotine
Nitrosopiperidine
Nitroso-3-piperidinol
Nitroso-4-piperidinol
Nitroso-4-piperidinone

Nitrosoanabasine
Nitrosothiomorpholine
Dinitrosopiperazine
Nitrosohexamethyleneimine
Nitrosoheptamethyleneimine
Nitrosooctamethyleneimine

C. Nitrosamides (5)

Methylnitrosourea
Methylnitrosourethane
Butylnitrosourethane

N-Ethyl-N'-nitro-N-nitrosoguanidine
Ethylnitrosocyanamide

Nose/Nasal Sinus: 35 Compounds

A. Aliphatic Nitrosamines (13)

Dimethylnitrosamine
Vinylmethylnitrosamine
n-Propylmethylnitrosamine
Allymethylnitrosamine
4-N-Methyl-N-nitrosamino)-1-
  (3-pyridyl)-1-butanone
Diethylnitrosamine
Vinylethylnitrosamine

Di-n-propylnitrosamine
2-Hydroxypropyl-propylnitrosamine
2-Oxopropyl-propylnitrosamine
Di-(2-hydroxypropyl)nitrosamine
Dibutylnitrosamine
4-Picolylethylnitrosamine

B. Cyclic Nitrosamines (21)

Nitrosonornicotine
Nitrosopiperidine

Nitrosothiomorpholine
Nitrosopiperazine

Table 3 (continued)
TISSUE SUSCEPTIBILITY TO CARCINOGENICITY OF 120 N-NITROSO COMPOUNDS[a]

B. Cyclic Nitrosamines (21) (continued)

Nitroso-2-methylpiperidine
Nitroso-3-methylpiperidine
Nitroso-4-methylpiperidine
Nitroso-2,6-dimethylpiperidine
Nitroso-3-piperidinol
Nitroso-4-piperidinol
Nitroso-4-piperidinone
Nitrosomorpholine

N-Nitroso-N'-methyl-piperazine
Carbethoxynitrosopiperazine
Dinitrosopiperazine
2,5-Dimethyl-dinitroso-piperazine
2,6-Dimethyl-dinitroso-piperazine
Nitrosohexamethyleneimine
Dinitrosohomopiperazine
Nitrosoheptamethyleneimine
2-Methyl-dinitroso-piperazine

C. Nitrosamides (1)

Ethylnitrosocyanamide

Lung: 29 Compounds

A. Aliphatic Nitrosamines (17)

Dimethylnitrosamine
Methoxymethyl-methylnitrosamine
1-Methoxyethyl-methylnitrosamine
Propylmethylnitrosamine
4-(N-methyl-N-nitrosamino)-1-(3-pyridyl)-1-butanone
Nitrosoephedrine
Undecylmethylnitrosamine
N[6]-Methylnitrosoadenosine
1-Methoxyethyl-ethylnitrosamine

Diethylnitrosamine
4-Picolyl-ethylnitrosamine
2-Hydroxypropyl-propylnitrosamine
Di(2-hydroxypropyl)nitrosamine
Dinitroso-N,N'-bis-(1-hydroxymethyl-propyl)-ethylenediamine
Dibutylnitrosamine
4-Hydroxybutyl-butylnitrosamine
Dipentylnitrosamine

B. Cyclic Nitrosamines (9)

Nitrosoazetidine
Nitrosopyrrolidine
Nitrosonornicotine
Nitrosopiperidine
Nitrosomorpholine

Dinitrosopiperazine
Nitrosohexamethyleneimine
Nitrosoheptamethyleneimine
Nitrosooctamethyleneimine

C. Nitrosamides (3)

Methylnitrosourea
Ethylnitrosourea

Methylnitrosourethane

Stomach: 26 Compounds

A. Aliphatic Nitrosamines (7)

Dimethylnitrosamine
Nitrososarcosine ethyl ester
Nitrosoephedrine

Diethylnitrosamine
Vinylethylnitrosamine
Butylethylnitrosamine
Dibutylnitrosamine

Table 3 (continued)
TISSUE SUSCEPTIBILITY TO CARCINOGENICITY OF 120 N-NITROSO COMPOUNDS[a]

B. Cyclic Nitrosamines (3)

Nitrosohydantoin
Nitrosopiperidine

Nitrosoheptamethyleneimine

C. Nitrosamides (16)

Methylnitrosoacetamide
Methylnitrosourea
Methylnitrosobiuret
1-Methyl-3-acetyl-1-nitrosourea
Methylnitrosourethane
Ethylnitrosourethane
Butylnitrosourea
Butylnitrosourethane

N-Ethyl-N′-nitro-N-nitrosoguanidine
N-Propyl-N′-nitro-N-nitrosoguanidine
N-Butyl-N′-nitro-N-nitrosoguanidine
N-i-Butyl-N′-nitro-N-nitrosoguanidine
N-Pentyl-N′-nitro-N-nitrosoguanidine
N-Hexyl-N′-nitro-N-nitrosoguanidine
Nitrosobenzthiazuram
Nitrosocarbaryl

Kidney: 22 Compounds

A. Aliphatic Nitrosamines (9)

Dimethylnitrosamine
Allymethylnitrosamine
Diethylnitrosamine
2-Hydroxyethyl-ethylnitrosamine
Butylethylnitrosamine

2-Oxopropyl-propylnitrosamine
Di-(2-hydroxypropyl)nitrosamine
Di-(2-acetoxypropyl)nitrosamine
Nitrosotrimethylhydrazine

B. Cyclic Nitrosamines (3)

Nitrosoimidazolidone
Nitrosohydantoin

Nitrosomorpholine

C. Nitrosamides (10)

Methylnitrosourea
1,3-Dimethylnitrosourea
Methylnitrosobiuret
Trimethylnitrosourea
Methylnitrosourethane

Ethylnitrosourea
Butylnitrosourea
N-Methyl-N′-nitro-N-nitrosoguanidine
Stretozotocin
Nitrosobenzthiazuram

GI Tract (Unspecified): 17 Compounds

A. Aliphatic Nitrosamines (1)

Acetoxymethyl-methylnitrosamine

B. Cyclic Nitrosamines (10)

Nitrosopiperidine
Nitroso-2-methylpiperidine
Nitroso-3-methylpiperidine
Nitroso-4-methylpiperidine
Nitroso-2,6-dimethylpiperidine

Dinitrosopiperazine
2-Methyl-dinitroso-piperazine
2,5-Dimethyl-dinitroso-piperazine
2,6-Dimethyl-dinitroso-piperazine
Dinitrosohomopiperazine

Table 3 (continued)
TISSUE SUSCEPTIBILITY TO CARCINOGENICITY OF 120 N-NITROSO COMPOUNDS[a]

C. Nitrosamides (6)

Methylnitrosourea
Ethylnitrosourea
Ethylnitrosourethane
Pentylnitrosourethane
Butylnitrosourea
N-Methyl-N'-nitro-N-nitrosoguanidine

Pharynx: 13 Compounds

A. Aliphatic Nitrosamines (6)

Vinylmethylnitrosamine
Benzylmethylnitrosamine
2-Methylbenzyl-methylnitrosamine
3-Methylbenzyl-methylnitrosamine
4-Methylbenzyl-methylnitrosamine
Diethylnitrosamine

B. Cyclic Nitrosamines (5)

Nitrosohydantoin
Nitroso-3-piperidinol
Nitroso-4-piperidinol
Nitroso-4-piperidinone
Nitrosoheptamethyleneimine

C. Nitrosamides (2)

Methylnitrosourea
Butylnitrosourethane

Trachea: 11 Compounds

A. Aliphatic Nitrosamines (4)

Propylmethylnitrosamine
Diethylnitrosamine
Dipropylnitrosamine
2-Hydroxypropyl-propylnitrosamine

B. Cyclic Nitrosamines (6)

Nitrosopyrrolidine
Nitrosonornicotine
Nitrosopiperidine
Nitrosohexamethyleneimine
Nitrosoheptamethyleneimine
Nitrosooctamethyleneimine

C. Nitrosamides (1)

Methylnitrosourea

Tongue: 11 Compounds

A. Aliphatic Nitrosoamines (3)

Nitrososarcosine ethyl ester
Vinylmethylnitrosamine
Dipropylnitrosamine

B. Cyclic Nitrosamines (8)

Nitrosohydantoin
Nitroso-3-piperidinol
Nitroso-4-piperidinol
Nitroso-4-piperidinone
Nitrosothiomorpholine
Nitrosohexamethyleneimine
Nitrosoheptamethyleneimine
Nitrosooctamethyleneimine

Table 3 (continued)
TISSUE SUSCEPTIBILITY TO CARCINOGENICITY OF 120 N-NITROSO COMPOUNDS[a]

C. Nitrosamides (0)

None reported

Pancreas: 7 Compounds

A. Aliphatic Nitrosamines (4)

Vinylethylnitrosamine
Di-(2-hydroxypropyl)nitrosamine
Di-(2-oxopropyl)nitrosamine
Di-(2-acetoxypropyl)nitrosamine

B. Cyclic Nitrosamines (1)

2,6-Dimethyl-nitrosomorpholine

C. Nitrosamides (2)

Methylnitrosourea
Methylnitrosourethane

Urinary Bladder: 7 Compounds

A. Aliphatic Nitrosamines (6)

4-Hydroxybutyl-methylnitrosamine
Dodecylmethylnitrosamine
4-Hydroxybutyl-ethylnitrosamine
3-Carboxypropyl-ethylnitrosamine
Dibutylnitrosamine
4-Hydroxybutyl-butylnitrosamine

B. Cyclic Nitrosamines (0)

None reported

C. Nitrosamides (1)

Methylnitrosourea

Peripheral Nerves: 7 Compounds

A. Aliphatic Nitrosamines (0)

None reported

B. Cyclic Nitrosamines (1)

Nitroso-2-methyl-piperidine

C. Nitrosamides (6)

Methylnitrosourea
1,3-Dimethylnitrosurea
Methylnitrosobiuret
Trimethylnitrosourea
1-Methyl-3-acetyl-1-nitrosourea
Ethylnitrosourea

Hematopoietic System: 7 Compounds

A. Aliphatic Nitrosamines (0)

None reported

Table 3 (continued)
TISSUE SUSCEPTIBILITY TO CARCINOGENICITY OF 120 N-NITROSO COMPOUNDS[a]

B. Cyclic Nitrosamines (1)

1-Nitrosohydantoin

C. Nitrosamides (6)

Methylnitrosourea
1,3-Dimethyl-nitrosourea
Ethylnitrosourea

Propylnitrosourea
Butylnitrosourethane
Butyl-dimethyl-nitrosourea

Ovary: 6 Compounds

A. Aliphatic Nitrosamines (0)

None reported

D. Cyclic Nitrosamines (1)

Nitrosomorpholine

C. Nitrosamides (5)

Methylnitrosourea
Methylnitrosourethane
Ethylnitrosourea

Propylnitrosourea
Butylnitrosourea

Other Tissues

Tissue	Aliphatic nitrosamines	Cyclic nitrosamines	Nitrosamides
Brain (5)	None reported	None reported	Methylnitrosourea 1,3-dimethylnitrosourea Methylnitrosobiuret Trimethylnitrosourea Ethylnitrosourea
Respiratory tract (unspecified) (5)	Vinylethylnitrosamine 2-Oxopropyl-propylnitrosamine Di-(2-hydroxypropyl) nitrosamine Diallylnitrosamine	None reported	None reported
Skin (4)	None reported	None reported	Methylnitrosourea Trimethyl-nitrosourea Ethylnitrosourea N-Methyl-N'-nitro-N-nitrosoguanidine
Spinal cord (3)	None reported	None reported	Methylnitrosourea 1,3-dimethylnitrosourea Trimethylnitrosourea
Olfactory (2)	None reported	Nitroso-2,5-dimethylpyrrolidine Nitrosopiperazine	None reported
Bronchus (2)	Diethylnitrosamine	None reported	Methylnitrosourea
Vagina (2)	Di-(2-acetoxypropyl) nitrosamine	None reported	Butyl-dimethyl-nitrosourea

Table 3 (continued)
TISSUE SUSCEPTIBILITY TO CARCINOGENICITY OF 120 N-NITROSO COMPOUNDS[a]

Tissue	Aliphatic nitrosamines	Cyclic nitrosamines	Nitrosamides
Uterus (2)	None reported	None reported	Methylnitrosourea Ethylnitrosourea
Ear Duct (2)	None reported	None reported	Methylnitrosourea Butylnitrosourea
Thyroid (2)	Di-(2-hydroxypropyl) nitrosamine	None reported	Propylnitrosourea
Mammary (2)	None reported	None reported	Propylnitrosourea Butylnitrosourea
Larynx (1)	None reported	Nitrosopiperidine	None reported
Thymus (1)	N[6]-Methylnitroso-adenosine	None reported	None reported
Adrenal (1)	None reported	None reported	Propylnitrosourea
Testis (1)	Nitrosopyrrolidine	None reported	None reported
Gall bladder (1)	Di-(2-acetoxypropyl) nitrosamine	None reported	None reported
Blood vessels (1)	None reported	Nitrosomorpholine	None reported
Jaw (1)	None reported	None reported	Methylnitrosourea
Bone (1)	None reported	None reported	2-Hydroxyethylnitrosourea

[a] Refer to Table 2 to determine applicable species and references.

targets. No other chemical class has so broad a scope of tissues for carcinogenic attack. Descriptions of the tumor types are available in the 1967 reviews of Magee and Barnes[1] and Druckrey and colleagues;[2] the carcinogenicity of these compounds is based not on shortening the induction time of spontaneous benign tumors, but on the production of malignant and metastatic tumors in animals for which the spontaneous occurrence of such tumors is rare or completely absent.

With the large number of compounds tested by bioassay, some structure-activity relationships seem apparent. Nitrosamines produce tumors primarily in the liver, kidney, esophagus, and respiratory tract; asymmetric methylalkylnitrosamines have strong carcinogenic action toward the esophagus regardless of route of administration, and di-n-butylnitrosamine and several of its derivatives are carcinogenic for the urinary bladder. Nitrosamides, on the other hand, are more specific for the nervous system and GI tract as well as the kidney.

A single administration of some nitroso compounds can result in subsequent tumor formation at high incidence. A single administration of dimethylnitrosamine[16] or methylnitrosourea[14] to newborn rats can produce high tumor incidence when the animals reach maturity. Even in adult animals, one dose can produce tumors; malignant genital tumors result from one administration of ethylnitrosourea to pregnant rats,[18] while a single treatment of an adult rat with methylnitrosourea is carcinogenic for the stomach, small and large intestine, kidney, skin, jaw, heart,[21] urinary bladder,[22] lung, and lymphoid tissue. One dose of dimethylnitrosamine to adult rats can induce kidney tumors[16] and even liver tumors if the tissue is growing rapidly, such as recovering from partial hepatectomy.[24]

By no means is the rat the only species susceptible to the carcinogenicity of N-nitroso compounds. Professor Schmahl at the Cancer Research Center in Heidelberg, Germany has paid special attention to species susceptibility and demonstrated the carcinogenicity of diethylnitrosamine to 22 species, including birds, amphibia, and fish as well as mammals.[25] From this observation alone it is difficult not to regard the human as a susceptible species as well.

The N-nitroso compounds can induce tumors transplacentally. Ivankovic and Druckrey[26] treated pregnant rats with ethylnitrosourea on the 15th day of gestation and observed that several months later, the offspring developed brain and spinal cord tumors; the same compound has also produced kidney tumors transplacentally.[27] Diethylnitrosamine given to rats on the 9th through 15th day of gestation induced tracheal papillomas in the offspring.[28]

Another property which characterizes many of the N-nitroso carcinogens is the rapidity with which they can induce tumors. For example, Lijinsky and co-workers[29] reported the formation of lung and esophageal tumors in only 7 and 16 weeks of treatment respectively with 200 mg N-nitrosoheptamethyleneimine.

IV. MUTAGENICITY

The mutagenicity of nitrosamines and nitrosamides has recently been reviewed by Montesano and Bartsch.[30] In several in vitro assay systems, the nitrosamides are mutagenic without the need for metabolic activation by mammalian enzymes; the "direct" action of these compounds is due presumably to the nonenzymatic production of reactive electrophiles which form covalent adducts with various macromolecules including DNA. The nitrosamines do not form electrophilic alkylating intermediates without mammalian enzymes and thus have failed to induce mutations in systems where these catalysts are lacking.

The correlation between positive results in mutagenicity and carcinogenicity assays so far has been high. Montesano and Bartsch[30] compared 47 N-nitroso compounds, 23 nitrosamides, and 24 nitrosamines. Of the 47 compounds, 38 (81%) were positive in both mutagenicity and carcinogenicity tests, and 3 (6%) were negative in both tests; thus the correlation between mutagenicity and carcinogenicity held for 41 of the 47 compounds. A total of five compounds (11%) were positive for carcinogenicity, but negative as mutagens, that is, five compounds (methylphenylnitrosamine, di-N,N'-bis-nitroso-N,N'-bis-(1-hydroxymethylpropyl)-ethylenediamine, trimethylnitrosourea, p-tolylsulfonylmethylnitrosamide, methylnitrososulfolan) failed to indicate their carcinogenicity in mutagenicity assays. Only one compound, N,N'-dinitroso-N,N' dimethylphthalamide, which has not produced animal tumors, has given positive results in mutagenicity assays. It is necessary to point out, as did Montesano and Bartsch,[30] that demonstration of mutagenicity of a compound cannot at this time be taken to imply the compound is also a carcinogen, for more studies need to be made on comparative mechanisms in mutagenesis and carcinogenesis.

V. TERATOGENICITY

The nitrosamides have been shown to be teratogenic. Methylnitrosurea given to rats on the 13th or 14th day of gestation produces fetal death and resorption and malformation of those fetuses that survive.[31] Ethylnitrosourea is teratogenic to rats[32] and pigs,[33] but dialkylnitrosamines did not have teratogenic effects in the rat.[32] The inactivity of the nitrosamines presumably was due to lake of metabolic activation of the agents by the fetal cells. A great deal more work needs to be done on the teratogenicity of the N-nitroso compounds.

VI. METABOLISM

The metabolism, especially the metabolic activation to reactive intermediates, of nitrosamines has been the subject of intensive study during the last 25 years and has recently been reviewed.[30,34-36] Generally it has been viewed that the carcinogenicity and perhaps the toxicity of the N-nitroso compounds can be explained in that their administration to animals results in covalent binding of a portion of the carcinogen to DNA, that this binding disturbs the genetic fidelity of DNA, and that a transformed cell may be produced which either dies or serves as the focus for development of a tumor. N-nitrosamides can break down nonenzymatically at physiological pH to produce such reactive fragments that will bind to DNA, and thus their metabolism has commanded little attention. N-nitrosamines are chemically stable at physiological pH and require enzyme-catalyzed activation to produce the reactive fragments that will bind to DNA; thus the nitrosamines have been the center of focus in metabolism studies on N-nitroso compounds.

Magee and Vandekar[9,37] demonstrated over 20 years ago that dimethylnitrosamine is oxidatively N-demethylated by enzymes bound to the endoplasmic reticulum in liver cells especially and in kidney and other tissues as well. The result of this oxidation was that one carbon of the nitrosamine was metabolized to formaldehyde and carbon dioxide,[9,37] and the other was converted eventually to an electrophilic reactant, presumably, a methonium ion, CH_3^+, which bound to nucleophilic sites on protein[38] and nucleic acids.[39] Other N-nitrosamines were shown to be metabolically activated to alkylating agents (see review by Lawley[40]).

Most of the evidence available today indicates that the oxidative N-demethylase which initiates the metabolic activation of dialkynitrosamines is a cytochrome P-450 monooxygenase.[9,12] The resultant methyl-hydroxymethylnitrosamine is thought to break down spontaneously to monomethylnitrosamine and formaldehyde; the amount of formaldehyde released in this decomposition is not sufficient to explain the toxicity of dimethylnitrosamine. It has be proposed[41] that the monomethylnitrosamine undergoes a protonic change to yield the diazohydroxide; these two intermediates are unstable and thought to decompose in a fraction of a second.[42] The manner in which the diazohydroxide breaks down is not known, but diazomethane, methyldiazonium ion, and diazotate have been suggested intermediates.[41,43] Lijinsky and co-workers[44] have shown that diazomethane is an unlikely intermediate in dimethylnitrosamine metabolism, as the transmethylation in rat liver nucleic acid alkylation resulting from dimethylnitrosamine administration involves no exchange of hydrogen atoms in deriving the methonium ion.

To account for metabolism of cyclic and higher dialkynitrosamines, Kruger[45] proposed a metabolic pathway similar to the β-oxidation of fatty acids. Enzymatic dehydrogenation between the α- and β-carbons of one alkyl chain and addition of water to the resultant double bond produces a β-hydroxylated nitrosamine because of the inductive effect of the nitroso group, analogous to the effect that activated carbonyl groups have in β-oxidation of fatty acids. The β-hydroxylated nitrosamine is further oxidized, yielding acetyl CoA and methylalkylnitrosamine. The nitrosamine apparently then undergoes α-oxidation, as above, producing either the methonium ion or the carbonium ion corresponding to the second alkyl group.

Arcos and co-workers[12] have demonstrated the existence of at least two enzymic forms of dimethylnitrosamine demethylase in rat liver microsomes, one active at low substrate concentrations and regarded as the critical enzyme in dimethylnitrosamine carcinogenesis and the other operative at high substrate concentrations. If the carcinogenic mechanism depends on metabolic production of an electrophile which alters the DNA template, then it is highly unlikely that microsomal production of reactive carbonium ions can explain alkylation in the nucleus, as the electrophiles are far too reactive to persist long enough to diffuse or be transported from the endoplasmic reticulum to the nuclear DNA. Argus and colleagues[46] have suggested the presence of a monooxygenase in the nuclear membrane which may activate nitrosamines at the site of DNA alkylation. Gold and Linder[47] have provided evidence to suggest the first product in dimethylnitrosamine metabolic activation in the cytoplasm, methylhydroxymethylnitrosamines, is more stable in the cell than predicted earlier[42] and may undergo further activation once inside the nucleus.

Other pathways for the metabolic activation of nitrosamines have been suggested,[47-54] and the complete elucidation of the process does not appear close at hand.

VII. MECHANISM OF ACTION

The carcinogenicity, and perhaps the toxicity, of the N-nitroso compounds is believed to be dependent upon the alkylation of macromolecules, especially DNA, by breakdown products of both nitrosamines and nitrosamides. Magee and Farber[39] in 1962 showed that administration of dimethylnitrosamine results in methylation of DNA in the 7 position of guanine; methylation of RNA and protein was also demonstrated.[38,39]

Swann and Magee[55] later showed that there was poor correlation between carcinogenicity and the formation of 7-alkylguanine in target organ DNA, and Ludlum[56] demonstrated that 7-methylguanylic acid in polyribonucleotides codes normally for RNA polymerase, making this methylation an unlikely basis for the somatic mutation proposed in chemical carcinogenesis. Loveless[57] pointed out that the O^6 position of DNA guanine is also quantitatively an important site of alkylation by N-nitroso compounds, and Lawley[40] and Singer[58] have shown several other sites of alkylation in DNA. Gerchman and Ludlum[59] found that O^6-methylguanine does not code as guanine, but as adenine in template for RNA polymerase, giving a biochemical basis for the proposed somatic mutation which is thought to be induced in animals treated with carcinogenic N-nitroso compounds. Goth and Rajewsky[60] and Craddock[24] have gone another step farther in demonstrating that not only is it important to alkylate DNA at a critical site, but that alkylation must persist long enough until the DNA replicates, building the error(s) into the daughter cells even though the parent carcinogen was removed (by metabolism and excretion) long before. Although the mechanism(s) for carcinogenicity by N-nitroso compounds has not yet been elucidated, the hypothesis that tumor initiation is a direct consequence of replication of DNA alkylated indirectly by N-nitroso compounds is an integral part of the generalized concept of chemical carcinogenesis enunciated by the Millers[61] several years ago; chemicals induce cancer by reacting covalently at nucleophilic sites on DNA and producing somatic mutations.

REFERENCES

1. **Magee, P. N. and Barnes, J. M.**, Carcinogenic nitroso compounds, *Adv. Cancer Res.*, 10, 163, 1967.
2. **Druckrey, H., Preussmann, R., Ivankovic, S., and Schmahl, D.**, Organotrope carcinogene Wirkungen bei 65 verschiedenen N-Nitroso-Verbindungen an BD-Ratten, *Z. Krebsforsch.*, 69, 103, 1967.
3. **Magee, P. N. and Swann, P. F.**, Nitroso compounds, *Br. Med. Bull.*, 25, 240, 1969.
4. **Magee, P. N.**, Toxicity of nitrosamines: their possible human health hazards, *Food Cosmet. Toxicol.*, 9, 207, 1971.
5. **Shank, R. C.**, Toxicology of N-nitroso compounds, *Toxicol. Appl. Pharmacol.*, 31, 361, 1975.

6. **Crosby, N. T. and Sawyer, R.,** N-nitrosamines: a review of chemical and biological properties and their estimation in foodstuffs, *Adv. Food Res.,* 22, 1, 1976.
7. **Barnes, J. M. and Magee, P. N.,** Some toxic properties of dimethylnitrosamine, *Br. J. Ind. Med.,* 11, 167, 1954.
8. **Heath, D. F. and Magee, P. N.,** Toxic properties of dialkylnitrosamines and some related compounds, *Bri. J. Ind. Med.,* 19, 276, 1962.
9. **Magee, P. N. and Vandekar, M.,** Toxic liver injury. The metabolism of dimethylnitrosamine *in vitro, Biochem. J.,* 70, 600, 1958.
10. **Brouwers, J. A. J. and Emmelot, P.,** Microsomal N-demethylation and the effect of the hepatic carcinogen dimethylnitrosamine on amino acid incorporation into the proteins of rat livers and hepatomes, *Exp. Cell Res.,* 19, 467, 1960.
11. **McLean, A. E. M. and Verschuuren, H. G.,** Effects of diet and microsomal enzyme induction on the toxicity of dimethylnitrosamine, *Br. J. Exp. Pathol.,* 50, 22, 1969.
12. **Arcos, J. C., Davies, D. L., Brown, C. E. L., and Argus, M. F.,** Repressible and inducible enzymic forms of dimethylnitrosamine-demethylase, *Z. Krebsforsch.,* 89, 181, 1977.
13. **Hard, G. C. and Butler, W. H.,** Cellular analysis of renal neoplasia: light microscope study of the development of interstitial lesions induced in the rat kidney by a single carcinogenic dose of dimethylnitrosamine, *Cancer Res.,* 30, 2806, 1970.
14. **Hard, C. C. and Butler, W. H.,** Toxicity of dimethylnitrosamine for the rat testis, *J. Pathol.,* 102, 201, 1970.
15. **Magee, P. N. and Barnes, J. M.,** The production of malignant primary hepatic tumors in the rat by feeding dimethylnitrosamine, *Br. J. Cancer,* 10, 114, 1956.
16. **Magee, P. N. and Barnes, J. M.,** The experimental production of tumors in the rat by dimethylnitrosamine (N-nitrosodimethylamine), *Acta Union Int. Contre Cancer,* 15, 187, 1959.
17. **Druckrey, H., Steinhoff, D., Preussmann, R., and Ivankovic, S.,** Carcinogenesis in rats by a single administration of methylnitrosourea and various dialkylnitrosamines, *Z. Krebsforsch.,* 66, 1, 1964.
18. **Druckrey, H. and Ivankovic, S.,** Erzeugung von Genitalkrebs bei Trachtigen Ratten, *Arzneim.-Forsch.,* 19, 1040, 1969.
19. **Leaver, D. D., Swann, P. F., and Magee, P. N.,** Induction of tumors in the rat by a single oral dose of N-nitrosomethylurea, *Br. J. Cancer,* 23, 177, 1969.
20. **Fort, L., Taper, H. S., and Brucher, J. M.,** Gastric carcinogenesis in rat induced by methylnitrosourea (MNU). Morphology and histochemistry of nucleases, *Z. Krebsforsch.,* 81, 51, 1974.
21. **Schreiber, D., Batka, H., Warzok, R., and Quentin, E.,** Induction of cardiac tumors in rats by nitrosomethylurea, *Zentralbl. Allg. Pathol. Anat.,* 115, 31, 1972.
22. **Hicks, M. and Wakefield, J. St. J.,** Rapid induction of bladder cancer in rats with nitrosomethylurea, *Chem.-Biol. Interact.,* 5, 139, 1972.
23. **Murthy, A. S. K., Vawter, G. F. and Bhaktaviziam, A.,** Neoplasms in Wistar rats after N-nitroso-N-methylurea injections, *Arch. Pathol.,* 96, 53, 1973.
24. **Craddock, V. M.,** Induction of liver tumors in rats by a single treatment with nitroso compounds given after partial hepatectomy, *Nature (London)* 245, 386, 1973.
25. **Schmahl, D.,** The role of nitrosamines in carcinogenesis — an overview, in *Safety Evaluation of Nitrosatable Drugs and Chemicals,* Parke, D. V., Ed., Pergamon Press, London, in press.
26. **Ivankovic, S. and Druckrey, H.,** Transplazentare Erzeugung Maligner Tumoren des Nervensystems. I. Athylnitroso-Harnstoff an BD IX-Ratten, *Z. Krebsforsch.,* 71, 320, 1968.
27. **Wrba, H., Pielsticker, D., and Mohr, U.,** Die Diaplazentacarcinogene Wirkung von Diathyl-Nitrosamin bei Ratten, *Naturwissenschaften,* 54, 47, 1967.
28. **Mohr, U., Althoff, J., and Authaler, A.,** Diaplacental effect of the carcinogen diethylnitrosamine in the Syrian golden hamster, *Cancer Res.,* 26, 2349, 1966.
29. **Lijinsky, W., Tomatis, L., and Wenyon, C. E.,** Lung tumors in rats treated with N-nitrosoheptamethyleneimine and N-nitrosooctamethyleneimine, *Proc. Soc. Exp. Biol. Med.,* 130, 945, 1969.
30. **Montesano, R. and Bartsch, H.,** Mutagenic and carcinogenic N-nitroso compounds: possible environmental hazards, *Mutat. Res.,* 32, 179, 1976.
31. **von Kreybig, T.,** Effect of a carcinogenic dose of methylnitrosourea on the embryonic development of the rat, *Z. Krebsforsch.,* 67, 46, 1965.
32. **Napalkov, N. P. and Alexandrov, V. A.,** On the effects of blastomogenic substances on the organism during embryogenesis, *Z. Krebsforsch.,* 71, 32, 1968.
33. **Ehrentraut, W., Juhla, H., Kupfer, G., Kupfer, M., Zintzsch, J., Rommel, P., Wahmer, M., Schnurrbusch, U., and Mockel, P.,** Experimentell erzeugte Missbildungen bei Schweinefeten durch intravenose Applikation von N-Athyl-N-nitrosoharnstoff, *Arch. Geschwulstforsch.,* 33, 31, 1969.

34. **Montesano, R. and Magee, P. N.**, Comparative metabolism *in vitro* of nitrosamine in various animal species including man, in *Chemical Carcinogenesis Essays,* IARC Scientific Publication No. 10, International Agency for Research on Cancer, Lyon, France, 1974, 39.
35. **Magee, P. N., Montesano, R., and Preussmann, R.**, N-Nitroso compounds and related carcinogens, in *Chemical Carcinogens,* Searle, C. E., Ed., ACS Monograph 173, American Chemical Society, Washington, D.C., 1976, 491.
36. **Pegg, A. E.**, Formation and metabolism of alkylated nucleosides: possible role in carcinogenesis by nitroso compounds and alkylating agents, *Adv. Cancer Res.,* 25, 195, 1977.
37. **Magee, P. N.**, Toxic liver injury. The metabolism of dimethylnitrosamine, *Biochem. J.,* 64, 676, 1956.
38. **Magee, P. N. and Hultin, T.**, Toxic liver injury and carcinogenesis. Methylation of proteins of rat liver slices by dimethylnitrosamine *in vitro, Biochem. J.,* 83, 106, 1962.
39. **Magee, P. N. and Farber, E.**, Toxic liver injury and carcinogenesis. Methylation of rat liver nucleic acids by dimethylnitrosamine *in vivo, Biochem. J.,* 83, 114, 1962.
40. **Lawley, P. D.**, Carcinogenesis by alkylating agents, in *Chemical Carcinogens,* Searle, C. E., Ed., ACS Monograph 173, American Chemical Society, Washington, D.C., 1976, 83.
41. **Druckrey, H., Preussmann, R., Schmahl, D., and Muller, M.**, Chemische Konstitution und Carcinogene Wirkung bei Nitrosaminen, *Naturwissenschaften,* 48, 134, 1961.
42. **Muller, E., Haiss, H., and Rundel, W.**, Investigations of diazomethanes. XII. Potassium methyldiazotate, a stabilized diazomethane, and monomethylnitrosamine, *Chem. Ber.,* 93, 1541, 1960.
43. **Rose, F. C.**, *Symposium on the Evaluation of Drug Toxicity,* Walpole, A. L. and Spinks, A., Eds. Churchill, London, 1958, 116.
44. **Lijinsky, W., Loo, J., and Ross, A. E.**, Mechanisms of alkylation of nucleic acids by nitrosodimethylamine, *Nature (London),* 218, 1174, 1968.
45. **Kruger, F. W.**, New aspects in metabolism of carcinogenic nitrosamines, in *Proc. 2nd Int. Symp. Princess Takamatsu Cancer Research Foundation, Topics in Chemical Carcinogenesis,* Nakahara, W., Takayama, S., Sugimura, T., and Odashima, S., Eds., University Park Press, Baltimore, 1972, 213.
46. **Argus, M. F., Hoch-Ligeti, C., Arcos, J. C., and Conney, A. H.**, Differential effects of B-naphthoflavone and pregnenolone-16α-carbonitrile on dimethylnitrosamine hepatocarcinogenesis, *J. Natl. Cancer Inst.,* 61, 441, 1978.
47. **Gold, B. and Linder, W. B.**, α-Hydroxynitrosamines: transportable metabolites of dialkylnitrosamines, *J. Am. Chem. Soc.,* 101, 6772, 1979.
48. **Lake, B. G., Philips, J. C., Heading, C. E., and Gangolli, S. D.**, Studies on the *in vitro* metabolism of dimethylnitrosamine by rat liver, *Toxicology,* 5, 297, 1976.
49. **Gangolli, S. D.**, Metabolic activation and detoxication of nitroso compounds, in *Safety Evaluation of Nitrosatable Drugs and Chemicals,* Parke, D. V., Ed., Pergamon Press, London, in press.
50. **Olah, G. A., Donovan, D. J., and Keefer, L. K.**, Carcinogen chemistry. I. Reactions of protonated dialkylnitrosamines leading to alkylating and aminoalkylating agents of potential metabolic significance, *J. Natl. Cancer Inst.,* 54, 465, 1975.
51. **Grilli, S. and Prodi, G.**, Identification of dimethylnitrosamine metabolites *in vitro, Gann,* 66, 473, 1975.
52. **Grilli, S., Tosi, M. R., and Prodi, G.**, Degradation of dimethylnitrosamine catalyzed by physical and chemical agents, *Gann,* 66, 481, 1975.
53. **Fahmy, O. G., Fahmy, M. J., and Wiessler, M.**, α-Acetoxy-dimethylnitrosamine: a proximate metabolite of the carcinogenic amine, *Biochem. Pharmacol.,* 24, 1145, 1975.
54. **Daugherty, J. P., Clapp, N. K., Zehfus, M. M., and Brock, S. E.**, Radioactive components in the acid-soluble fraction of mouse liver cytosol after dimethylnitrosamine (methyl-^{14}C) administration, *Gann,* 68, 697, 1977.
55. **Swann, P. F. and Magee, P. N.**, The alkylation of nucleic acids of the rat by N-methyl-N-nitrosourea, dimethylnitrosamine, dimethylsulfate and methylmethanesulfonate, *Biochem. J.,* 110, 39, 1968.
56. **Ludlum, D. B.**, The properties of 7-methylguanine-containing templates for ribonucleic acid polymerase, *J. Biol. Chem.,* 245, 477, 1970.
57. **Loveless, A.**, Possible relevance of O^6-alkylation of deoxyguanosine to the mutagenicity and carcinogenicity of nitrosamines and nitrosamides, *Nature (London),* 233, 206, 1969.
58. **Singer, B.**, The chemical effects of nucleic acid alkylation and their relationship to mutagenesis and carcinogenesis, *Prog. Nucleic Acid Res. Mol. Biol.,* 152, 19, 1975.
59. **Gerchman, L. L. and Ludlum, D. B.**, The properties of O^6-methylguanine in template for RNA polymerase, *Biochim. Biophys. Acta,* 308, 310, 1973.
60. **Goth, R., and Rajewsky, M. F.**, Molecular and cellular mechanisms associated with pulse-carcinogenesis in the rat nervous system by ethylnitrosourea: ethylation of nucleic acids and elimination rates of ethylated bases from the DNA of different tissues, *Z. Krebsforsch.,* 82, 37, 1974.

61. Miller, J. A., Carcinogenesis by chemicals: an overview, G. H. A. Clowes Memorial Lecture, *Cancer Res.*, 30, 559, 1970.
62. Druckrey, H., Preussmann, R., Ivankovic, S., Schmidt, C. H., Mennel, H. D., and Stahl, K. W., Selective induction of bladder cancer in rats with dibutyl- and N-butyl-N-(4-hydroxybutyl)-nitrosamine, *Z. Krebsforsch.*, 66, 280, 1964.
63. Druckrey, H., Preussmann, R., Blum, G., Ivankovic, S., and Afkham, J., Erzeugung von Karzinomen der Speiserohre duroh unsymmetrische Nitrosamine, *Naturwissenschaften*, 50, 100, 1963.
64. Schmahl, D., Enstehung, Wachstum und Chemotherapie maligner Tumoren, *Arzneim.-Forsch.*, 13, Beiheft., 1963.
65. Lijinsky, W., Lee, K. Y., Tomatis, L., and Butler, W. H., Nitrosoazetidine. A potent carcinogen of low toxicity, *Naturwissenschaften*, 54, 518, 1967.
66. Garcia, H., Keefer, L., Lijinsky, W., and Wenyon, C. E. M., Carcinogenicity of nitrosothiomorpholine and 1-nitrosopiperazine in rats, *Z. Krebsforsch.*, 74, 179, 1970.
67. Wiessler, M. and Schmahl, D., Zur Carcinogenen Wirkung von N-Nitroso-Verbindungen. II. Mitteilung. S(+) und R(−)-N-Nitroso-2-Methyl-piperidin, *Z. Krebsforsch.*, 79, 118, 1973.
68. Druckrey, H., Preussmann. R., Afkham, J., and Blum, G., Erzeugung von Lungenkrebs durch Methylnitrosourethan bei intravenoser Gabe an Ratten, *Naturwissenschaften*, 49, 451, 1962.
69. Goodall, C. M., Lijinsky, W., and Tomatis, L., Tumorigenicity of N-nitrosohexamethyleneimine, *Cancer Res.*, 28, 1217, 1968.
70. Zak, F. G., Holzner, J. H., Singer, E. J., and Popper, H., Renal and pulmonary tumors in rats fed dimethylnitrosamine, *Cancer Res.*, 20, 96, 1960.
71. Druckrey, H., Steinhoff, D., Preussmann, R., and Ivankovic, S., Cancer production by single dosage of N-methyl-N-nitrosourea and several dialkylnitrosamines, *Naturwissenschaften*, 50, 735, 1963.
72. Druckrey, H., Ivankovic, S., Mennel, H. D., and Preussman, R., Selective production of carcinomas of the nasal cavity in rats by N,N'-dinitrosopiperazine, nitrosopiperidine, nitrosomorpholine, methylallylnitrosamine, dimethylnitrosamine, and methylvinylnitrosamine, *Z. Krebsforsch.*, 66, 138, 1964.
73. Takayama, S. and Oota, K., Malignant tumors induced in mice fed with N-nitrosodimethylamine, *Gann*, 54, 465, 1963.
74. Toth, B., Magee, P. N., and Shubik, P., Carcinogenesis study with dimethylnitrosamine administered orally to adult and subcutaneously to newborn BALB/c mice, *Cancer Res.*, 24, 1712, 1964.
75. Tomatis, L., Magee, P. N., and Shubik, P., Induction of liver tumors in the Syrian golden hamster by feeding dimethylnitrosamine, *J. Natl. Cancer Inst.*, 33, 341, 1964.
76. Mohr, U., Haas, H., and Hilfrich, J., The carcinogenic effects of dimethylnitrosamine and nitrosomethylurea in European hamsters (*Cricetus cricetus* L.), *Br. J. Cancer*, 29, 359, 1974.
77. Homberger, F., Handler, A. H., Soto, E., Hseuh, S. S., VanDongen, C. G., and Russfield, A. B., Adenocarcinoma of glandular stomach following 3-methylcholanthrene, N-nitrosodiethylamine or N-nitrosodimethylamine feeding in carcinogen-susceptible inbred Syrian hamster, *J. Natl. Cancer Inst.*, 57, 141, 1976.
78. Koppang, N., The hepatotoxic and carcinogenic actions of dimethylnitrosamine (DMNA), *Acta Pathol. Microbiol. Scand. Suppl.*, 215, 30, 1970.
79. Halver, J. E., Johnson, C. L., and Ashley, L. M., Dietary carcinogens induce fish hepatoma, *Fed. Proc. Fed. Am. Soc. Exp. Biol.*, 21, 390, 1962.
80. Wiessler, M. and Schmahl, D., On the carcinogenic action of N-nitrosocompounds. 6th communication. Methoxymethyl-methylnitrosamine, 1-(methoxy)-ethyl-ethylnitrosamine, methoxymethyl-ethylnitrosamine, 1-(methoxy)-ethyl-methylnitrosamine, and N-nitrosoxazolidine, *Z. Krebsforsch.*, 88, 25, 1976.
81. Wogan, G. N., Paglialunga, S., Archer, M. C., and Tannenbaum, S. R., Carcinogenicity of nitrosation products of ephedrine, sarcosine, folic acid and creatine, *Cancer Res.*, 35, 1981, 1975.
82. Wiessler, M. and Schmahl, D., On the carcinogenic action of N-nitroso-compounds. 5th Communication: Acetoxymethyl-methyl-nitrosamine, *Z. Krebsforsch.*, 85, 47, 1976.
83. Joshi, S. R., Rice, J. M., Wenk, M. L., Roller, P. P., and Keefer, L. K., Selective induction of intestinal tumors in rats by methyl(acetoxymethyl) nitrosamine, an ester of the presumed reactive metabolite of dimethylnitrosamine, *J. Natl. Cancer Inst.*, 58, 1531, 1977.
84. Reznick, G., Mohr, U., and Kruger, F. W., Carcinogenic effect of di-*n*-propylnitrosamine, β-hydroxypropyl-*n*-propylnitrosamine, and methyl-*n*-propylnitrosamine on Sprague Dawley rats, *J. Natl. Cancer Inst.*, 54, 937, 1975.
85. Hecht, S. S., Chen, C.-h. B., Ohmori, T., and Hoffmann, D., Comparative carcinogenicity in F344 rats of the tobacco-specific nitrosamines, N'-nitrosonornicotine and 4-(N-methyl-N-nitrosamino)-1-(3-pyridyl)-1-butanone, *Cancer Res.*, 40, 298, 1980.
86. Eisenbrand, G., Preussmann, R., and Schmahl, D., Carcinogenicity of N-nitroso-ephedrine in rats, *Cancer Lett.*, 5, 103, 1978.

87. **Okada, M., Suzuki, E., and Hashimoto, Y.**, Carcinogenicity of N-nitrosamines related to N-butyl-N-(4-hydroxybutyl) nitrosamine and N,N-dibutylnitrosamine in ACI/N rats, *Gann,* 67, 825, 1976.
88. **Lijinsky, W., Taylor, H. W., Mangino, M., and Singer, G. M.**, Carcinogenesis of nitrosomethylundecylamine in Fischer rats, *Cancer Lett.,* 5, 209, 1978.
89. **Okada, M., Suzuki, E., and Mochizuki, M.**, Possible important role of urinary N-methyl-N-(3-carboxypropyl) nitrosamine in induction of bladder tumors in rats by N-methyl-N-dodecylnitrosamine, *Gann,* 67, 771, 1976.
90. **Althoff, J. and Lijinsky, W.**, Urinary bladder neoplasms in Syrian hamsters after administration of N-nitroso-N-methyl-*n*-dodecylamine, *Z. Krebsforsch.,* 90, 227, 1977.
91. **Druckrey, H., Preussmann, R., Schmahl, D., and Blum, G.**, Carcinogenic effect of N-methyl-N-nitrosoaniline, *Naturwissenschaften,* 48, 722, 1961.
92. **Schweinsberg, F., Schott-kollat, P., and Burkle, G.**, Change of toxicity and carcinogenicity of N-methyl-N-nitrosobenzylamine in rats by methyl-substitution in the phenyl residues, *Z. Krebsforsch.,* 88, 231, 1977.
93. **Anderson, L. M., Giner-Sorolla, A., Breenbaum, J., and Good, R. A.**, A new carcinogen — N[6](methylnitroso)adenosine, *Proc. Am. Assoc. Cancer Res.,* 18, 79, 1977.
94. **Schmahl, D., Preussman, R., and Hamperl, H.**, Hepatocarcinogenic effect of diethylnitrosamine administered orally to rats, *Naturwissenschaften,* 49, 89, 1960.
95. **Druckrey, H., Schildbach, A., Schmahl, D., Preussman, R., and Ivankovic, S.**, Quantitative Analyse der carcinogenen Wirkung von Diathylnitrosamin, *Arzneim.-Forsch.,* 13, 841, 1963.
96. **Schmahl, D., Thomas, C., and Konig, K.**, Experiments with diethylnitrosamine on cancer production in mice, *Naturwissenschaften,* 50, 407, 1963.
97. **Shvemberger, I. N.**, Induction of malignant esophagus and stomach tumors in C_3HA mice with N-nitrosodiethylamine, *Vopr. Onkol.,* 11, 74, 1965.
98. **Hoffmann, F. and Graffi, A.,** *Arch. Geschwulstforsch.,* 23, 274, 1964; as cited by **Magee, P. N. and Barnes, J. M.,** *Adv. Cancer Res.,* 10, 163, 1967.
99. **Herrold, K. M. and Durham, L. J.**, Induction of tumors in the Syrian hamster with diethylnitrosamine (N-nitrosodiethylamine), *Cancer Res.,* 23, 773, 1963.
100. **Dontenwill, W. and Mohr, U.**, Carcinomas of the respiratory tract after treatment of golden hamsters with diethylnitrosamines, *Z. Krebsforsch.,* 64, 305, 1961.
101. **Herrold, K. M.**, Induction of olfactory neuroepithelial tumors in Syrian hamsters by diethylnitrosamine, *Cancer,* 17, 114, 1964.
102. **Druckrey, H. and Steinhoff, D.,** *Naturwissenschaften,* 49, 497, 1962; as cited by **Magee, P. N. and Barnes, J. M.,** *Adv. Cancer Res.,* 10, 163, 1967.
103. **Argus, M. F. and Hoch-Ligeti, C.**, Induction of malignant tumors in the guinea pig by oral administration of diethylnitrosamine, *J. Natl. Cancer Inst.,* 30, 533, 1963.
104. **Schmahl, D. and Thomas, C.**, Carcinogenic effect of diethylnitrosamine on rabbit liver, *Naturwissenschaften,* 52, 165, 1965.
105. **Rapp, H. J., Carleton, J. H., Crisler, C., and Nadel, E. M.**, Induction of malignant tumors in the rabbit by oral administration of diethylnitrosamine, *J. Natl. Cancer Inst.,* 34, 453, 1965.
106. **Graw, J. J., Berg, H., and Schmahl, D.**, Carcinogenic and hepatotoxic effects of diethylnitrosamine in hedgehogs, *J. Natl. Cancer Inst.,* 53, 589, 1974.
107. **Schmahl, D., Thomas, C., and Scheid, G.**, Carcinogenic effect of diethylnitrosamine in the dog, *Naturwissenschaften,* 51, 466, 1964.
108. **Schmahl, D., Osswald, H., and Mohr, U.**, Hepatotoxic and carinogenic effect of diethylnitrosamine in pigs, *Naturwissenschaften,* 54, 341, 1967.
109. **O'Gara, R. W. and Kelly, M. G.**, Induction of hepatomas in monkeys given N-nitrosodiethylamine (DENA), *Proc. Am. Assoc. Cancer Res.,* 6, 50, 1965.
110. **Kelly, M. G., O'Gara, R. W., Adamson, R. H., Gadekar, K., Botkin, C. C., Reese, W. H., and Kerber, W. T.**, Induction of hepatic cell carcinomas in monkeys with N-nitrosodiethylamine, *J. Natl. Cancer Inst.,* 36, 323, 1966.
111. **Halver, J. A.,** *Res. Rep. U.S. Fish Wildl. Serv.,* 160, 22, 1963; as cited by **Magee, P. N., Montesano, R., and Preussmann, R.,** *Chemical Carcinogens,* Searle, C. E., Ed., ACS Monograph 173, American Chemical Society, Washington, D.C., 1976, 491.
112. **Stanton, M. F.**, Diethylnitrosamine-induced hepatic degeneration and neoplasia in the aquarium fish, *Brachydanio rerio, J. Natl. Cancer Inst.,* 34, 117, 1965.
113. **Thomas, C. and Schmahl, D.**, Morphology of kidney tumors in the rat, *Z. Krebsforsch.,* 66, 125, 1964.
114. **Althoff, J., Grandjean, C., Russel, L., and Pour, P.**, Vinylethylnitrosamine: a potent respiratory carcinogen in Syrian hamsters, *J. Natl. Cancer Inst.,* 58, 439, 1977.
115. **Schmahl, D., Thomas, C., and Scheid, G.**, Carcinogenic action of ethylbutylnitrosamine in mice, *Naturwissenschaften,* 50, 717, 1963.

116. Pour, P., Kruger, F. W., Cardesa, A., Althoff, J., and Mohr, U., Carcinogenic effect of di-*n*-propylnitrosamine in Syrian golden hamsters, *J. Natl. Cancer Inst.*, 51, 1019, 1973.
117. Pour, P., Althoff, J., Cardesa, A., Kruger, F. W., and Mohr, U., Effect of beta-oxidized nitrosamine on Syrian golden hamster. II. 2-Oxopropyl-*n*-propylnitrosamine, *J. Natl. Cancer Inst.*, 52, 1869, 1974.
118. Mohr, U., Reznick, G., and Pour, P., Carcinogenic effects of diisopropanolnitrosamine in Sprague-Dawley rats, *J. Natl. Cancer Inst.*, 58, 361, 1977.
119. Konishi, Y., Denda, A., Kondo, H., and Takahashi, S., Lung carcinomas induced by oral administration of N-bis-(2-hydroxypropyl)nitrosamine in rats, *Gann*, 67, 773, 1976.
120. Kruger, F. W., Pour, P., and Althoff, J., Induction of pancreas tumors by di-iso-propanolnitrosamine, *Naturwissenschaften*, 61, 328, 1974.
121. Rao, M. S. and Reddy, J. K., Induction of malignant vascular tumors of the liver in guinea pigs treated with 2,2,′-dihydroxy-di-*n*-propylnitrosamine, *J. Natl. Cancer Inst.*, 58, 387, 1977.
122. Gingell, R., Wallcave, L., Nagel, D., Kupper, R., and Pour, P., Metabolism of the pancreatic carcinogens N-nitroso-bis-(2-oxopropyl)amine and N-nitroso-bis(2-hydroxypropyl)amine in Syrian hamsters, *J. Natl. Cancer Inst.*, 57, 1175, 1976.
123. Pour, P., Althoff, J., Kruger, F. W., and Mohr, U., A potent pancreatic carcinogen in Syrian hamsters: N-nitroso-bis(2-oxopropyl)amine, *J. Natl. Cancer Inst.*, 58, 1449, 1977.
124. Pour, P., Althoff, J., Gingell, R., Kupper, R., Kruger, F., and Mohr, U., N-Nitroso-bis-(2-acetoxypropyl)amine as a further pancreatic carcinogen in Syrian golden hamsters, *Cancer Res.*, 36, 2877, 1976.
125. Biancifiori, C., Montesano, R., and Bolis, G. B., Indagini sulla cancerogenesi da sodio nitrito elo etambitolo in topi Balb/c/cb/se, *Lav. Anat. Pat. Perugia*, 35, 45, 1975.
126. Althoff, J., Grandjean, C., and Gold, B., Dialkylnitrosamine: a potent respiratory carcinogen in Syrian golden hamsters, *J. Natl. Cancer Inst.*, 59, 1569, 1977.
127. Druckrey, H., Preussmann, R., Schmahl, D., and Muller, M., Erzeugung von Blasenkrebs an Ratten mit N,N-Dibutylnitrosamin, *Naturwissenschaften*, 49, 19, 1962.
128. Fujii, K., Odashima, S., and Okada, M., Induction of tumors by administration of N-dibutylnitrosamine and derivatives to infant mice, *Br. J. Cancer*, 35, 610, 1977.
129. Reznik-Schuller, H. and Mohr, U., Ultrastructure of N-nitrosodibutylamine-induced tumors of the nasal cavity in the European hamster, *J. Natl. Cancer Inst.*, 57, 401, 1976.
130. Reznik, G., Mohr, U., and Kmoch, N., Carcinogenic effects of different nitrosocompounds in Chinese hamsters: N-dibutylnitrosamine and N-nitrosomethylurea, *Cancer Lett.*, 1, 183, 1976.
131. Druckrey, H. and Preussmann, R., Production of lung cancer in rats by means of subcutaneous injection of N,N-diamylnitrosamine, *Naturwissenschaften*, 49, 111, 1962.
132. Greenblatt, M. and Lijinsky, W., Nitrosamine studies: neoplasms of liver and genital mesothelium in nitrosopyrrolidine-treated MRC rats, *J. Natl. Cancer Inst.*, 48, 1687, 1972.
133. Greenblatt, M. and Lijinsky, W., Failure to induce tumors in Swiss mice after concurrent administration of amino acids and sodium nitrite, *J. Natl. Cancer Inst.*, 48, 1389, 1972.
134. Dontenwill, W., Experimental studies on the organotropic effect of nitrosamines in the respiratory tract, *Food Cosmet. Toxicol.*, 6, 571, 1968.
135. Lijinsky, W. and Taylor, H. W., The effect of substituents on the carcinogenicity of N-nitrosopyrrolidine in Sprague-Dawley rats, *Cancer Res.*, 36, 1988, 1976.
136. Singer, G. M. and Taylor, H. W., Carcinogenicity of N-nitroso-nornicotine in Sprague-Dawley rats, *J. Natl. Cancer Inst.*, 57, 1275, 1976.
137. Lijinsky, W., Interaction with nucleic acids of carcinogenic and mutagenic N-nitroso compounds, *Prog. Nucleic Acid Res. Mol. Biol.*, 17, 247, 1976.
138. Sander, J. and Burkle, G., Induction of malignant tumors in rats by oral administration of 2-imidazolidine and nitrite, *Z. Krebsforsch.*, 75, 301, 1971.
139. Pelfrene, A., Mirvish, S. S., and Garcia, H., Carcinogenic action of ethylnitrosocyanamide, 1-nitrosohydantoin and ethylnitrosourea in the rat, *Proc. Am. Assoc. Cancer Res.*, 16, 117, 1975.
140. Lijinsky, W. and Taylor, H. W., Carcinogenicity of methylated nitrosopiperidines, *Int. J. Cancer*, 16, 318, 1975.
141. Mirvish, S. S. and Kaufman, L., A study of nitrosamines and S-carboxyl derivatives of cysteine as lung carcinogens in adult SWR mice, *Int. J. Cancer*, 6, 69, 1970.
142. Takayama, S., Induction of tumors in ICR mice with N-nitrosopiperidine, especially in forestomach, *Naturwissenschaften*, 56, 142, 1969.
143. Dontenwill, W. and Mohr, U., The organotropic effects of nitrosamines, *Z. Krebsforsch.*, 65, 166, 1962.
144. Althoff, J., Wilson, R., Cardesa, A., and Pour, P., Comparative studies of neoplastic response to a single dose of nitroso compounds. III. The effect of N-nitrosopiperidine and N-nitrosomorpholine in Syrian golden hamsters, *Z. Krebsforsch.*, 81, 251, 1974.
145. O'Gara, R. W., Adamson, R. H., and Dolgard, D. W., Induction of tumors in subhuman primates by two nitrosamine compounds, *Proc. Am. Assoc. Cancer Res.*, 11, 60, 1970.

146. **Wiessler, M. and Schmahl, D.,** The carcinogenic action of N-nitroso-compounds. Second Communication. S(+) and R(−)-N-nitroso-2-methyl-piperidine, *Z. Krebsforsch.,* 79, 118, 1973.
147. **Boyland, E., Roe, F. J. C., Gonod, J. W., and Mitchley, B. C. V.,** The carcinogenicity of nitrosoanabasine, a possible constituent of tobacco smoke, *Br. J. Cancer,* 18, 265, 1964.
148. **Shank, R. C. and Newberne, P. M.,** Dose-response study of the carcinogenicity of dietary sodium nitrite and morpholine in rats and hamsters, *Food Cosmet. Toxicol.,* 14, 1, 1976.
149. **Bannasch, P. and Muller, M. A.,** Lichtmikroskopische Untersuchungen uber die Wirkung von N-Nitrosomorpholin auf die Lebervon Ratte und Maus, *Arzneim.-Forsch.,* 14, 805, 1964.
150. **Lijinsky, W. and Taylor, H. W.,** Increased carcinogenicity of 2,6-dimethylnitrosomorpholine compared with nitrosomorpholine in rats, *Cancer Res.,* 35, 2123, 1975.
151. **Mohr, U., Reznik, G., Emminger, E., and Lijinsky, W.,** Induction of pancreatic duct carcinomas in the Syrian hamster with 2,6-dimethylnitrosomorpholine, *J. Natl. Cancer Inst.,* 58, 429, 1977.
152. **Lijinsky, W. and Taylor, H. W.,** Carcinogenicity of methylated dinitrosopiperazine in rats, *Cancer Res.,* 35, 1270, 1975.
153. **Love, L. A., Lijinsky, W., Keefer, L. K., and Garcia, H.,** Chronic oral administration of 1-nitrosopiperazine at high doses to MRC rats, *Z. Krebsforsch.,* 89, 69, 1977.
154. **Schmahl, D. and Thomas, C.,** Induction of lung and liver tumors in mice with N,N'-dinitrosopiperazine, *Z. Krebsforsch.,* 67, 11, 1965.
155. **Mirvish, S. S. and Garcia, H.,** 1-Nitroso-5,6,-dihydrouracil: induction of liver cell carcinomas and kidney adenomas in the rat, *Z. Krebsforsch.,* 79, 304, 1973.
156. **Goodall, C. M., Lijinsky, W., and Tomatis, L.,** Tumorigenicity of N-nitrosohexamethyleneimine, *Cancer Res.,* 28, 1217, 1968.
157. **Schmahl, D.,** Zur carcinogen Wirkung von N-Nitrosohexamethyleninmin, *Naturwissenschaften,* 55, 653, 1968.
158. **Althoff, J., Pour, P., Cardesa, A., and Mohr, U.,** Comparative studies of neoplastic response to a single dose of nitroso compounds. I. The effect of N-nitrosohexamethyleneimine in Syrian golden hamsters and Swiss mice, *Z. Krebsforsch.,* 78, 78, 1972.
159. **Lijinsky, W., Tomatis, L., and Wenyon, C. E. M.,** Lung tumors in rats treated with N-nitrosoheptamethyleneimine and N-nitrosooctamethyleneimine, *Proc. Soc. Exp. Biol. Med.,* 130, 945, 1969.
160. **Lijinsky, W., Ferraro, A., Montesano, R., and Wenyon, C. E. M.,** Tumorigenicity of cyclic nitrosamines in Syrian golden hamsters, *Z. Krebsforsch.,* 74, 185, 1970.
161. **Druckrey, H., Preussmann, R., Schmahl, D., and Muller, M.,** Production of a gastric carcinoma by nitrosamines in rats, *Naturwissenschaften,* 48, 165, 1961.
162. **Druckrey, H., Ivankovic, S., and Preussmann, R.,** Selective induction of malignant tumors in the brain and spinal cord of the rat by N-methyl-N-nitrosourea, *Z. Krebsforsch.,* 66, 389, 1965.
163. **Frei, J. V.,** Toxicity, tissue changes, and tumor induction in inbred Swiss mice by methylnitrosamine and -amide compounds, *Cancer Res.,* 30, 11, 1970.
164. **Terracini, B. and Testa, M. C.,** Carcinogenicity of a single administration of N-nitrosomethylurea: a comparison between newborn and 5-week-old mice and rats, *Br. J. Cancer,* 24, 588, 1970.
165. **Graffi, A. and Hoffmann, F.,** A strong carcinogenic effect of methylnitrosourea on the mouse skin in the drop test, *Acta Biol. Med. Ger.,* 16, K1, 1966.
166. **Graffi, A. and Hoffmann, F.,** *Acta Biol. Med. Ger.,* 17, K33, 1966; as cited by **Magee, P. N., Montesano, R., and Preussmann, R.,** *Chemical Carcinogens,* Searle, C. E., Ed., ACS Monograph 173, American Chemical Society, Washington, D.C., 1976, 491.
167. **Terracini, B. and Stramignoni, A.,** Malignant lymphomas and renal changes in Swiss mice given nitrosomethylurea, *Eur. J. Cancer,* 3, 435, 1967.
168. **Kelly, M. G., O'Gara, R. W., Yancey, S. T., and Botkin, C.,** Carcinogenicity of 1-methyl-1-nitrosourea in newborn mice and rats, *J. Natl. Cancer Inst.,* 41, 619, 1968.
169. **Joshi, V. V. and Frei, J.,** Gross and microscopic changes in the lymphoreticular system during genesis of malignant lymphoma induced by a single injection of methylnitrosourea in adult mice, *J. Natl. Cancer Inst.,* 44, 379, 1970.
170. **Joshi, V. V. and Frei, J.,** Effects of dose and schedule of methylnitrosourea on incidence of malignant lymphoma in adult female mice, *J. Natl. Cancer Inst.,* 45, 335, 1970.
171. **Eckert, H. and Seidler, E.,** Tumorigenic effect of methylnitrosourea in mice, *Arch. Geschwulstforsch.,* 38, 7, 1971.
172. **Denlinger, R. H., Koestner, A., and Wechsler, W.,** Induction of neurogenic tumors in C3HeB/FeJ mice by nitrosourea derivatives: observations by light microscopy, tissue culture, and electron microscopy, *Int. J. Cancer,* 13, 559, 1974.
173. **Graffi, A., Hoffmann, F., and Schutt, M.,** N-Methyl-N-nitrosourea as a strong topical carcinogen when painted on skin of rodents, *Nature (London),* 214, 611, 1967.

174. Herrold, K. M., Carcinogenic effect of N-methyl-N-nitrosourea administered subcutaneously to Syrian hamsters, *J. Pathol. Bacteriol.*, 92, 35, 1966.
175. Herrold, K. M., Upper respiratory tract tumors induced in Syrian hamsters by N-methyl-N-nitrosourea, *Int. J. Cancer*, 6, 217, 1970.
176. Druckrey, H., Ivankovic, S., Bucheler, J., Preussman, R., and Thomas, C., Cancer of the stomach and pancreas in quinea-pigs, induced by methylnitroso-urea and -urethane, *Z. Krebsforsch.*, 71, 167, 1968.
177. Osske, G., Schreiber, D., Schneider, J., and Janisch, W., Enzymhistochemische Untersuchengen an experimentellen Glioblastomen des kaninchens, *Eur. J. Cancer*, 5, 525, 1969.
178. Schreiber, D., Janisch, W., Warzok, R., and Tausch, H., Experimental induction of brain and spinal cord tumors in rabbits by methylnitrosourea (MNU), *Z. Ges. Exp. Med.*, 150, 76, 1969.
179. Stavrou, D., Morphology and histochemistry of experimental brain tumours in rabbits, *Z. Krebsforsch.*, 73, 98, 1969.
180. Kleinhues, P., Zulch, K. J., Matsumoto, S., and Radke, U., Morphology of malignant gliomas induced in rabbits by systemic application of N-methyl-N-nitrosourea, *Z. Neurol.*, 198, 65, 1970.
181. Warzok, R., Schneider, J., Schreiber, D., and Janisch, W., Experimental brain tumours in dogs, *Experientia*, 26, 303, 1970.
182. Stavrou, D. and Haglid, K. G., Experimentell induzierte Tumoren des peripheren Nervensystems beim Hund, *Naturwissenschaften*, 59, 317, 1972.
183. Druckrey, H., Landschutz, C. H., Preussmann, R., and Ivankovic, S., Stomach cancer and neurogenic tumors induced by oral administration of methyl-nitrosobiuret (MNB) in rats, *Z. Krebsforsch.*, 75, 229, 1971.
184. Ivankovic, S., Druckrey, H., and Preussmann, R., Induction of tumors of the peripheral and central nervous system by trimethylnitrosourea in the rat, *Z. Krebsforsch.*, 66, 541, 1965.
185. Maekawa, A., Odashima, S., and Nakadate, M., Induction of tumors in stomach and nervous-system of ACI-N rat by continuous oral-administration of 1-methyl-3-acetyl-1-nitrosourea, *Z. Krebsforsch.*, 86, 195, 1976.
186. Schoental, R., Carcinogenic action of diazomethane and of nitroso-N-methyl urethane, *Nature (London)*, 188, 420, 1960.
187. Schoental, R. and Magee, P. N., Induction of squamous carcinoma of the lung and of the stomach and oesophagus by diazomethane and N-methyl-N-nitroso-urethane, respectively, *Br. J. Cancer*, 16, 92, 1962.
188. Druckrey, H., Preussman, R., Afkham, J., and Blum, G., Production of lung cancer in rats by intravenous administration of methylnitrosourethan, *Naturwissenschaften*, 49, 451, 1962.
189. Schoental, R., Induction of tumours of the stomach in rats and mice by N-nitroso-N-alkylurethanes, *Nature (London)*, 199, 190, 1963.
190. Lijinsky, W., Garcia, H., Keefer, L., Loo, J., and Ross, A. E., Carcinogenesis and alkylation of rat liver nucleic acids by nitrosomethylurea and nitrosoethylurea administered by intraportal injection, *Cancer Res.*, 32, 893, 1972.
191. Druckrey, H., Ivankovic, S., and Gimmy, J., Cancerogenic effects of methyl- and ethyl-nitrosourea (MNU and ENU) at single intracerebral and intracarotidal injection in newborn and young BD-rats, *Z. Krebsforsch.*, 79, 282, 1973.
192. Druckrey, H., Schagen, B., and Ivankovic, S., Induction of neurogenic malignancies by one single dose of ethyl-nitrosourea (ENU) given to newborn and juvenile BDIX-strain rats, *Z. Krebsforsch.*, 74, 141, 1970.
193. Hadjiolov, D., Thymic lymphoma and myeloid leukemia in the rat induced with ethylnitrosourea, *Z. Krebsforsch.*, 77, 98, 1972.
194. Frei, J. V., Tumor induction by low molecular weight alkylating agents, *Chem. Biol. Interact.*, 3, 117, 1971.
195. Lombard, L. S. and Vesselinovitch, S. D., Pathogenesis of renal tumors in mice treated with ethylnitrosourea, *Proc. Soc. Am. Assoc. Cancer Res.*, 12, 55, 1971.
196. Rice, J. M. and Davidson, J. K., Spontaneous regression of chemically induced malignant lymphoma in Swiss mice, *Cancer Res.*, 31, 2008, 1971.
197. Vesselinovitch, S. D., Lombard, L. S., Mihailovich, N., Itze, L., and Rice, J. M., Broad spectrum carcinogenicity of ethylnitrosourea in the newborn and infant mice, *Proc. Soc. Am. Assoc. Cancer Res.*, 12, 56, 1971.
198. Searle, C. E. and Jones, E. L., Tumours of the nervous system in mice treated neonatally with N-ethyl-N-nitrosourea, *Nature (London)*, 240, 559, 1972.
199. Vesselinovitch, S. D., Itze, L., Mihailovich, N., Rao, K. V. N., and Manojlovski, B., Role of hormonal environment, partial hepatectomy, and dose of ethylnitrosourea in renal carcinogenesis, *Cancer Res.*, 33, 339, 1973.
200. Pelfrene, A., Mirvish, S. S., and Gold, B., Induction of malignant bone tumors in rats by 1-(2-hydroxyethyl)-1-nitrosourea, *J. Natl. Cancer Inst.*, 56, 445, 1976.

201. **Schoental, R.,** Induction of intestinal tumours by N-ethyl-N-nitrosourethane, *Nature (London)*, 208, 300, 1965.
202. **Ogui, T., Nakadate, M., and Odashima, S.,** Induction of tumors in female Donryu rats by a single administration of 1-propyl-1-nitrosourea, *Gann*, 67, 121, 1976.
203. **Maekawa, A., Kamiya, S. and Odashima, S.,** Tumors of upper digestive-tract of ACI-N rats given N-propyl-N-nitrosourethan in drinking water, *Gann*, 67, 549, 1976.
204. **Okada, K., Teratani, M., Takahashi, A., Uchimo, H., Kajihara, M., and Ohkita, T.,** Rat myeoloid-leukemia model of human disease — hematological characteristics of BNU induced W-FU rat leukemia L1504 and L1005, *Acta Haematol. J.*, 40, 72, 1977.
205. **Takizawa, S.,** Influence of sex hormones on kidney tumors induced in rats with N-butylnitrosourea, *Gann*, 67, 33, 1976.
206. **Odashima, S., Hashimoto, Y., Ogiu, T., and Maekawa, A.,** Carcinogenic effect of 1-butyl-1-nitrosourea on female Sprague-Dawley rats, *Gann*, 66, 615, 1975.
207. **Takeuchi, M., Maekawa, A., Tada, K., and Odashima, S.,** Leukemias and vaginal tumors induced in female Donryu rats by continuous administration of 1-butyl-3-,3-dimethyl-1-nitrosourea in drinking water, *J. Natl. Cancer Inst.*, 56, 1177, 1976.
208. **Takeuchi, M., Kamiya, S. and Odashima, S.,** Induction of tumors of the forestomach, esophagus, pharynx, and oral cavity of the Donryu rat given N-butyl-N-nitrosourethan in the drinking water, *Gann*, 65, 227, 1974.
209. **Fujii, M.,** Carcinogenic effect of N-butyl-N-nitrosourethan on CDF_1 mice, *Gann*, 67, 231, 1976.
210. **Schoental, R.,** Carcinogenic activity of N-methyl-N-nitroso-N'-nitroguanidine, *Nature (London)*, 209, 726, 1966.
211. **Sugimura, T. and Fugimura, S.,** Tumour production in glandular stomach of rat by N-methyl-N'-nitro-N-nitrosoguanidine, *Nature (London)*, 216, 943, 1967.
212. **Koyama, Y., Omori, K., Hirota, T., Sano, R., and Ishihara, K.,** Leiomyosarcomas of the small intestine induced in dogs by N-methyl-N'-nitro-N-nitrosoguanidine, *Gann*, 67, 241, 1976.
213. **Sugimura, T., Tanaka, N., Kawachi, T., Kogure, K., Fugimura, S., and Shimosato, Y.,** Production of stomach cancer in dogs by N-methyl-N'-nitro-N-nitrosoguanidine, *Gann*, 62, 67, 1971.
214. **Takayama, S., Kuwabora, N., Azama, Y., and Sugimura, T.,** Skin tumors in mice painted with N-methyl-N'-nitro-N-nitrosoguanidine and N-ethyl-N'-nitro-N-nitrosoguanidine, *J. Natl. Cancer Inst.*, 46, 973, 1971.
215. **Sasajima, K., Kawachi, T., Sano, T., Sugimura, T., Shimosata, Y., and Shirota, A.,** Esophageal and gastric cancers with metastases induced in dogs by N-ethyl-N'-nitro-N-nitrosoguanidine, *J. Natl. Cancer Inst.*, 58, 1789, 1977.
216. **Sekizuka, H., Doi, H., Sunagawa, M., Nagai, S., Kojima, S., Hiraide, H., Hoshi, K., and Murakami, T.,** Induction of esophageal cancer associated with gastric cancer in a dog by N-ethyl-N'-nitro-N-nitrosoguanidine, *Gann*, 66, 683, 1975.
217. **Kurihara, M., Shirakabe, H., Izumi, T., Miyasaka, K., Yamaya, F., Maruyama, T., and Yasui, A.,** Adenocarcinomas of the stomach induced in Beagle dogs by oral administration of N-ethyl-N'-nitro-N-nitroso-guanidine, *Z. Krebsforsch.*, 90, 241, 1977.
218. **Horton, L., Fox, C., Corrin, B., and Sonksen, P. H.,** Streptozotocin-induced renal tumors in rats, *Br. J. Cancer*, 36, 692, 1977.
219. **Berman, L. D., Hayes, J. A. and Sibay, T. M.,** Effect of streptozotocin in the Chinese hamster (*Cricetulus griseus*), *J. Natl. Cancer Inst.*, 51, 1287, 1973.
220. **Ungerer, O., Eisenbrand, G., and Preussmann, R.,** The reaction of nitrite with pesticides. Formation, chemical properties and carcinogenic activity of the N-nitroso compound of the herbicide N-methyl-N'-(2-benzothiazolyl)urea (Benzthiazuron), *Z. Krebsforsch.*, 81, 217, 1974.
221. **Eisenbrand, G., Schmahl, D., and Preussmann, R.,** Carcinogenicity in rats of high doses of N-nitrosocarbaryl, a nitrosated pesticide, *Cancer Lett.*, 1, 281, 1976.
222. **Eisenbrand, G., Ungerer, O., and Preussmann, R.,** The reaction of nitrite with pesticides. II. Formation, chemical properties and carcinogenic activity of the N-nitroso derivative of N-methyl-1-naphthyl carbamate (carbaryl), *Food Cosmet. Toxicol.*, 13, 365, 1975.

Chapter 7

HUMAN RESPONSES TO N-NITROSO COMPOUNDS

Ronald C. Shank

TABLE OF CONTENTS

I.	Introduction	220
II.	Acute Toxicity	220
III.	Chronic Toxicity	221
	References	225

I. INTRODUCTION

In spite of what appears widespread occurrence of N-nitroso compounds in the environment, there are only a few reports of human responses to these poisons and carcinogens. This is probably due to the low levels of environmental exposures precluding acute toxicity and the multitude of target organs making extremely difficult epidemiologic association of human cancers to long term exposures. What human data are available do indicate, however, that man appears to respond to nitrosamines qualitatively much like other mammalians respond.

II. ACUTE TOXICITY

The first documented human poisoning by an N-nitroso compound appears to be a report by Freund in 1937 in which he describes the case of two young chemists involved in the production of dimethylnitrosamine for use as an anticorrosive agent. The first case report concerns a 29-year-old male assigned to work on the development and manufacture of the nitrosamine. After an undisclosed period of exposure, chiefly by inhalation, the chemist developed headache, upper right quadrant pain, nausea, sensation of weakness in the upper abdomen, and occasional abdominal cramps. He associated his illness with breathing fumes of dimethylnitrosamine and reported to the company physician. He continued to work irregularly for approximately 2 weeks, until he felt so ill that he quit his job. He was hospitalized for further observation and was suffering from cramps, a distended abdomen, and exhaustion. The abdomen was surgically punctured (paracentesis) to withdraw fluid which had accumulated in the peritoneum (approximately 2.5 qt). The patient apparently was told his "peritonitis" was of tuberculosis origin, and he entered a sanatorium where he remained for several months. There was no evidence of tuberculosis, but his headaches, abdominal pain, and weakness persisted. With continued rest at home, his condition slowly improved.

Approximately 16 months after the first chemist became ill, a second chemist, age 26 years, resumed the production of dimethylnitrosamine for the company. Several days after starting this work, he spilled approximately a liter of dimethylnitrosamine which he cleaned up with an ordinary mop and rag. That night he became ill and could not work the following day. Six days after the accidental spill, he began to have upper abdominal pain, cramps, and a feeling of distention; in another 2 days, his abdomen was more tender and the distention worsened, and he was hospitalized 6 days later. The patient was slightly icteric, there was fluid in the pleural cavities, and his liver was enlarged. A total of 5 paracenteses, each removing between 3.5 to 6 ℓ of fluid, were performed over the following 4 weeks. An exploratory laparotomy was performed, revealing an enlarged purplish liver and spleen; the patient died shortly after the operation, 48 days after the dimethylnitrosamine spill. Autopsy showed acute degeneration of liver parenchyma. Interviews of other workers in the laboratory, who also had some contact with dimethylnitrosamine, indicated they all had some history of dizziness, faintness, headache, and weakness.

Hamilton and Hardy,[2] in their book on industrial toxicology, mentioned 2 cases of dimethylnitrosamine poisoning found by Hamilton in the course of a survey of a large automobile factory 12 years after this report from Freund.[1] The first case was described as becoming violently ill, but recovered, while the second subject was less affected but died of an infection following a paracentesis. In the second case, an autopsy revealed a cirrhotic liver, with areas of regeneration.

The title to Freund's report bore little resemblance to nitrosamine poisoning, and it was a study by Barnes and Magee[3] in 1954 that drew widespread attention to human poisoning by dimethylnitrosamine. Approximately 10 months after introducing dimethylnitrosamine into a research laboratory of a large industrial firm in Great Britain, one man working in the laboratory died of bronchopneumonia, and cirrhosis was an incidental finding at autopsy. A second worker, with a long history of dyspepsia, underwent a laparotomy 5 months after the introduction of the nitrosamine. At that time, no liver abnormalities were reported, but 6 months later, while being examined for an incisional hernia which developed in the operational scar, a hardened liver was discovered. Liver function tests confirmed liver damage, suspected to be cirrhosis. Exposure to the nitrosamine ended, and 3 months later liver function had improved. A third man in the laboratory apparently was asymptomatic. Barnes and Magee went on to study in controlled experiments the toxic properties of dimethylnitrosamine in rats, mice, guinea pigs, rabbits, and dogs and published the classic description of hepatotoxicity by this compound (see Chapter 6).

Acute poisoning by one other N-nitroso compound has been reported. Watrous[4] claimed that chemists who had spilled N-nitrosomethylurethane on their hands experienced severe itching, uniform erythema, and marked swelling of the skin which became cyanotic. The symptoms lasted 5 days after which extensive desquamation of the skin followed.

Five chemists in another laboratory were exposed to N-nitrosomethylurethane.[5] One man spilled some of the compound which splashed onto his feet and face; reddening of the skin and conjuctiva occurred at the point of contact and persisted for over a month. Respiratory difficulties also developed, but gradually subsided. The other workers were less affected.

Recently, two cases in forensic toxicology offered further support to the potential role of dimethylnitrosamine as a human toxicant and carcinogen. In an apparent murder attempt, an adult female was exposed to dimethylnitrosamine in her food over a period of several weeks; the victim died with liver cirrhosis and hepatic lesions, suggestive of preneoplastic change.[23] In an unrelated murder attempt, several people drank a beverage alleged to contain dimethylnitrosamine by deliberate addition; two of the people died days later with massive liver failure. Liver from one of the victims, a 23-year-old man, was assayed for methylated DNA, as it was considered unlikely that the parent toxin would have persisted long enough in the liver to be detected after the time of death. Dimethylnitrosamine administration to animals leads to methylation of liver DNA (see Chapter 7), and detection of aberrant methylation in the DNA of the victim's liver could be taken as presumptive evidence of exposure to a strong methylating intermediate. The DNA from the victim's liver was found to contain amounts of 7-methylguanine and O^6-methylguanine, not constituents of normal mammalian DNA, consistent with exposure to a probable lethal dose of dimethylnitrosamine.[24]

III. CHRONIC TOXICITY

All studies related to the possible chronic toxicity of N-nitroso compounds in humans have focused on the carcinogenicity of these agents. One of the first reports of a possible association was a geographical study, suggesting that esophageal cancer in Africa could be related to drinking locally distilled spirits containing dimethylnitrosamine or similar compounds.[6] Interviews with staff members of hospitals in part of Zaire (Katanga Province in Congo Republic), part of Mozambique, and all of Malawi, Rhodesia, and Zambia suggested an uneven geographical distribution for esophageal cancer; however, little or no data were available to support the clinical diagnosis to determine the sizes of the population at risk, confirm places of residence of reported cases, or to account for variation in the power of recall by the physicians during the interview.[7]

It appeared that hospitals reporting the largest numbers of esophageal cancer cases were located in areas where certain distilled spirits were consumed. The distillate was prepared from fermented sugar and corn husks with herbs added after fermentation. Eight samples of the beverage were analyzed for polycyclic hydrocarbons by spectrofluorometry, and N-nitrosamines by polarography and thin-layer chromatography (TLC); none of the samples contained detectable amounts of benzpyrene or other polycyclic hydrocarbons, but all eight give positive results for nitrosamines by both tests.[6] The analyses have not been confirmed by mass spectroscopy.

In studies on cancer of the uterine cervix in southern Africans, Harington and co-workers[8] examined pooled specimens of human cervical and vaginal discharge matter for the presence of nitrosamines. Samples (about 0.2 to 0.5 mℓ per patient) were collected from 100 women attending an antenatal clinic in Johannesberg and dichloromethane extracts were pooled for gas chromatography (GC). Analysis indicated the presence of dimethylnitrosamine, the identity of which was supported by mass spectroscopy (MS). The amounts of nitrosamine present were not known, but the report implies a possible association between dimethylnitrosamine formation in the vaginal vault and human cervical cancer in Africa.

It has been suggested that cancer of the stomach in certain provinces in Chile (which has the highest age-adjusted mortality rate for stomach cancer in the world) may result from exposure to nitrosamines in areas where environmental nitrate concentrations are high.[9,10] The predominantly agricultural areas in Chile use the greatest amounts of nitrate fertilizers and these same areas have the highest death rates for stomach cancer.[10,11] Presumably the use of nitrates agriculturally is equitable to human exposure to nitrates and carcinogenic nitrosamines; however, no scientific evidence has been presented to support this proposed relationship. The suggested etiology is that nitrates applied to soils and crops enter the human diet though meat, vegetables, and water; after reduction to nitrite, nitrosation of secondary and tertiary amines takes place to form the carcinogens which initiate the stomach cancer. In northern provinces of Chile in which the great natural nitrate deposits are located, esophageal, but not stomach, cancer is prevalent and apparently a similar etiology could explain this particular risk; i.e., nitrate yielded the nitrite which indirectly nitrosated secondary amines producing nitroso compounds whose carcinogenic activities were specific to the esophagus.[9]

A more direct measure of nitrate exposure was done in two English towns, and the town with the greater exposure was found to report an increased death rate from stomach cancer.[12] Estimates of nitrate consumption, based on nitrate contents of food and drinking water, were made for the residents of Paddington and Worksop in England, and urinary nitrate concentrations were determined directly. In Paddington the nitrate intake was estimated to be approximately 57 to 64 mg on a daily basis, and the urinary nitrate concentration was 1.0 mmol (62 mg)/ℓ; in Worksop the intake was about 143 mg/day, and the nitrate in the urine was 2.6 mmol (161 mg)/ℓ. The higher nitrate levels in Worksop were attributed to greater nitrate concentrations in the drinking water supplied to that town. Analysis of death records for Worksop and several other towns in which nitrate concentrations in drinking water were not elevated revealed that death rates for stomach cancer in Worksop were 27% higher (statistically significant at the 5% level) than in 8 of the 9 other towns studied. This association was not apparent when all cancer sites were considered. As in the other studies, the suggested etiology for the increased risk from gastric cancer is that the high nitrate concentrations in drinking water are responsible for the formation of carcinogenic N-nitroso compounds specific for the stomach; however, direct scientific evidence to associate nitrosamines to gastric cancer in Worksop is not available.

Other cancer investigators have also discussed the possible role of nitrosamines in human cancers. Weisburger and co-workers[13,14] have discussed the impact of home refrigeration on the conversion of nitrate to nitrite and the formation of nitrosamines in leftover cooked foods. In their considerations of the possible etiologies for cancer of the gallbladder, bile ducts, and salivary glands in workers in the rubber industry, Mancuso and Brennan[15] point out the several N-nitroso compounds used in this industry and call for epidemiological studies to be carried out to determine whether these compounds, and others, have a causal role in the induction of these cancers in rubber industry workers.

There have been two studies in which attempts were made to measure directly dietary consumption of nitrosamines and relate this intake to cancer rates in defined populations. One study, carried out by the International Agency for Research on Cancer (World Health Organization), focused on esophageal cancer in 15 specific areas of Iran, where this cancer in unusually prevalent, and in nearby areas where esophageal cancer rates are not high.[16] The sampling unit in the study was the village, restricted to rural areas to avoid the rapid changing life styles found in towns and cities. Dietary surveys were made twice during each of three seasons. Information collected in each village included a census, household characteristics, such as ethnic group, birth place, marital status, cosanguinity, occupational history, food and smoking habits (especially whether household members smoked or chewed opium and tobacco, consumed tea, hot drinks, alcohol, seeds, nuts, and dried fruit), crops raised and animals owned, pregnancy history including special diets, and esophageal cancer history within the family. Six randomly selected households were surveyed over 5 days for quantitative food consumption; information of food handling and use was obtained, and in two villages located in high esophageal cancer incidence areas and two in low incidence areas, samples of the main foodstuffs were collected and analyzed for volatile (but not nonvolatile) nitrosamines, polycyclic aromatic hydrocarbons, aflatoxins, nitrate, and nitrite; analyses for nitrate and nitrite were also done on drinking water samples. A medical examination was performed on each adult member of dietary survey households, and samples of blood, urine, and feces were collected for laboratory analysis. Cancer rates were determined through the Caspian cancer registry.[17]

GC analyses failed to detect nitrosamines in excess of 5 μg/kg food sample (see Table 1); apparently only dimethylnitrosamine was detected, rather frequently at the 1 μg/kg level, but the frequency of contamination was not significantly greater in the high cancer incidence area. Nitrate and nitrite intakes did not differ between the high and low incidence areas, and aflatoxin was found in only a single food sample. While the results fail to associate nitrosamines with human esophageal cancer, the investigators point out that several esophageal carcinogens are nonvolatile N-nitroso compounds and would not have been detected by the methods used in the study. Now that a simple method for the specific quantitative determination of nonvolatile nitrosamines exists,[18] it might be valuable to expand the Caspian littoral dietary survey to include these possible contaminants.

The second study on the association of dietary nitrosamines and human esophageal cancer was conducted in China in the Anyang region.[19] Between 1969 and 1971, an interdisciplinary coordinating group conducted an epidemiologic survey to determine the geographic distribution of esophageal cancer, measure mortality rates for this disease, obtain evidence to associate environmental factors, and explore preventive measures against esophageal cancer.

Population statistics were based on brigade or commune registries, and diagnosis was made radiologically in 66.48% of the cases; less than one third of the cases were diagnosed on clinical symptoms and medical histories, and only 5.99% were confirmed histopathologically. The average age/sex adjusted mortality rate for esophageal cancer over the

Table 1
DIMETHYLNITROSAMINE IN FOOD SAMPLES IN IRAN[16]

DMN[a] conc (µg/kg)	Percentage of samples analyzed	
	High incidence area[b]	Low incidence area[b]
1—5	12	6
1	62	50
None detectable	26	44

[a] DMN = dimethylnitrosamine.
[b] Incidence of esophageal cancer.

entire 181 counties and cities in the survey was 37.39 per 100,000 population; the highest mortality rate in any single county was 139.80 per 100,000 (Oupi County) which was 97 times higher than the lowest rate measured, 1.43 per 100,000 in Tse-yuan County.

Linhsien County in Honan Province began a cancer registry in 1959 and over a 12-year period measured a county-wide average death rate, adjusted for age and sex, of 99.76 per 100,000; the adjusted incidence was 108.56 per 100,000. In a total population of 10,264 people 30 years and older, an examination of 7212 people (83.1%) found 118 cases of esophageal cancer, giving a point prevalence of 379 per 100,000 population. The report claims that the number of deaths from esophageal cancer accounted for about 20% of all deaths (not just cancer deaths) in Linhsien county, a truly remarkable incidence calling for immediate efforts in preventive medicine.

A comparative study was also conducted on the prevalence of pharyngeal and esophageal cancer in chickens in Linhsien and in Fanhsien where the incidence of human esophageal cancer is low. Inspection of 18,774 chickens in Linhsien uncovered 33 cases of pharyngeal/esophageal cancer (confirmed by pathology), and in Fanhsien, two cases were found among 11,399 chickens. Thus, the prevalence of this form of cancer in chickens in Linhsien, where human esophageal cancer is prevalent, is 175.78 per 100,000, compared to 17.55 per 100,000 in Fanhsien, where the prevalence of human esophageal cancer is 6.5 times lower than in Linhsien. This striking parallelism between cancers in certain populations and their domestic animals suggests an environmental factor in the etiology of the disease.

Three environmental factors were reported as possibly associated with human esophageal cancer in this region of China. The first was nitrosamines. Analysis of 124 food samples from Linhsien by TLC and GC showed that 29 samples contained dimethylnitrosamine, diethylnitrosamine, and/or methylbenzylnitrosamine (a known esophageal carcinogen);[20] only 1 sample of 86 from Fanhsien contained nitrosamine (identity was not specified). Since publication of this report, these findings have been confirmed by high resolution MS.[21]

The second environmental factor studied was trace elements in foodstuffs and drinking water. Chemical and spectrographic analyses indicated that in areas where esophageal cancer rates were high, drinking waters were lower in cobalt and nickel contents than in areas with low cancer rates; molybdenum contents in foodstuffs were also low in the high esophageal cancer rate areas; numerical data were not supplied.

The third factor that may have a role in esophageal carcinogenesis in northern China was the "white ground mushroom", *Geotrichum candidum* Link, which is fact is not a mushroom, but rather a mold. This dietary component in Linhsien was cultured in the laboratory, and a liquid product was placed in the drinking water of rats, resulting 10 to 15 months later in papilliform tumors of the anterior stomach in 3 of the 11 animals.

The fungal liquid also demonstrated possible promoter properties in the induction of upper gastric tumors by methylbenzylnitrosamine.

While this epidemiological study is not conclusive, it is the only evidence available so far which associates to any degree environmental nitrosamines with a human cancer.

Other studies are currently in progress. The International Agency for Research on Cancer (IARC) Directory of Ongoing Research in Cancer Epidemiology[22] lists seven current studies on the possible role N-nitroso compounds may have in the etiology of human cancers. Two of the studies, one in Colorado and the other in Israel, are focusing on associations between nitrate levels in drinking water and mortality rates for cancer of the stomach (Colorado and Israel) and esophagus, urinary bladder, and liver (Colorado only). In Thailand a study is analyzing the association between the ingestion of fermented salted fish and cancer of the liver. A study in the Federal Republic of Germany is concentrating upon dietary habits of vegetarians, nitrosamine precursors in foods, and all cancer sites. A study in Egypt is looking at the occurrence of nitrosamines in urine and bladder cancer in patients with schistosomiasis (bilharziasis). Another group is conducting a nonconcurrent prospective epidemiological study of a chart of workers in the U.S. exposed to dimethylnitrosamine in the manufacture of 1,1-dimethylhydrazine, a rocket propellant. The next few years promise to provide much new information on the possible associations between environmental N-nitroso compounds and human cancers.

REFERENCES

1. **Freund, H. A.**, Clinical manifestations and studies in parenchymatous hepatitis, *Ann. Intern. Med.*, 10, 1144, 1937.
2. **Hamilton, A. and Hardy, H.**, *Industrial Toxicology*, 2nd ed., Hoeber, New York, 1949, 301.
3. **Barnes, J. M. and Magee, P. N.**, Some toxic properties of dimethylnitrosamine, *Br. J. Ind. Med.*, 11, 167, 1954.
4. **Watrous, R. M.**, Health hazards of the pharmaceutical industry, *Br. J. Ind. Med.*, 4, 111, 1947.
5. **Wrigley, F.**, Toxic effects of nitrosomethylurethane, *Br. J. Ind. Med.*, 5, 26, 1948.
6. **McGlashan, N. D., Walters, C. L., and McLean, A. E. M.**, Nitrosamines in African alcoholic spirits and oesophageal cancer, *Lancet*, 2, 1017, 1968.
7. **McGlashan, N. D.**, Oesophageal cancer and alcoholic spirits in central Africa, *Gut*, 10, 643, 1969.
8. **Harington, J. S., Nunn, J. R., and Irwig, L.**, Dimethylnitrosamine in the human vaginal vault, *Nature (London)*, 241, 49, 1973.
9. **Zaldivar, R.**, Geographic pathology of oral, esophageal, gastric, and intestinal cancer in Chile, *Z. Krebsforsch.*, 75, 1, 1970.
10. **Armijo, R. and Coulson, A. N.**, Epidemiology of stomach cancer in Chile. The role of nitrogen fertilizer, *Int. J. Epidemiol.*, 4, 301, 1975.
11. **Zaldivar, R. and Wetherstrand, W. H.**, Further evidence of a positive correlation between exposure to nitrate fertilizers (NaNO$_3$ and KNO$_3$) and gastric cancer death rates: nitrites and nitrosamines, *Experentia*, 31, 1354, 1975.
12. **Hill, M. J., Hawksworth, G., and Tattersall, G.**, Bacteria, nitrosamines and cancer of the stomach, *Br. J. Cancer*, 28, 563, 1973.
13. **Weisburger, J. H. and Raineri, R.**, Assessment of human exposure and response to N-nitroso compounds: a new view on the etiology of digestive tract cancers, *Toxicol. Appl. Pharmacol.*, 31, 369, 1975.
14. **Marquardt, H., Rufino, F., and Weisburger, J. H.**, Mutagenic activity of nitrite-treated foods: human stomach cancer may be related to dietary factors, *Science*, 196, 1000, 1977.
15. **Mancuso, T. F. and Brennan, M. J.**, Epidemiological considerations of cancer of the gallbladder, bile ducts, and salivary glands in the rubber industry, *J. Occup. Med.*, 12, 333, 1970.
16. Joint Iran-International Agency for Research on Cancer Study Group, Esophageal cancer studies in the Caspian littoral of Iran: results of population studies — a prodrome, *J. Natl. Cancer Inst.*, 59, 1127, 1977.

17. **Mahboubi, E., Kmet, J., and Cook, P. J.,** Oesophageal cancer studies in the Caspian littoral of Iran: the Caspian cancer registry, *Br. J. Cancer,* 28, 197, 1973.
18. **Fine, D. H., Ross, R., Rounbehler, D. P., Silvergleid, A., and Song, L.,** Analysis of nonionic nonvolatile N-nitroso compounds in foodstuffs, *J. Agric. Food Chem.,* 24, 1069, 1976.
19. Coordination Group for Research on Etiology of Esophageal Cancer in North China, The epidemiology and etiology of esophageal cancer in North China (in Chinese), *Chinese Med. J.,* 1, 167, 1975.
20. **Druckrey, H., Preussmann, R., Blu, G., Ivankovic, S., and Afkham, J.,** Erzeugung von Karzinomen der Speiserohre durch Unsymmetrische Nitrosamine, *Naturwissenschaften,* 50, 100, 1963.
21. **Li, M.-K.,** Department of Chemical Etiology and Carcinogenesis, Cancer Institute, Chinese Academy of Medical Sciences, Peking, personal communication, 1978.
22. International Agency for Research on Cancer, *Directory of On-going Research in Cancer Epidemiology,* International Agency for Research on Cancer, Lyon, 1977.
23. **Schmahl, D.,** The role of nitrosamines in carcinogenesis — an overview, in *Safety Evaluation of Nitrosatable Drugs and Chemicals,* Parke, D. V., Ed., Pergamon Press, London, in press.
24. **Herron, D. C. and Shank, R. C.,** Methylated purines in human liver DNA after probable dimethylnitrosamine poisoning, *Cancer Res.,* in press.

Chapter 8
N-NITROSO COMPOUNDS: ASSESSMENT OF CARCINOGENIC RISK

Ronald C. Shank

TABLE OF CONTENTS

I. Introduction 228

II. Estimate of Exposure 228
    A. Estimation from Ingested Nitrate and Nitrite 229
    B. Estimation from Ambient Air and Certain Foods 230
    C. Estimation from Nitrosamine Blood Levels 230

III. Dose-Response Relationships 232

IV. Extrapolations from Animal Data to Human Risk 237
    A. Mathematical Models and Their Consistency with the Biology of Cancer 237
    B. Extrapolation According to a Typical Mathematical Model ... 241

V. Risk Vs. Benefit 242

References 243

I. INTRODUCTION

The assessment of risk made for environmental mycotoxins (see Chapter 4) was limited to consideration of a single compound, aflatoxin B_1, because its known exposure to humans and its potency as a hepatocarcinogen in laboratory animals are much greater than for any other known mycotoxin. Consideration cannot be limited to a single compound in assessing the risk presented by N-nitroso compounds because several are likely to occur in the human environment, although current data are available primarily for only dimethylnitrosamine and N-nitrosopyrrolidine; also, the more than 120 known carcinogenic N-nitroso compounds cover a wide range of potencies, with no single compound standing out as exceptionally more potent than the other. Also, the mycotoxin assessment focussed essentially on the risk of liver cancer, although other cancer sites were acknowledged. In assessing N-nitroso compound risks, all vital tissues must be considered, for as a class of compounds, these agents are able to induce primary cancers in all these tissues.

Finally, the mycotoxin risk assessment was facilitated by the existence of four comparable epidemiological studies on the association between dietary aflatoxins and human liver cancer, whereas for the N-nitroso compounds no evidence is currently available to associate these agents with any form of human cancer. Nevertheless, because of their potency, their activity in so many tissues in a variety of experimental animals, and their occurrence in the human environment, an attempt to assess the risk these compounds present to human health seems warranted. Two such risk assessments have already been made: the first in 1976 by an ad hoc study group appointed by the U.S. Environmental Protection Agency[1] and the second by a panel appointed by the National Research Council of the U.S. National Academy of Sciences;[2] the latter report included nitrates and nitrites in the risk assessment. The conclusions of the above reports and the conclusion of this chapter are the same, namely, that with the limited data on exposure and the many uncertainties involved in extrapolating from responses in laboratory animals given large doses to humans exposed to low concentrations, it is not possible at this time to make a scientifically reliable estimate of risk for N-nitroso compounds and human cancer. On the other hand, the question on whether a risk is likely can be approached.

II. ESTIMATE OF EXPOSURE

The occurrence of N-nitroso compounds in the environment was discussed in detail in Chapter 5. The major source of human exposure to these compounds appears to be in vivo formation from ingested precursors; it seems that occupational exposure to these carcinogens rarely occur, and where they do occur, they can be controlled or avoided altogether. The principal source of one precursor of the N-nitroso compounds, nitrate, is vegetables and certain other foods, unless the drinking water comes from private wells contaminated with larger amounts of nitrate. Most nitrite in the saliva and gastric juice is derived from the ingested nitrate.

Human exposures to N-nitroso compounds cannot be accurately measured, but can be estimated three ways:

1. Using available estimates for nitrate and nitrite exposure and assuming a fraction of this results in in vivo nitrosamine formation.
2. Using available information on dimethylnitrosamine levels in ambient air and certain foods.
3. Using the data of Fine and co-workers[3] on human blood levels of nitrosamines following ingestion of a defined meal.

A. Estimation from Ingested Nitrate and Nitrite

The average daily intake of nitrate in the U.S. was estimated by White[4] to be about 100 mg/day; if the drinking water is from shallow private wells, the nitrate intake may be increased by as much as 2200 mg/day,[2] although this would be unusual. In their studies on the relationship between ingestion of nitrate and appearance of nitrite in human saliva, Spiegelhalder and co-workers[5] determined that the average increase in nitrite in saliva was 20 ppm for every 100 mg nitrate ingested; ingestion of 470 mg of nitrate was estimated to result in an average salivary nitrite concentration of about 100 ppm, and assuming a salivary flow rate of 50 mℓ/hr, about 40 mg nitrite would enter the stomach through the saliva over a 5-hr period. Similarly then, 100 to 2300 mg of ingested nitrate, if taken at one time, would be expected to result in average salivary nitrite concentrations of about 20 to 500 ppm nitrite or 8 to 200 mg nitrite entering the stomach within 5 hr. Assuming that this represents the total 24-hr ingestion of nitrite plus another 2 mg nitrite from cured meats, the daily nitrite exposure is estimated to range from 10 mg (perhaps "typical") to 202 mg (for persons drinking contaminated water).

It is far more difficult to make the next step and estimate the rate of nitrosation of secondary and tertiary amines in the stomach in order to estimate the amounts of carcinogenic N-nitroso compounds formed. It is known that the rate of formation of nitrosamines is proportional to the square of the nitrite concentration, but the relationship between salivary nitrite concentration and output to actual variation in nitrite concentrations in stomach contents is not known. The pH and amounts of amines, catalysts, and inhibitors are also important but unknown variables determining the extent to which ingested nitrite is used in in vivo nitrosation reactions.

An experiment by Mysliwy and co-workers[6] investigated in vivo nitrosation in the stomach of dogs. Intubation of 725 μmol of sodium nitrite and 10 μmol of pyrrolidine resulted in as much as 1.4 μmol N-nitrosopyrrolidine in the stomach at any one time. This indicates that at least 0.2% of administered nitrite was recovered as the nitrosamine at any one time. Using this figure as an estimate of a lower bound for in vivo nitrosation, the National Research Council Panel on Nitrates[2] assumed that the extent of formation of nitrosamines (as dimethylnitrosamine) in the human stomach was about 5% of the nitrite ingested.

Using the range of nitrite ingestion of 10 to 202 mg, or 217 to 4391 μmol nitrite, a 5% utilization of the nitrite to carcinogenic nitrosamines would produce 11 to 220 μmol nitrosamine if sufficient amine precursor were available. This assumes that all the daily nitrite is present in the stomach at one time, even though the salivary nitrite enters the stomach throughout the day and night; thus, this estimate approaches an upper bound or maximum.

If the nitrosamine burden is expressed in terms of dimethylnitrosamine, then the range of 11 to 220 μmol nitrosamine becomes 814 to 16,280 μg dimethylnitrosamine per day. For a 70-kg body weight adult, this intake becomes 12 to 233 μg/kg body weight per day; expressed in terms of dietary concentrations assuming a daily consumption of 2 kg food for an adult, the daily intake is equivalent to a diet containing 407 to 8140 parts per billion dimethylnitrosamine. Such an estimate indicates that nitrate and nitrite contribute to the nitrosamine body burden to a much greater extent than do food-borne nitrosamines (see Table 6, Chapter 5), even if a kilogram of cured meat is ingested each day. This assumes, of course, that volatile nitrosamines account for most of the total nitrosamines present in foods, for data are not available on nonvolatile nitrosamine concentrations in foods.

B. Estimation from Ambient Air and Certain Foods

Limited information is available on human exposure to airborne nitrosamines. Studies focussing on dimethylamine manufacturing plants[7] and a chemical factory using dimethylnitrosamine in the synthesis of a rocket propellant[8] have shown that these point sources can contaminate industrial air and even the air supply in the urban area in the vicinity of these point sources. The average dimethylnitrosamine concentrations in the air at the site where the carcinogen was reduced to 1,1-dimethylhydrazine was about 12 $\mu g/m^3$, about 1 $\mu g/m^3$ in adjacent neighborhoods, and about 0.1 $\mu g/m^3$ in more distant urban areas. Secondary amine manufacturing plants produced much smaller amounts of dimethylnitrosamine as airborne contaminants, and in fact, a U.S. Environmental Protection Agency survey of municipal air supplies near sources of amine emissions has shown only a few samples to contain detectable nitrosamines[9]. It would seem, then, that except for certain chemical plant workers, exposure to detectable levels of airborne nitrosamines is rare. The National Research Council report[2] on nitrates estimated that a worker in a rocket propellant plant using dimethylnitrosamine breathing 10 m^3 of air during the working period would inhale approximately 116 μg dimethylnitrosamine; during the nonworking hours, he might inhale another 10 m^3 of air containing 0.1 $\mu g/m^3$. Thus, over a 24-hr period, assuming 100% of the inhaled nitrosamine is absorbed, the daily exposure would be 117 $\mu g/day$ 5 days out of the week and 1 $\mu g/day$ over the weekend, for an average exposure of 84 μg dimethylnitrosamine per day; for residents of the city living not in the immediate neighborhood of the chemical plant, the exposure would be expected not to exceed 1 $\mu g/day$. These exposures can be compared to those expected from the consumption of certain foods.

Fried bacon, sausages, fish, and cheese appear to be the largest dietary sources of preformed nitrosamines. Assuming a resident of the above "city" ate over a week's period 250 g each of fried bacon, frankfurters, assorted cured meats, fish, and cheese, and using average values of nitrosamine concentrations in these foods, calculated by multiplying the midpoint of the range by the fraction of analyzed samples with positive results, it is estimated that the resident consumes approximately 4.9 μg total nitrosamines per day; the calculations are outlined in Table 1. This would represent about 0.9 $\mu g/day$ for dimethylnitrosamine, mainly from fish, and 4.0 $\mu g/day$ N-nitrosopyrrolidine, 90% of which is estimated to come from fried bacon. Whether it is justified toxicologically to add these two burdens is wholly undetermined. Such a diet would also include small amounts of diethylnitrosamine (0.046 $\mu g/day$ from cured meats) and N-nitrosopiperidine (0.003 $\mu g/day$ from fried bacon). These two nitrosamines would account for approximately 1/100 of the total nitrosamines in this dietary model.

These values have no scientific merit other than to help to a small extent in putting into perspective various sources of exposure. For people living in atmospheres not contaminated by certain chemical plants, the dietary model suggests that nitrite-treated foods are the major source of exposure to preformed nitrosamines. On the other hand, the model suggests a burden two to four orders of magnitude less than that from dietary sources of nitrosamine precursors with nitrosamine formation occurring in vivo after ingestion.

C. Estimation from Nitrosamine Blood Levels

In 1977 Fine and co-workers[3] measured the concentration of dimethylnitrosamine in the blood of a human subject before and after the volunteer ingested a beer and a large (800-g) sandwich made of bread, bacon, spinach, and tomato. Analysis of the lunch components by gas chromatography (GC) and thermal energy analysis (TEA) indicated the meal contained 1.6 μg preformed dimethylnitrosamine and an unspecified amount of nitrosopyrrolidine. Blood taken before the meal contained 2 μg dimethylnitrosamine

Table 1
CALCULATED EXPOSURE TO NITROSAMINES THROUGH CONSUMPTION OF SELECTED FOODS

Daily nitrosamine ingestion
(μg/person/day)

Food product	Dimethylnitrosamine[a] A × B × C = D				N-nitrosopyrrolidine[a] A × B × C = D			
	A	B	C	D	A	B	C	D
Fried bacon	$\dfrac{5-0}{2}$	$\dfrac{50}{50}$	$\dfrac{250}{1000} \cdot \dfrac{1}{7}$	0.1	$\dfrac{200-1+1}{2}$	$\dfrac{50}{50}$	$\dfrac{250}{1000} \cdot \dfrac{1}{7}$	3.6
Frankfurters	$\dfrac{84-11+11}{2}$	$\dfrac{3}{34}$	$\dfrac{250}{1000} \cdot \dfrac{1}{7}$	0.1				
Cured meats	$\dfrac{35-2+2}{2}$	$\dfrac{29}{80}$	$\dfrac{250}{1000} \cdot \dfrac{1}{7}$	0.2	$\dfrac{105-13+13}{2}$	$\dfrac{17}{80}$	$\dfrac{250}{1000} \cdot \dfrac{1}{7}$	0.4
Fish	$\dfrac{26-4+4}{2}$	$\dfrac{22}{23}$	$\dfrac{250}{1000} \cdot \dfrac{1}{7}$	0.5				
Individual totals				0.9				4.0
Grand total								4.9

Note: Data from Table 6, Chapter 5.

[a] A = midpoint of the range of concentrations, μg/kg
B = fraction of analyzed samples containing nitrosamine
C = portion of a kilogram food eaten over a 1-week period, on a daily basis
D = daily nitrosamine ingestion, μg/person/day = A × B × C

in the total blood volume (assumed to be 5640 mℓ). The dimethylnitrosamine level in the blood rose to a maximum of 4.35 μg for total blood volume 35 min after the lunch was eaten, and fell to below initial levels 162 min after lunch. Diethylnitrosamine was also detected; the total blood contained 0.51 μg before lunch and a maximum of 2.6 μg 65 min after lunch. Nitrosopyrrolidine, present in the sandwich, was not detected in the blood samples. No estimation was made of how much nitrosamine was absorbed from the blood into the tissues and how much was lost into urine and exhaled air. Estimation for minimum in vivo nitrosamine formation would indicate that at least 0.75 μg of dimethylnitrosamine and 2.1 μg diethylnitrosamine formed in the body after eating the lunch. If it is assumed that the nitrosamine formation in this subject is typical for what occurs in the general population and that the lunch represents an approximation of what the daily nitrosamine burden may be, then the daily exposure due to diet would be approximately 2.8 μg per person per day.

III. DOSE-RESPONSE RELATIONSHIPS

A few studies on dose-response relationships in nitrosamine carcinogenesis have been done and provide limited information in making risk assessments. The classical studies were carried out in the early 1960s by Druckrey and colleagues in which the length of tumor induction time and the magnitude of total carcinogenic doses which were directly proportional to total carcinogenic response were compared to the level of daily doses provided in the drinking water of rats. Several detailed experiments[10] on diethylnitrosamine indicated that the relationship could be described as the product of the daily dose, "d", and the average induction time, described by the following equation

$$d \cdot t^{2.3} = \text{constant}$$

For example, a daily dose of 0.15 mg diethylnitrosamine per kg body weight induced liver tumors in 27 of 30 surviving rats, with an average induction time of about 609 days. The lowest dose tested, 0.075 mg/kg, produced liver tumors in all four survivors, with an average induction time of about 830 days. Such a relationship invites one to use the above formula to calculate the daily dose that would produce tumors with an induction time about equal to the lifespan of the species. Assuming the rat has a lifespan of 3 years (1095 days), the daily dose that would be expected to induce liver tumors at or near the end of natural life for the rat would be approximately 0.039 mg/kg body weight, this would be equivalent to a dietary level of about 0.5 ppm diethylnitrosamine. One is tempted to conclude then that diets containing slightly less than 0.5 ppm diethylnitrosamine would induce liver tumors in rats so slowly that the animals would live out their lifespans before the tumors developed. One might even wish to make this kind of analysis in extrapolating to safe human exposures for carcinogens, safe because people would die of other causes before the exposures could produce tumors. One serious problem which is often overlooked in dose-response carcinogenicity studies is the relationship between dose levels and target organs, e.g., in the Druckrey studies,[10] administration of one or a few high oral doses of diethylnitrosamine to rats produced kidney tumors, and intermediate doses yielded only liver tumors, while both liver and esophageal tumors were obtained with low daily doses. Also, whether the Druckrey equation remains valid for low doses which produce low (less than 1%) tumor incidences has not yet been established experimentally.

In 1967 Terracini et al.[11] reported on a multidose feeding study on dimethylnitrosamine in rats. Female Porton rats, 4 to 6 weeks old, were fed chow diets containing 0, 2, 5, 10, 20 or 50 ppm dimethylnitrosamine for 120 weeks. They observed a dose-related tumor response which is illustrated in Figure 1. Rather than comparing induction times for

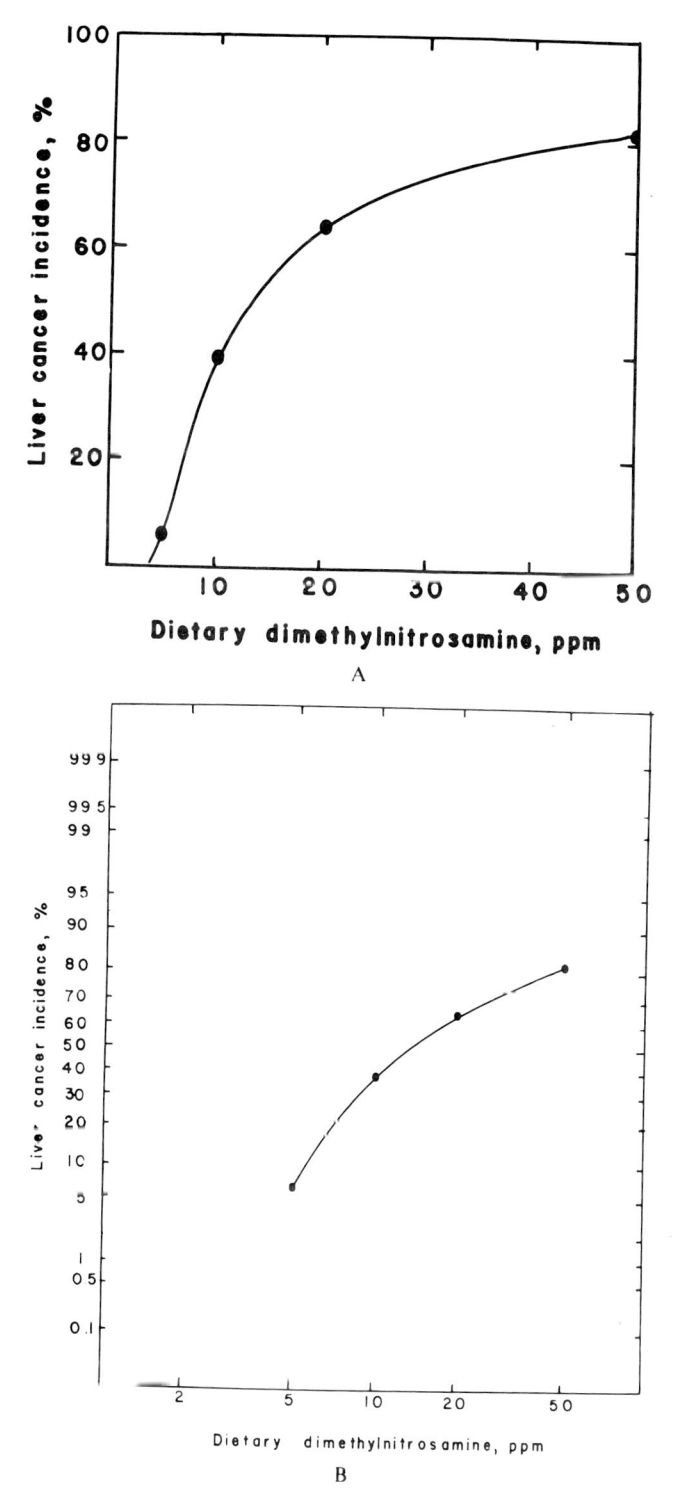

FIGURE 1. Incidence of liver cancer in rats fed diets containing various doses of dimethylnitrosamine. (A) Dose and tumor response plotted on linear coordinates; (B) log dose plotted vs. tumor response in probits expressed as probability.[11]

each dose, this relationship correlates liver cancer incidence over the lifespan of the population and daily dose of dimethylnitrosamine; whether expressed on linear (Figure 1A) or log-probit (Figure 1B) coordinates, it is apparent that the incidence of liver cancer at low doses is less than would be expected from a linear extrapolation from the portion of the curve determined at the higher doses. Figure 1A implies a no-observable tumor incidence in rats would occur when the dietary level of dimethylnitrosamine fell below about 4 ppm, and in Figure 1B, a linear extension tangential to the curve at the lowest dose suggests a dietary nitrosamine concentration of 2 ppm would be expected to produce a liver cancer incidence of about 0.01% of the rat population during the lifespan. To measure an incidence of 0.01% in a laboratory experiment with a high probability of detection (say, 80%) would require testing at least 16,000 animals at the required dose (30,000 animals for a 95% probability).[12] The above rat experiment used less than 100 animals per dose (few studies use more than 100 rats per dose) and therefore would not be expected to detect such a low incidence.

A dose-response study on dimethylnitrosamine has also been done in the mouse by Vesselinovitch.[13] He injected 7-day-old (C57B1 × C3H)F$_1$ mice i.p. every 3 days with 1, 2, or 4 μg dimethylnitrosamine per injection for six injections; the 15 to 31 surviving animals per dose were killed when 66 weeks old, and the incidence of tumors was measured. Liver and lung tumors were most common, and some hemangiomas occurred in the high-dose animals; no liver or lung tumors or hemangiomas were seen in the control animals. Male mice developed more liver and lung tumors than did female mice, and different dose-response curves were observed for the three tumor sites. Figures 2A and B illustrate the dose-response for dimethylnitrosamine and liver cancer in both males and females and lung adenoma incidence in male mice (lung tumors in females were seen only at the highest dose). The liver and lung tumor response in males was linear over the range of doses tested when the data were expressed in terms of log dose and probit incidence (see Figure 2B); in females liver cancer incidence fell off faster for lower doses than would be expected from a linear extrapolation from high doses. The does-response curves are nonlinear when the data are expressed in arithmetical terms (see Figure 2A).

Clapp and co-workers determined dose-response relationships in mice for dimethylnitrosamine[14] and diethylnitrosamine.[15] Dimethylnitrosamine was added to the drinking water of male RF mice so that the cumulative doses were 0, 87, 89, 170, and 243 mg/kg body weight; daily doses varied from 0.4 to 1.8 mg/kg.[14] For diethylnitrosamine the cumulative doses were 0, 57, 213, 321, 572, 780, and 943 mg/kg body weight, with daily doses ranging from 2 to 115 mg/kg.[15] The data are summarized in Table 2. The lack of clear dose-response relationships in these experiments may be due, in part, to the occurrence of lung and liver tumors spontaneously in this strain of mouse and the discontinuity between the daily exposure rates and durations.

Montesano and Saffioti[16] reported a positive dose-response correlation for tumor induction by diethylnitrosamine in the upper respiratory tract of hamsters. A total of 36 Syrian golden hamsters were given 12 weekly s.c. injections of 0.5, 1, 2, or 4 mg carcinogen per animal each time. The data are summarized in Table 3.

Montesano and Saffioti expressed their results in terms of the probability of an animal developing a tumor and the length of time the animal was on the experiment; as the dose of diethylnitrosamine increased, the sooner the animal was expected to develop a tumor with a given probability. Figure 3 presents the data in the log-probit mode; here, the dose-response curve for diethylnitrosamine and tracheal cancer is linear over the entire dose range. For cancer of the nasal cavities, the relationship falls faster for lower doses than would be anticipated from a linear extrapolation from the high dose relationship; the relationship does not appear to be linear for cancer of the larynx.

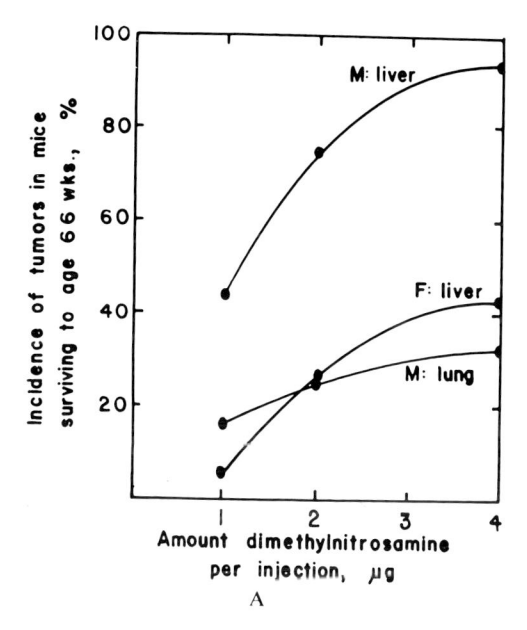

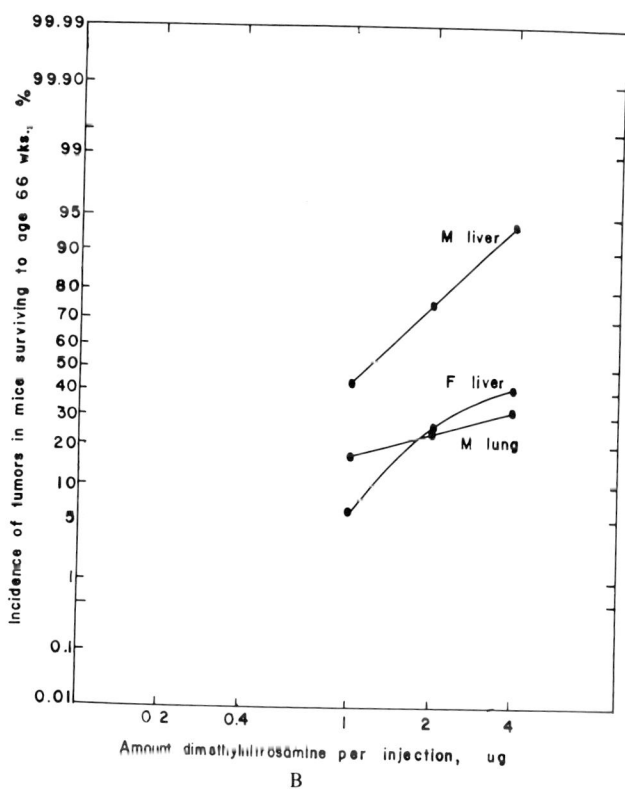

FIGURE 2. Incidence of liver and lung tumors in mice given six i.p. injections of various doses of dimethylnitrosamine. (A) dose and tumor responses plotted on linear coordinates; (B) log dose plotted vs. tumor response in probits expressed as probability.[13]

Table 2
DOSE-RESPONSE RELATIONSHIPS BETWEEN DIMETHYLNITROSAMINE AND DIETHYLNITROSAMINE INGESTION AND TUMORS AT VARIOUS SITES IN MICE[14,15]

Tumor Incidence in Dimethylnitrosamine-Treated Mice[a]

Cumulative dose (mg/kg body weight)	Daily dose (mg/kg body weight)	Lung adenomas	Hepatocellular carcinomas	Liver hemangiosarcomas
0	0	97/250 (39)	7/174 (4)	1/120 (1)
87	1.8	82/83 (99)	0/68 (0)	9/68 (13)
89	0.4	13/17 (76)	2/10 (20)	0/17 (0)
170	0.43	40/47 (85)	1/47 (2)	24/47 (51)
243	0.91	91/92 (99)	0/93 (0)	89/93 (96)

Tumor Incidence in Diethylnitrosamine-Treated Mice[a]

		Lung adenomas	Liver tumors	Stomach tumors
0	0	63/155 (41)	6/139 (4)	0
57	6	27/32 (84)	4/24 (17)	20/32 (62)
213	2	48/62 (77)	17/36 (47)	57/62 (92)
321	6	15/22 (68)	14/16 (87)	20/20 (100)
572	11.5	46/57 (81)	55/56 (98)	56/57 (98)
780	3.5	70/115 (61)	74/103 (72)	115/124 (93)
943	6	24/42 (74)	37/42 (90)	42/42 (100)

[a] Tumor-bearing animals divided by number of animals tested (percentage).

Table 3
INCIDENCE OF UPPER RESPIRATORY TRACT TUMORS IN HAMSTERS TREATED WITH DIETHYLNITROSAMINE[16]

Cumulative dose (mg)	Weekly dose (mg)	Number of animals with tumors out of 36 tested per dose (%)		
		Trachea	Nasal cavities	Larynx
6	0.5	29 (81)	6 (17)	6 (17)
12	1.0	32 (89)	25 (69)	18 (50)
24	2.0	34 (94)	26 (72)	17 (47)
48	4.0	35 (97)	27 (75)	26 (72)

No dose-response data of inhaled nitrosamines are available. One study carried out by Moiseev and Benemanshii[17] exposed Wistar rats and Balb/C mice to air concentrations of dimethylnitrosamine of 0.005 and 0.219 mg/m^3 continuously for 17 months for the mice and 25 months for the rats; control animals were included in the experiment, and each group contained 31 to 40 male rats, 30 to 51 female rats, 33 to 47 male mice, and 30 to 68 female mice. Animals exposed to the lower dose developed approximately the same number of tumors as did the controls. All animals at the higher dose developed more lung and kidney tumors, and all except male mice developed more liver tumors than controls.

Although a dose-response relationship cannot be drawn from these data, the "no-effect" level and the "effect" level can be expressed in terms of body burdens. Assuming

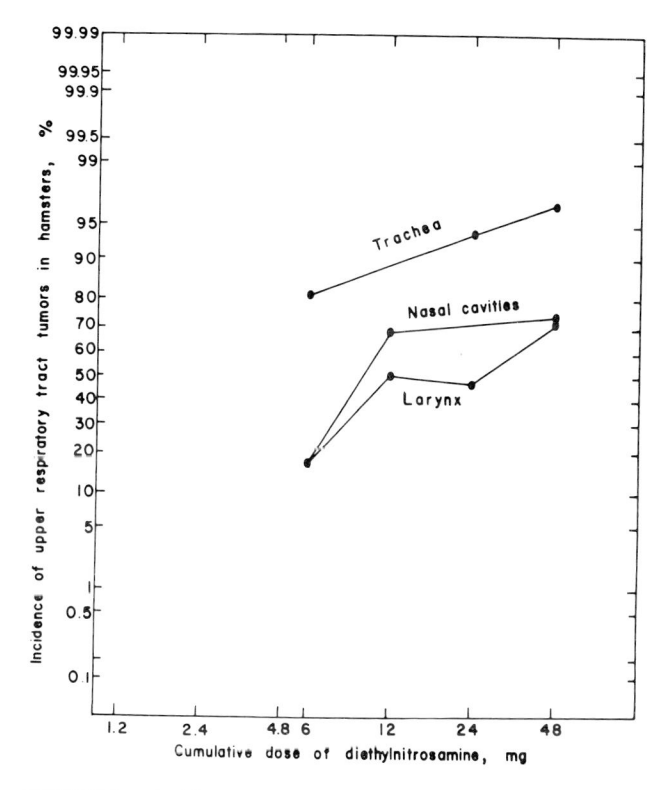

FIGURE 3. Incidence of upper respiratory tract tumors in hamsters given 12 s.c. injections of various doses of diethylnitrosamine; log dose is plotted vs. tumor response in probits expressed as probability.[16]

a rat breathes an average of 73 mℓ/min and a mouse, 24 mℓ/min, an air concentration of 0.005 mg dimethylnitrosamine per cubic meter would result in an exposure of about 0.5 μg per rat (0.2 μg per mouse) per day if 100% of the inhaled nitrosamine is absorbed; similarly, an air concentration of 0.219 mg dimethylnitrosamine per cubic meter is an exposure of about 23 μg per rat and 8 μg per mouse.

IV. EXTRAPOLATIONS FROM ANIMAL DATA TO HUMAN RISK

Chapter 4 outlined several of the problems, limitations, and assumptions involved in extrapolating to human risk from animal data, where relatively small numbers of rodents (assumed to be appropriate models for man) are exposed to doses greatly in excess of what might be expected to occur in the human environment. Many of the limitations could be avoided in the mycotoxin risk assessment because of a relatively strong data base for both animal and human exposure and response. These limitations cannot be avoided in the case of the risk presented by environmental N-nitroso compounds because nothing is known about the carcinogenicity of these chemicals in humans.

A. Mathematical Models and Their Consistency with the Biology of Cancer

Several mathematical models have been proposed for use in extrapolating from results of high-dose animal experiments to expected response in human populations exposed to carcinogens at concentrations several orders of magnitude lower than the animal doses. Faced with a paucity of experimental, epidemiological, and clinical data and having no scientifically elucidated mechanisms of carcinogenesis to use as guides, most of these models are necessarily based on several assumptions, the validities of which are undeterminable at this time. The major assumptions are

1. The carcinogen acts directly on a single cell producing heritable alterations, as opposed to possible indirect action which might alter an organism's susceptibility to cancer.
2. The dose-response relationship for small exposures is linear and predictable from experimental data obtained from large exposures, perhaps only a single datum point.
3. Species differences when not known can be ignored in spite of possible major consequences resulting from differences in metabolism, DNA repair activity, immunosuppression competence, etc.
4. Cancer arises from a single cell mutation and, hence, thresholds ("no-effect doses") cannot be assumed.
5. The carcinogen acts independently of other environmental agents, so that potentiation, synergism, additive toxicity/carcinogenicity, and antagonism do not have to be considered.

The models are carefully considered attempts to meet the needs of regulatory agencies in answering the demands for immediate control of carcinogens in the environment. Nevertheless, the models are forced to go beyond the biology of the phenomenon they attempt to describe, and for this reason, the confidence is low in the numbers (extrapolated low-dose tumor responses) the models produce.

Basically, there are three types of cancer risk extrapolation models: the one-hit model, the probit-log dose model, and the threshold model. In the one-hit model, the assumption is made that the observed tumor is the result of a mutagenic event having taken place in a single somatic cell. Although it may be consistent with the somatic mutation theory of carcinogenesis to trace the development of a malignant tumor backward to a mutagenic event in a single cell, this argument cannot be used in the opposite direction, namely that an obligatory result of a single transformation is a cancer; this is inconsistent with scientific observations in chemistry and biology, as will be explained below. The probit-log dose model[18,19] was developed empirically and is consistent with a great many experimental studies in pharmacology and toxicology, at least in dose ranges where responses in animals and humans are observable; the dose-response relationship in this model is unknown for extremely large dose ranges involving more than one species (the probit-log dose analysis for aflatoxin carcinogenicity in humans in Chapter 4 appears to be one of the first attempts to use this model where human data are available). The third model, the threshold model,[20] is based on the assumption that there is a dose of carcinogen to which an individual (extended to a population) can be exposed without the carcinogenic process going to completion and resulting in a tumor; this model lacks mathematical development presumably due to the many complex interactions that could influence the carcinogenic process.

A basic principle in pharmacology and toxicology is that the intensity of a biological response is proportional to the amount of active compound interacting at the target site and the rate at which the interaction occurs. Factors affecting this interaction are

1. The extent and rate of absorption of the toxin
2. The extent of distribution of the toxin within the body
3. The rates of metabolic alterations of the toxin
4. The rate of excretion of the compound and its metabolites

One of the major developments in chemical carcinogenesis has been the recent realization that most carcinogens exist in the environment in inactive states and that they become active only at the site of tumor induction. Much evidence indicates that these carcinogens must be converted at the target site to an electrophilic agent which interacts with DNA, creating a heritable change ultimately expressed as a cancerous cell. There are several competing mechanisms governing the success with which a carcinogen transforms the target cell and produces a tumor. These competing mechanisms are not taken into account in linear extrapolations from effects at high doses to effects at low doses. This concept of activation is of critical importance in extrapolating from animal carcinogenicity testing data to assessing risks to humans. The metabolism of a test compound can be elucidated in a variety of animal species, as well as in human tissues, to find the species which most closely matches the human in the metabolism of that particular compound. Carcinogenicity testing, then, should be done in the animal species which is biologically and biochemically most relevant to predicting the activity of the compound in the human.

There are several illustrations of the importance of metabolism in chemical carcinogenesis. Acetylaminofluorene (2-AAF) is carcinogenic only to those species capable of hydroxylating the nitrogen atom, which in turn becomes esterified with acetic acid. N-Acetoxy-AAF degrades, losing the acetate moiety to produce a strong electrophilic derivative of AAF.[21,22] A similar formation of an electrophilic derivative of the hepatocarcinogenic aminoazo dye, N-methyl-4-aminoazobenzene, has also been described.[23] Aflatoxin B_1, the most potent chemical carcinogen known,[24] diethylstibestrol,[25] and benzo(a)pyrene[26,27] are activated to carcinogens by metabolic epoxidation and subsequent decomposition to strong electrophilic derivatives. The N-nitroso compounds appear to be active by forming positively charged carbonium ions.[28-30] Even carcinogenic metals ionize to electrophilic reagents.

These positively charged compounds are highly reactive and bind strongly and immediately to nearby sites on large molecules which are especially dense in electrons (nucleophilic sites). These sites exist in protein, ribonucleic acid, and deoxyribonucleic acid (DNA), the genetic material of a cell. Such sites are $>$S: in methionine; -S:$^-$ in cysteine; $>$N: as N-1 and N-3 of histidine, N-3 and N-7 of guanine, N-1, N-3, and N-7 of adenine, and N-3 of cytosine; $>$C: as C-3 of tyrosine and C-8 of guanine; $>$C-O:$^-$ as O-6 of guanine.

The ultimate metabolic pathway for a toxin or carcinogen usually represents the final product of several competing pathways; the extent to which various metabolic products are formed depends, in part, on how much substrate is available for metabolic conversion. When toxin is present at low concentrations, metabolism proceeds according to the reactions with the fastest rates. As toxin concentration increases, pathways with rapid rats can become saturated, making substrate available to slow reactions, forming additional products. For example, it has been suggested that at low concentrations vinyl chloride metabolism may follow a pathway of detoxication and facilitation of urinary excretion, while at high concentrations, this pathway may become saturated, leaving sufficient vinyl chloride or one of its intermediary metabolites in liver cells to be metabolized via a minor pathway leading to the formation of the putative carcinogen.[31] When young adult male Sprague Dawley rats were exposed to vinyl chloride at air concentrations of 100 ppm, 50% of the absorbed toxin was metabolized in 86 min and involved hydroxylation to 2-chloroethanol; alcohol dehydrogenase converts 2-chloroethanol to chloroacetaldehyde which reacts rapidly with sulfhydryl groups of glutathione and cysteine. At concentrations greater than 220 ppm, 50% of the toxin is metabolized in 261 min, suggesting saturation of the above pathway. In addition at high concentrations of vinyl chloride, monochloracetic acid appears as a metabolite; Hefner et al.[31] suggest that this acid may be derived from further oxidation of chloroacetaldehyde remaining after saturation of the above sulfhydryl groups and also derived from decomposition of a putative vinyl chloride epoxide, a compound likely to produce a stong electrophilic product.

Thus, at doses of vinyl chloride high enough to deplete liver stores of glutathione and other sources of free sulfhydryl groups, compounds which make up part of the normal chemical line of defense in the liver, chloroacetaldehyde and perhaps vinyl chloride epoxide may be produced in sufficient quantity near DNA to react covalently with the macromolecule to initiate a mutagenic and carcinogenic event.

Gehring and Young[32] have recently summarized several instances demonstrating this important phenomenon of dose-dependent pharmacokinetics. The renal uptake and excretion of the herbicide 2,4,5-trichlorophenoxyacetic acid (2,4,5-T) is governed by the organic acid secretory process of the kidney; at a dose of 5 mg/kg, 2,4,5-T is readily excreted by the kidney, but the secretory process becomes saturated at doses approaching 100 mg/kg, resulting in a disproportionately larger volume of distribution in the body, involving tissues undisturbed at lower doses and leading to biotransformation products not seen at lower doses.

Saturation of the metabolic oxidation pathway for 1,4-dioxane is cited by Gehring and Young[32] as an example of dose-dependent pharmacokinetics where rates of metabolism and routes of excretion become dose-sensitive. Other chemicals cited by Gehring and co-workers[32-34] which show dose-dependent (nonlinear) pharmacokinetics include

Acetaminophene	Ethylene glycol
Amphetamine	Furosemide
Aniline	Isonicotinic acid hydrazide
Benzo(*a*)pyrene	2-Naphthylamine
Bishydroxycoumarin	Salicylamide
Bromobenzene	Styrene
Carbon disulfide	Sulfobromophthalein

The principal implication that dose-dependent pharmacokinetics has on carcinogen metabolism is that it makes probable that at low doses these compounds could be metabolized only via detoxication pathways, while at higher doses, upon saturation of the usual detoxication reactions (in effect reaching the "metabolic threshold" suggested by Gehring), the carcinogens become activated. The kind of dose-tumor response curve one would obtain under such a range of dose then would be shallow or flat for low doses with a sharp steep slope at higher doses.

Two other mechanisms in chemical carcinogenesis are also important to consider here: (1) repair of the damage done to DNA by the electrophilic reactant (ultimate carcinogen) and (2) immunologic repression of tumor growth.

Immunologic repression refers to the ability of the body to recognize a malignant cell as foreign and initiate the immunologic system to isolate and destroy that cell. If only very few malignant cells are formed and the immunologic system is intact, the probability for immunologic suppression of tumor growth is high. When large amounts of carcinogen are administered, the immunologic system may become overloaded, or it may become repressed, resulting in the escape of malignant cells from this protective mechanism and the establishment of an irreversible tumor.[35] The significance of such overloading of the immunologic system is that the number of resultant tumors would not be directly proportional to exposure to carcinogen until overloading was complete, i.e., the dose-response (tumor) curve would be nonlinear, with fewer tumors at low doses than expected from a linear extrapolation from high-dose responses.

Most mammalian cells are able to recognize alteration of DNA (supposedly the carcinogenic event) and reverse it, restoring the fidelity of the genetic template; this is known as DNA repair.[36-38] When an electrophilic carcinogen binds strongly to the DNA molecule, the binding is presumed to alter "permanently" the genetic template of the cell. This alteration can be recognized by the cell which enzymatically examines the template, locates the bound carcinogen or the site where it had been bound (e.g., apurinic sites), excises the portion of the DNA molecule damaged by the carcinogen and repairs the excision site with normal DNA complement, thus restoring the original genetic template; this repaired cell produces daughter cells which are completely normal, even though the parent cell once was transformed by a carcinogen.

Pegg[39] has reported the results of a study he made on the effect of dosage of dimethylnitrosamine on the alteration of DNA template in target cells. Most researchers in this field believe there is strong evidence to support methylation of DNA at the O^6-position of guanine as one of the carcinogenic initiation mechanisms (see Chapter 6). The proposed ultimate carcinogen from dimethylnitrosamine is the positively charged methonium ion, and the critical adduct with DNA is said to be O^6-methylguanine. This methylated base in DNA is enzymatically removed in liver by the repair system. Pegg[39] showed that for doses above 2.5 mg dimethylnitrosamine per kilogram body weight in the rat, there was a linear relationship between O^6-methylguanine levels in liver DNA and dose of the nitrosamine. Below 2.5 mg/kg, however, O^6-methylguanine levels were not directly proportional to dosage, but were less than would be expected from a linear extrapolation from the dose-response (DNA adduct) curve at higher doses. Evidence was presented indicating that enzymatic removal of O^6-methylguanine, repair of DNA, was more efficient at lower doses and became saturated, overloaded, and/or partially inhibited at high doses of carcinogen. Changes in metabolic activation of dimethylnitrosamine in going from low to high doses may also have had an effect on the levels of O^6-methylguanine in the liver DNA. Such results offer strong evidence to support consideration of nonlinear pharmacokinetics for cellular interactions of carcinogens at low doses.

Recent reports from Maher and co-workers[40,41] have demonstrated a nonlinear dose-response curve for the cytotoxicity, mutagenicity, and DNA binding of the putative carcinogenic metabolite of benzo(*a*)pyrene in human cells. At high doses of 7,8-dihydrodiol-9,10-epoxybenzo(*a*)pyrene in human fibroblasts, the typical linear relationships were observed between epoxide concentration and binding to DNA, mutation rate, and cell death; at low doses, however, DNA binding appeared to be repaired before cells divided, and thus mutants did not occur.

If we are to give any credence to interactions between ultimate carcinogens (and mutagens) and DNA, then we can only conclude that the growing body of evidence for dose-dependency in metabolic activation and DNA repair is strong support for nonlinearity in the pharmacokinetics describing the dose-response curves at low doses and that linear extrapolations from high-dose effects are likely to overestimate or underestimate the magnitude of the response at low doses. Nevertheless, there are times when one is forced to make a decision on whether a given carcinogen in the environment presents a real and serious risk to human health; at such times, the decision maker usually resorts to some form of mathematical model, much like those used below to "evaluate" the dose-response data for dimethylnitrosamine carcinogenicity in the rat.

B. Extrapolation According to a Typical Mathematical Model

With an understanding of the biological implications of the assumptions upon which the mathematical models are based, one can now proceed to see how such models are being used to estimate risk. To permit comparison of models in a situation relevant to

N-nitroso compounds and human cancer, use is made of the dose-response data of Terracini et al.[11] for dimethylnitrosamine feeding to female rats.

In a paper by Guess and colleagues,[42] the dimethylnitrosamine data were used to illustrate a statistical model for low-dose extrapolations of animal carcinogenicity data to estimate human risk. Using this linear, multistage model, the authors estimated that the excess risk of developing liver cancer over the risk of other "background" causes of liver cancer would be in the order of 1 in 10 for a dietary level of eight parts dimethylnitrosamine per million and between 1 in 100 and 1 in 10,000 for 0.4 ppm in the diet. The excess risk approximates 1 in 100 million somewhere between dietary levels of 10^{-3} and 10^{-5} parts dimethylnitrosamine per million. The range of 0.4 to 8 ppm is considered here, for this is the range estimated to represent human exposures based on ingestion of nitrate and nitrite (see Section II.A.1 of this chapter).

The incidence of primary liver cancer in the U.S. is approximately two cases per 100,000 people per year, or about 140 cases per 100,000 on a lifetime basis (assuming an average lifespan of 70 years). The mathematical model predicts, using the upper confidence limits, that dimethylnitrosamine in the U.S. dietary, if the lowest estimate (test meal for a human subject)[3] of approximately 0.4 ppb is used, would present an excess (additional) lifetime risk of approximately 1 in 1 million, or an increase of approximately 0.07%. On the other hand, if the estimates of dietary burden calculated from ingestion of nitrate and nitrite are used, the model predicts liver cancer rates in humans greatly in excess of known rates due to all causes. The model cannot be faulted for poor estimates of exposure to N-nitroso compounds, but then it should not be used if there is little confidence in the exposure estimates currently available which is certainly the case for environmental nitrosamines.

V. RISK VS. BENEFIT

In that it is concluded that the major environmental sources of N-nitroso compounds are indirectly the nitrate and nitrite and various secondary and tertiary amines in the human dietary and pharmacopoeia, it is necessary to offer some balance to the problem and point out the benefit to man in using nitrite as a food additive. At this writing, the U.S. Food and Drug Administration and Department of Agriculture are considering total or partial bans on the use of nitrate and nitrite as intentional additives to human foods in compliance with the Delaney clause in the Food and Drug Act.

Since the early 1900s, it has been known that a characteristic flavor and color of meat products cured with salt was attributed to the presence of nitrite in the curing mix. In 1925 the Department of Agriculture formally authorized the direct addition of nitrite to meats as part of the curing process; nitrate is added as reserve source of the more unstable nitrite.

The antimicrobial effect of nitrite was demonstrated in the 1920s, and it was soon substantiated that nitrite was particularly effective in inhibiting the growth and toxin production of the bacterium, *Clostridium botulinum*. This microorganism, a frequent contaminant of foods and foodstuffs, under anaerobic conditions found in some vacuum-packed foods, produces a protein, which is the most potent poison known to man. The particular poisoning that results, botulism, often escapes early diagnosis and has a high fatality rate, although some patients survive if given antitoxin early in the course of toxicosis or if maintained on a mechanical respirator until the nervous system has recovered, a period of several weeks.

It wasn't until 1973 that the amounts of nitrite needed to prevent botulinum toxin production in various vacuum-packed products were determined. In the absence of another readily available and inexpensive agent that could be used in lieu of nitrite to prevent toxin production, the benefit to human health offered by the use of nitrite in certain processed foods seems undeniable. The benefit is difficult to quantitate, however, because one cannot calculate how many people would suffer botulinum poisoning if the nitrite were not used. Toxin production is not a necessary consequence of failure to add nitrite to certain food products, and even if the toxin is formed, it can be destoyed by thorough cooking. Nevertheless, since the food industry essentially has always had the benefit of using nitrite, save a few well-publicized exceptions, data are not available on how large a scale the botulism problem could be. Botulism has frequently resulted from underprocessed home packed foods, and if one can extrapolate over a similar problem on the scale of the commercial food industry, one would conclude that the problem is likely to be just as large as a nitrite-related cancer risk, or even larger.

Too few data are available to the scientist to permit quantitative argument of the risks and benefits to using nitrite food additives. The only recommendation that seems justifiable at this point is to recognize that there are both risks and benefits to consider and that nitrite usage could reasonably be restricted to levels necessary to control botulism, but higher usage levels would appear unwarranted.

REFERENCES

1. U.S. Environmental Protection Agency, Assessment of Scientific Information on Nitrosamines. A Report on an ad hoc Study Group of the Executive Committee of the Science Advisory Board, U.S., Environmental Protection Agency, Washington, D.C., 1976.
2. National Research Council, Nitrates: An Environmental Assessment, Panel on Nitrates of the Coordinating Committee for Scientific and Technical Assessments of Environmental Pollutants, Commission on Natural Resources, National Academy of Sciences, Washington, D.C., 1978, 723.
3. Fine, D. H., Ross, R., Roundbehler, D. P., Silvergleid, A., and Song, L., Formation *in vivo* of volatile N-nitrosamines in man after ingestion of cooked bacon and spinach, *Nature, (London)*, 265, 753, 1977.
4. White, J. W., Jr., Relative significance of dietary sources of nitrate and nitrite, *J. Agric. Food Chem.*, 23, 886, 1975; corrections, *J. Agric. Food Chem.*, 24, 202, 1976.
5. Spiegelhalder, B., Elsenbrand, G., and Preussmann, R., Influence of dietary nitrate on nitrite content of human saliva: possible relevance to *in vivo* formation of N-nitroso compounds, *Food Cosmet. Toxicol.*, 14, 545, 1976.
6. Mysliwy, T. S., Wick, E. L., Archer, M. C., Shank, R. C., and Newberne, P. M., Formation of N-nitrosopyrrolidine in a dog's stomach, *Br. J. Cancer*, 30, 279, 1974.
7. Fine, D. H., Rounbehler, D. P., Belcher N. M., and Epstein, S. S., N-nitroso compounds: detection in ambient air, *Science*, 192, 1328, 1976.
8. Fine, D. H., Rounbehler, D. P., Rounbehler, A., Silvergleid, A., Sawicki, E., Krost, K., and DeMarrais, G. A., Determination of dimethylnitrosamine in air, water and soil by thermal energy analysis: measurements in Baltimore, Md., *Environ. Sci. Technol.*, 11, 581, 1977.
9. Bachmann, J., personal communication, 1977.
10. Druckrey, H., Schildbach, A., Schmahl, D., Preussman, R., and Ivankovic, S., Quantitative Analyse der carcinogenen Wirkung von Diathylnitrosamine, *Arzneim.-Forsch.*, 13, 841, 1963.
11. Terracini, B., Magee, P. N., and Barnes, J. M., Hepatic pathology in rats on low dietary levels of dimethylnitrosamine, *Br. J. Cancer*, 21, 559, 1967.
12. Saracci, R., Problems of low-frequency phenomena in toxicological studies, in *The Problems of Species Difference and Statistics in Toxicology*, De, S. D., Baker, C., Tripod, J., and Jacob, J., Eds., Proc. Eur. Soc. for the Study of Drug Tox., Vol. 11, Excerpta Medica Foundation, Amsterdam, 1970, 100.
13. Vesselinovitch, S. D., The sex-dependent difference in the development of liver tumors in mice administered dimethylnitrosamine, *Cancer Res.*, 29, 1024, 1969.
14. Clapp, N. K. and Toya, R. E., Effect of cumulative dose and dose rate on dimethylnitrosamine oncogenesis in RF mice, *J. Natl. Cancer Inst.*, 45, 495, 1970.
15. Clapp, N. K., Craig, A. W., and Toya, R. E., Sr., Diethylnitrosamine oncogenesis in RF mice as influenced by variations in cumulative dose, *Int. J. Cancer*, 5, 119, 1970.

16. **Montesano, R. and Saffiotti, U.**, Carcinogenic response of the respiratory tract of Syrian Golden hamsters to different doses of diethylnitrosamine, *Cancer Res.*, 28, 2197, 1968.
17. **Moiseev, G. E. and Benemanskii, V.V.**, The carcinogenic activity of low concentrations of nitroso dimethylamine on inhalation, *Vopr. Onkol.*, 21, 107, 1975.
18. **Mantel, N. and Bryan, W. R.**, "Safety" testing of carcinogenic agents, *J. Natl. Cancer Inst.*, 27, 455, 1961.
19. **Mantel, N., Bohidar, N. R., Brown, C. C., Ciminera, J. L., and Tukey, J. W.**, An improved Mantel-Bryan procedure for "safety" testing of carcinogens, *Cancer Res.*, 35, 865, 1975.
20. **Cornfield, J.** Carcinogenic risk assessment, *Science*, 198, 693, 1977.
21. **Kriek, E., Miller, J. A., Juhl, U., and Miller, E. C.**, 8-(N-2-Fluorenyl-acetamido)-guanosine, an arylamidation reaction product of guanosine and the carcinogen N-acetoxy-N-2-fluorenylacetamide in neutral solution, *Biochemistry*, 6, 177, 1967.
22. **Miller, E. C., Lotlikar, P. D., Miller, J. A., Butler, B. W., Irving, C. C., and Hill, J. T.**, Reactions *in vivo* of some tissue nucleophiles with the glucuronide of N-hydroxy-2-acetylaminofluorene, *Mol. Pharmacol.*, 4, 147, 1968.
23. **Miller, J. A.**, Carcinogenesis by chemicals: an overview — G.H.A. Clowes Memorial Lecture, *Cancer Res.*, 30, 559, 1970.
24. **Swenson, D. H., Miller, E. C., and Miller, J. A.**, Aflatoxin B_1-2,3-oxide: evidence for its formation in rat liver *in vivo* and by human liver microsomes *in vitro*, *Biochem. Biophys. Res. Commun.*, 60, 1036, 1974.
25. **Metzler, M. and McLachlan, J. A.**, Diethylstilbestrol: its metabolism in fetal, neonatal and adult mice and electrophilic reactivity of metabolites, 1st Int. Congr. Toxicology, April 1, 1977, Toronto, Canada.
26. **Grover, P. L. and Sims, P.**, Interaction of the K-region epoxides of phenanthrene and dibenz(a,b)anthracene with nucleic acids and histones, *Biochem. Pharmacol.*, 19, 2251, 1970.
27. **Lehr, R. E. and Jerina, D. M.**, Relationships of quantum mechanical calculations, relative mutagenicity of benzo(a)anthracenediol epoxides and "bay region" concept of aromatic hydrocarbon carcinogenicity, *J. Toxicol. Environ. Health*, 2, 1259, 1977.
28. **Druckrey, H., Preussmann, R., Ivankovic, S., and Schmahl, D.**, Organotrope carcinogene Wirkungen bei 65 verschiedenen N-Nitroso-Verbindungen an BD-Ratten, (Organ-specific carcinogenic action of 65 different N-nitroso compounds in BD rats), *Z. Krebsforsch*, 69, 103, 1967.
29. **Magee, P. N.**, Toxicity of nitrosamines: their possible human health hazards, *Food Cosmet. Toxicol.*, 9, 207, 1971.
30. **Shank, R. C.**, Toxicology of N-nitroso compounds, *Toxicol. Appl. Pharmacol.*, 31, 361, 1975.
31. **Hefner, R. E., Jr., Watanabe, P. G., and Gehring, P. J.**, Preliminary studies of the fate of inhaled vinyl chloride monomer in rats, *Ann. N.Y. Acad. Sci.*, 246, 135, 1975.
32. **Gehring, P. J. and Young, J. D.**, Application of pharmacokinetic principles in practice, 1st Int. Congr. Toxicology, March 31, 1977, Toronto, Canada.
33. **Watanabe, P. G., McGowan, G. R., and Gehring, P. J.**, Fate of (14C) vinyl chloride after single oral administration in rats, *Toxicol Appl. Pharmacol.*, 36, 339, 1976.
34. **Gehring, P. J., Watanabe, P. G., and Young, J. D.**, The relevance of dose-dependent pharmacokinetics in the assessment of carcinogenic hazard of chemicals, in *Cold Spring Harbor Conf. Cell Proliferation*, 1977, Cold Spring Harbor Laboratory, Cold Spring Harbor, New York, 1977, 205.
35. **Penn, I.**, Malignancies associated with immunosuppressive or cytotoxic therapy, *Surgery*, 83, 492, 1978.
36. **Goth, R. and Rajewsky, M. F.**, Persistence of O^6-methylguanine in rat brain DNA, *Proc. Natl. Acad. Sci. U.S.A.*, 71, 639, 1974.
37. **Goth, R. and Rajewsky, M. F.**, Molecular and cellular mechanisms associated with pulse-carcinogenesis in the rat nervous system by ethylnitrosourea; ethylation of nucleic acids and elimination rates of ethylated bases from the DNA of different tissues, *Z. Krebsforsch*, 82, 37, 1974.
38. **Ramanathan, R., Rajalakshmi, S., Sarma, D. S. R., and Farber, E.**, Non-random nature of *in vivo* methylation by dimethylnitrosamine and the subsequent removal of methylated products from rat liver chromatin DNA, *Cancer Res.*, 36, 2073, 1976.
39. **Pegg, A. E.**, Alkylation of rat liver DNA by dimethylnitrosamine: effect of dosage on O^6-methylguanine levels, *J. Natl. Cancer Inst.*, 58, 681, 1977.
40. **Maher, V. M., Dorney, D. J., Mendrala, A., Harvey, R. G., and McCormick, J. J.**, Comparing in normal diploid human fibroblasts the cytotoxicity and mutagenicity of the 7,8-dihydrodiol-9,10-oxide ("anti" isomer) and "K-region" epoxide of benzo(a)pyrene, *Proc. Am. Assoc. Cancer Res.*, 18, 190 (Abstr.) March 1977, Denver, Colo.
41. **McCormick, J. J., Heflich, R. H., and Maher, V. M.**, Evidence of excision repair in human cells of DNA damage caused by the 7,8-dihydrodiol-9,10-oxide ("anti" isomer) and the 4,5-oxide of benzo(a)pyrene, *Proc. Am. Assoc. Cancer Res.*, 18, 190 (Abstr.) March 1977, Denver, Colo.
42. **Guess, H., Crump, K., and Peto, R.**, Uncertainty estimates for low-dose-rate extrapolations of animal carcinogenicity data, *Cancer Res.*, 37, 3475, 1977.

INDEX

A

2-AAF, see Acetylaminofluorene
Acanthamoeba, II: 79
Acetate incorporation, and trichothecenes, II: 44
C-Acetate incorporation, and aflatoxins, II: 23
Acetoxyscirpendiol, II: 33, 37
4-Acetoxyscirpendiol, II: 33
Acetylaminofluorene (2-AAF), I: 239
Acetylation, II: 45
Acetylchaetoglobosin A, II: 53
Acetylcholine, and cytochalasins, II: 66, 71
Acetylcholinesterase, II: 83
Acetyl CoA synthesis, and aflatoxins, II: 24
3-Acetyldeoxynivalenol, II: 33, 37, 40
O-Acetyl-dihydrosterigmatocystin (O-AcDHSTG), II: 139, 141, 143
Acetyl T-2 toxin, II: 33
Achlya, II: 75
Achlya ambisexualis, II: 64, 68
α_1-Acid glycoprotein, II: 20
Acid phosphatase, and aflatoxins, II: 24
Acinetobacter calcoaceticus, II: 30
O-AcSTG, see O-Acetylsterigmatocystin
ACTH, see Adrenocorticotropin
Actin, and cytochalasins, II: 72, 77—80, 90, 91, 93, 94
F-Actin, and cytochalasins, II: 78, 79
G-Actin, and cytochalasins, II: 78, 94
Actin-binding protein, and cytochalasins, II: 79
Actinomyces, and ochratoxins, II: 130
Actinomycin D, II: 21, 22
Action potential, II: 72
Actomyosin, and cytochalasins, II: 71, 78, 79
Actomyosin contraction, and cytochalasins, II: 68
Acute cardiac beriberi, I: 114
Acute hepatic disease, and aflatoxins, I: 52
Acylation, and cytochalasins, II: 91
Adduct formation
  and N-nitroso compounds, I: 207
  and patulin, II: 124—125
  and penicillic acid, II: 124
Adenine, II: 87
Adenocarcinomas, II: 5
  and sterigmatocystins, II: 139
Adenomas
  and ochratoxins, II: 131
  and sterigmatocystins, II: 139
Adenosine, II: 87
Adenyl cyclase, and cytochalasins, II: 76
ADH, see Alcohol dehydrogenase
Adhesion, II: 92
  and cytochalasins, II: 59, 60
Adipocytes, II: 88
  and cytochalasins, II: 73, 84, 85, 89, 90
Adrenal cells, and cytochalasins, II: 76
Adrenal gland
  and aflatoxins, I: 53, 54
  and N-nitroso compounds, I: 197, 206

and ochratoxins, I: 78
and trichothecenes, I: 61
Adrenocoritotropin (ACTH), and cytochalasins, II: 76
Aerobacter aerogenes, II: 89
AFB, see Aflatoxin B
AFG, see Aflatoxin G
AFL, see Aflatoxicol
Aflatoxicol (AFL), II: 10, 27
Aflatoxicol (AFL) dehydrogenase activity, II: 10
Aflatoxicol H$_1$ (AFLH$_1$), II: 10, 27
Aflatoxicol M$_1$ (AFLM$_1$), I: 55, 58, 123, 143; II: 8, 10, 12—16, 19, 22, 23, 27
Aflatoxin B$_1$ (AFB$_1$), see also Aflatoxins, I: 6, 19, 20, 24, 52—58, 63, 66, 71, 79—81, 86, 90, 121, 123, 126, 128, 130, 132—135, 142—150, 228, 239; II: 1, 4—8, 11—27, 37, 138, 140—142, 156
  cytotoxicity of, II: 4
  excretion of, II: 14—16
  hydroxylation of, II: 8, 11
  macromolecular complexes of, II: 12
  metabolic transformations of, II: 9, 17
  tissue distribution of, II: 14—16
C-Aflatoxin B$_1$ (AFB$_1$), II: 15
Aflatoxin B$_1$ (AFB$_1$)-2,3-dichloride, II: 12
Aflatoxin B$_1$ (AFB$_1$)-dihydrodiol, II: 11
Aflatoxin B$_1$ (AFB$_1$)-glutathione conjugate, II: 14
Aflatoxin B$_1$ (AFB$_1$) hydroxylase, II: 13
Aflatoxin B$_2$ (AFB$_2$), see also Aflatoxins, I: 19, 24, 28; II: 4—8, 10—12, 15, 17, 19—21, 24—26, 142
Aflatoxin G$_1$ (AFG$_1$), see also Aflatoxins, I: 6, 24, 53—55, 57, 66, 90; II: 4—8, 17, 19—21, 23, 25—27, 142
Aflatoxin G$_2$ (AFG$_2$), see also Aflatoxins, I: 24, 28, 132; II: 4—8, 12, 17, 20, 25—27, 142
Aflatoxin glutathione conjugate, II: 11
Aflatoxin M$_1$ (AFM$_1$), I: 6, 10, 19, 20, 24
Aflatoxin M$_2$ (AFM$_2$), I: 24, 27
Aflatoxin P$_1$ (AFP$_1$), II: 8—10, 12, 13, 15, 27
Aflatoxin Q$_1$ (AFQ$_1$), I: 55, 58; II: 10, 12—15, 27
Aflatoxins, see also specific aflatoxins, I: 5, 7—11, 18—22, 52—58; II: 1, 2, 156
  and C acetate incorporation, II: 23
  and acetyl CoA synthesis, II: 24
  and acid phosphatase, II: 24
  and acute hepatic disease, I: 52
  and adrenal gland, I: 53, 54
  and agglutinin, II: 26
  and albumin, II: 24, 26
  and amino acid incorporation, II: 19
  and amino acid transport, II: 26
  and aminoacyl-tRNA, II: 20
  and *Aspergillus*, I: 123
  and *Aspergillus clavatus*, I: 130
  and *Aspergillus flavus*, I: 123, 131, 134
  and *Aspergillus ochraceous*, I: 130, 131

and baboons, I: 66
and *Bacillus subtilis*, II: 7
and beans, I: 121, 122, 124
and bile, I: 133; II: 8, 14, 16
and bile duct hyperplasia, I: 21
biological activity of, II: 4-6
and biological tests, I: 28-29
and birds, II: 10
and blood, I: 19-21, 131; II: 15, 16, 25-26
and bone marrow, I: 21
and brain, I: 133
and breast milk, I: 134
and brine shrimp, II: 4
and calcium, II: 26
and capillary fragility, II: 24
and carcinogenicity, I: 8, 10-12, 19, 52, 55-58, 68, 120-130; II: 5-8, 10, 11, 13, 14, 16, 17, 19, 21, 24, 27
and cassava, I: 121, 122, 124, 130
and cell-mediated immunity, II: 26
and chang liver cells, II: 16
chemistry of, II: 4
and chicken embryos, II: 8, 10
and chickens, I: 20-22, 52; II: 4, 5, 8, 10, 11, 16, 24-26
and cholesterol, II: 23, 25
and cholesterol esters, II: 23
and chromosomal abnormalities, II: 5
and CO_2, II: 14
and cointoxication, I: 129-130
and colon, I: 128
and complement, II: 26
and confirmatory tests, I: 28
and copper, II: 26
and corn, I: 21
and corticosterone, II: 22
and cows, I: 19-20, 52; II: 8, 10, 15
and *Cyclops fuscus*, II: 4
and cytochalasin E, I: 131
and cytochalasins, I: 131; II: 48
and cytochromes, II: 23
and cytosol proteins, II: 12
and dairy products, I: 27-28
and degranulation, II: 22
and O-demethylation, II: 14
and deoxyguanosine, II: 12
detection of, I: 24-29
and dietary factors, II: 13, 14
distribution of, II: 6-16
and DNA, II: 11-14, 17, 18, 26, 27
and DNAse activity, II: 24
DNA-attacking ability of, II: 6
and DNA binding, II: 6, 7, 9, 11, 12, 14, 19, 27
and DNA damage, II: 6, 7
and DNA repair, II: 7, 13, 17
and DNA synthesis, I: 52, 53, 57; II: 7, 12, 16, 17, 19, 20, 22, 23
and dogs, I: 52, 53
dose-response relationships of, I: 144-145, 148
and *Drosophila*, II: 4
and ducks, I: 20-22, 52, 53, 58, 66, 71, 72; II: 4, 5, 7, 8, 17, 25

and eggs, I: 21, 22; II: 16
and electron transport, II: 23
epidemiology of, I: 10-11
and epoxidation, II: 11-13, 17, 27
and erythrocytes, II: 26
and *Escherichia coli*, II: 7
excretion of, I: 57; II: 6-16
and fat pads, II: 24
and fatty acids, II: 23
and feces, I: 133; II: 8, 14-16
and fibroblasts, II: 7
and fish, I: 57-58, 124; II: 4, 5, 7, 10
and fowl, I: 52-53
and *Fusarium*, I: 123
and gall bladder, I: 53
and garlic, I: 124
and globulin, II: 26
and glucose incorporation, II: 25
and glucuronide conjugates, II: 6-8, 15
and glutathione, II: 11, 13
and glycine methyltransferase, II: 24
and glycogen, II: 25
and goats, I: 52; II: 8
and guinea pigs, I: 53, 66; II: 8, 25, 26
and guppies, I: 58
and hamsters, I: 66; II: 5, 7, 8, 11, 13, 17
and heart, I: 19, 52, 131, 133; II: 23
and HeLa cells, II: 17, 18, 26
and helical polysomes, II: 7, 21
and hepatic disease, I: 52, 134
and hepatitis, I: 134
and hepatocellular carcinomas, II: 5
and hepatocyte, II: 8
and hormonal status, II: 13
and humans, I: 7-8, 57, 120-130; II: 5, 7, 8, 10, 24
and hydroxylation, II: 6, 13, 27
and immune system, I: 20
and immunoglobulin A, II: 26
and immunoglobulin G, II: 26
and infection, I: 129
and initiation factor, II: 21
and intestines, II: 24
and iron, II: 26
and Japanese killifish, II: 5
and kidney, I: 19, 20, 52, 56, 131, 133; II: 6-8, 13, 15-17, 19-21, 23, 24, 26
and leucine, II: 7, 26
and C-leucine, II: 19
and lipid synthesis, II: 23-24
and liver, I: 18-21, 52-58, 71, 72, 120-130, 133, 134; II: 5, 7, 8, 10-13, 15-23, 25, 26
and liver fluke, I: 129
and lung, I: 131; II: 6, 17
and lymph nodes, I: 131
and lysosomal enzymes, II: 24-25
and lysosomes, II: 23, 24, 27
macromolecular binding of, II: 11-13
and magnesium, II: 26
and maize, I: 121-123
and malate dehydrogenase, II: 25
and marmosets, I: 58

and membrane, II: 22
and membrane transport, II: 17, 26
and metabolic activation, II: 6, 11, 25, 27
and metabolism, I: 54, 57; II: 6-16, 18
metabolites of, II: 8-11
and mice, I: 55, 66; II: 5, 6, 8, 10, 13, 16, 17, 19-21, 24
and microsomal activity, II: 14
and microsomal enzymes, II: 12
and microsomes, II: 10, 11, 13, 19, 22, 25
and milk, I: 20, 27-28; II: 15
and mink, II: 16, 24
and mitochondria, II: 7, 23
and mitochondrial function, II: 23, 27
and mitosis, I: 52
and mixed function oxidase, II: 20
and monkeys, I: 52-54, 58, 66, 132, 133; II: 8, 10, 15-17, 19, 20, 25
and monosomes, II: 20
and mRNA synthesis, II: 18
and muscle, I: 19, 21; II: 8, 15, 16, 24
and mutagenicity, I: 54; II: 5, 7-9, 11, 12, 27, 37
and NADH, II: 25
and NADP, II: 25
and NADPH, II: 8, 25
and *Neurospora crassa*, II: 7
and N-nitroso compounds, I: 223
and nuclear membrane, II: 12
and nuclei, II: 7, 8, 11, 13, 16, 18, 19, 22, 24, 25
and nucleic acid synthesis, II: 17
and nucleolar polymerase, II: 18
and nucleolar RNA, II: 19
and nucleoplasmic polymerase, II: 18
and nucleoside transport, II: 27
and nutrition, I: 127-129
occurrence of, II: 4
and ochratoxin, I: 123, 130
and onions, I: 124
and orotic acid, II: 7
and C-orotic acid, II: 17
and ovariectomy, II: 13
and ovary, II: 5
and pancreas, I: 21, 52
pathways of, II: 8-11
and peanut butter, I: 122, 123
and peanuts, I: 18, 21, 52, 57, 58, 121, 122, 127, 130, 134; II: 124
and peas, I: 121, 122
and penicillic acid, I: 130
and *Penicillium*, I: 123
and peppers, I: 124
and peptidyl-tRNA, II: 20
and phosphate incorporation, II: 17, 23
and phospholipids, II: 23
and phosphorylation, II: 8
and physiochemical properties, I: 25
and pigs, I: 18-19, 22, 52; II: 8, 16
and plasma, II: 25
and plasma proteins, II: 20
and polynucleotides, II: 11

and polysome disaggregation, II: 21
and polysomes, II: 20-22
and protein, II: 11-13, 19-21, 25, 26
and protein binding, II: 6-8, 11-13, 25
and protein deficiency, II: 14, 23
and protein synthesis, II: 7, 12, 16, 19-23, 25, 26
and rabbits, II: 5, 10, 17, 20, 21, 26
and rainbow trout, I: 55, 57, 66; II: 5, 7
and rats, I: 18, 52-57, 66-68, 127-129; II: 5, 7, 8, 10-26
and *rec* assay, II: 6-7
and respiration, II: 23
and Reye's syndrome, I: 131-133
and ribonuclease, II: 18
and ribosomal subunits, II: 20, 21
and ribosomes, II: 20, 21
and rice, I: 123, 124, 130, 131
and risk assessment, I: 142, 149-151
and RNA, II: 11, 18, 26, 27
and mRNA, II: 20, 21
and rRNA, II: 12, 13, 17, 18
and tRNA, II: 12, 19, 20, 25
and RNA binding, II: 6, 7, 9, 11-13, 19, 27
and RNA polymerase, II: 18
and RNA synthesis, II: 7, 16, 19, 20, 22, 26, 27
and salmon, I: 58
and *Salmonella typhimurium*, II: 7
and sampling procedure, I: 24, 26
and serum, II: 26
and serum protein, II: 20
and sesame seeds, I: 124
and sheep, II: 8, 15
and shrews, I: 58
and shrimp, I: 124
and skin, II: 5, 6, 17
and smooth microsomes, II: 22
and spermatozoa, I: 19
and spleen, I: 19, 21, 52, 131; II: 26
and sterigmatocystins, I: 123, 130; II: 138
and steroid binding, II: 22
and stomach carcinomas, II: 5
structure, I: 25, 121
structure-activity relationships in, II: 7-8, 26-27
and sulfate conjugates, II: 8, 15
and template activity, II: 18
and teratogenicity, I: 55
and termination, II: 21
and testis, II: 23
and testosterone, II: 14, 22
and thymidine, I: 53, 56, II: 7
and H-thymidine, II: 17
and thymidine transport, II: 17
and thymus, I: 20, 131; II: 13
and tissue, II: 25-26
and toxicity, I: 19, 52-58, 66, 67, 130-135; II: 6-9, 11-13, 15-17, 19, 21, 23, 24, 27
and trachea, II: 4
and transcription, II: 27
and translation, II: 27
and trichothecenes, I: 15
and triglycerides, II: 25

and tRNA methylase, II: 24
and *Saccharomyces cerevisiae*, II: 7
and T-2 toxins, I: 130
and trout, I: 55, 57, 58, 66, 148
and turkeys, I: 52, 53; II: 4
and turkey X disease, I: 20
and urea, II: 26
and uridine transport, II: 18
and urine, I: 57, 123, 134; II: 8, 14-16
and uterus, II: 22
and vitamin A, II: 26
and zearalenone, I: 65
and zebra fish, II: 4, 5, 7
AFLH, see Aflatoxicol H
AFLM, see Aflatoxicol M
AFM, see Aflatoxin M
AFP, see Aflatoxin P
AFQ, see Aflatoxin Q
Africa, I: 120, 125, 127, 130, 142, 221, 222
Agglutination, II: 81
and cytochalasins, II: 73, 75, 82
Agglutinin
and aflatoxins, II: 26
and cytochalasins, II: 82
Aggregation, and cytochalasins, II: 59-60, 79, 81
AHH, see Aryl hydrocarbon hydroxylase
Air analysis, and N-nitroso compounds, I: 158
Airborne nitrosamines, I: 230
Alamine, II: 75
Alberta, I: 116
Albumin, II: 20, 63
and aflatoxins, II: 24, 26
and cytochalasins, II: 65
and ochratoxins, II: 132, 134, 136
and sterigmatocystins, II: 141
Alcohol, and N-nitroso compounds, I: 223
Alcohol dehydrogenase (ADH), II: 43
and penicillic acid, II: 126
Aldolase, and penicillic acid, II: 126
Alimentary toxic aleukia (ATA), I: 6, 60, 112-114; II: 30
and blood, I: 113
and bone marrow, I: 113
and *Fusarium*, I: 112
and mucous membranes, I: 113
and skin, I: 113
Aliphatic nitrosamines, I: 200-206
Allium cepa, II: 37, 122
Alloantigens, and cytochalasins, II: 81
Alterations in RNA metabolism, II: 17-18
Alternaria, I: 2, 3
Alveolar macrophages, and cytochalasins, II: 60, 69, 84, 92
α-Amanitin, and penicillic acid, II: 125
Ames assay, II: 6-8, 10, 12, 37, 123
and ochratoxins, II: 131
and sterigmatocystins, II: 139, 143
Amine alkaloids, I: 109, 110
Amino acid alkaloids, I: 110, 111
and ergot alkaloids, I: 109
Amino acid incorporation
and aflatoxins, II: 19

and cytochalasins, II: 74
Amino acids
and cytochalasins, II: 88-89
and ochratoxins, II: 136
and penicillic acid, II: 125
Amino acid transport
and aflatoxins, II: 26
and cytochalasins, II: 75, 88
Aminoacylation, and ochratoxins, II: 135
Aminoacyl-tRNA, II: 42
and aflatoxins, II: 20
α-Aminobutyric acid (GABA), and cytochalasins, II: 67
α-Aminoisobutyric acid, II: 88
and penicillic acid, II: 125
and zearalenone, II: 150
Amino sugar transport, and cytochalasins, II: 74
Amphibians
and cytochalasins, II: 72
and patulin, II: 122
α-Amylase, II: 64
Anabolic activity of zearalenone, I: 17; II: 148-149
Anchovies, and N-nitroso compounds, I: 171
Anguidine, II: 32
Animal data vs. human risks, I: 237-242
Anthraquinones, I: 114
Antiactin, and cytochalasins, II: 80
Antibodies
and cytochalasins, II: 64, 68
and ochratoxins, II: 136
Antiestrogens, and zearalenone, II: 149
Antigens
and cytochalasins, II: 64, 67-69, 81
and ochratoxins, II: 136
and penicillic acid, II: 125
Antisera, and cytochalasins, II: 82
AOAC, see Association of Official Analytical Chemists
Apple juice, and risk assessment, I: 142
Arborization, II: 56, 57, 92, 94
Arginine, and penicillic acid, II: 125
Artery
and risk assessment, I: 149
and sterigmatocystin, I: 100
Aryl hydrocarbon hydroxylase (AHH), II: 13
Ascites, and cytochalasins, II: 54, 75
Ascorbic acid, and N-nitroso compounds, I: 164
Asia, I: 120, 127, 130, 142
Aspergillus, II: 122, 123
and aflatoxins, I: 123
and Balkan (endemic) nephropathy, I: 116
and ochratoxins, I: 22, 58; II: 130
and sterigmatocystins, II: 138
Aspergillus amstelodami, II: 138
Aspergillus chevalieri, II: 138
Aspergillus clavatus, I: 5, 131; II: 49, 50, 122
and aflatoxins, I: 130
Aspergillus flavus, I: 2-5, 11, 18, 21, 24; II: 4, 138
and aflatoxins, I: 123, 131, 134
Aspergillus glaucus, I: 3
Aspergillus nidulans, I: 63; II: 138

Aspergillus niger, I: 131
Aspergillus ochraceous, I: 3, 4, 22, 33; II: 123
   and aflatoxins, I: 130, 131
   and ochratoxins, II: 130
Aspergillus oryzae, I: 5
Aspergillus parasiticus, I: 5
Aspergillus ruber, II: 138
Aspergillus rugulosus, I: 63; II: 138
Aspergillus versicolor, I: 4, 63, 64; II: 138
Aspertoxin, II: 139, 143
   and sterigmatocystins, II: 138
Associate Referee on Confirmational Methods, I: 28
Association of Official Analytical Chemists (AOAC), I: 24, 29
ATA, see Alimentary toxic aleukia
ATP, II: 23
   and cytochalasins, II: 58, 73, 76, 78, 80, 81, 84, 91
   and ochratoxins, II: 132, 133
ATPase, II: 23
   and cytochalasins, II: 78, 79, 83
   and penicillic acid, II: 126, 127
Atrazine, I: 176
Attachment, and cytochalasins, II: 59
Auckland, I: 132
Australia, I: 131
Autolysis, and ochratoxins, II: 135
Axoplasmic transport, and cytochalasins, II: 72

B

Baboons, and aflatoxins, I: 66
Baccharin, II: 30
Bacillus cereus, and ochratoxins, II: 130
Bacillus megaterium, and ochratoxins, II: 130
Bacillus subtilis, II: 6, 37, 54, 92, 96, 122-124
   and aflatoxins, II: 7
   and ochratoxins, II: 131, 135
   and *rec* assay, II: 56, 102
   and sterigmatocystins, II: 139, 142, 143
   and zearalenone, II: 147
Bacon, and N-nitroso compounds, I: 169-171, 173, 230, 231
Bacteria
   and aflatoxins, II: 6
   and cytochalasins, II: 54, 60, 69
   and ochratoxins, II: 130, 135
   and patulin, II: 122, 125
   and penicillic acid, II: 123, 125
   and psoralens, II: 107, 108, 114, 116
   and trichothecenes, II: 31, 39
Balkan (endemic) nephropathy, I: 115-116
   and *Aspergillus*, I: 116
   and barley, I: 116
   and blood, I: 115
   and kidney, I: 115, 116
   pathological changes in, I: 8
   and *Penicillium*, I: 116
   and risk assessment, I: 142
   and skin, I: 116
   and urine, I: 115

   and wheat, I: 116
Baltimore, I: 164-167
Bangkok, I: 129, 131
Barley, I: 13
   and Balkan (endemic) nephropathy, I: 116
   and ochratoxins, I: 59
   and zearalenone, I: 18
Bass, and N-nitroso compounds, I: 176
Bean hull toxicosis, II: 30
Beans, and aflatoxins, I: 121, 122, 124
Beer, and N-nitroso compounds, I: 173, 176, 230
Beets, and N-nitroso compounds, I: 174
Belle, West Virginia, I: 165, 166
7,8-Benzoflavone, II: 16, 140
Benzo(a)pyrene, I: 239, 241
Bergapten, II: 106
BF detection method, I: 26, 27
BHK cells, II: 60
   and cytochalasins, II: 65
Bile, II: 6
   and aflatoxins, I: 133; II: 8, 14, 16
   and penicillic acid, II: 124
   and zearalenone, II: 148
Bile duct
   and aflatoxins, I: 21
   and sterigmatocystins, II: 139
Bilirubin, II: 26
Biological confirmatory tests, I: 36
Biological detection methods, I: 32-33
Biological effects of psoralens, II: 107-109
Biological tests, and aflatoxins, I: 28-29
Bipolaris, I: 63; II: 138
Birds, see also specific birds
   and aflatoxins, I: 52-53; II: 10
   and ochratoxins, II: 130
   and sterigmatocystins, II: 138
Bishydroxycoumarin, I: 116-118
Bladder
   and cytochalasins, II: 71, 76, 88
   and N-nitroso compounds, I: 170, 188-192, 204, 206, 225
   and trichothecenes, I: 62; II: 30
Blastogenesis, and cytochalasins, II: 68
Blebbing, II: 56
Blebs, II: 57
Blood
   and aflatoxins, I: 19-21, 131; II: 15, 16, 25-26
   and alimentary toxic aleukia, I: 113
   and Balkan (endemic) nephropathy, I: 115
   and cyclochlorotine, II: 102
   and cytochalasins, II: 68, 71, 85
   and luteoskyrin, II: 96
   and N-nitroso compounds, I: 159, 171, 197, 198, 204, 223, 228, 230, 232
   and ochratoxins, I: 22; II: 131, 133, 134
   and penicillic acid, II: 124
   and psoralens, II: 109
   and zearalenone, II: 149
Blood clot retraction, and cytochalasins, II: 71, 92
Blood platelets, and cytochalasins, II: 66
Blood urea, I: 8

Blood vessels, and N-nitroso compounds, I: 194
"Blue-eye" fungi, I: 3
Bone
   and N-nitroso compounds, I: 197, 206
   and ochratoxins, II: 130
Bone marrow
   and aflatoxins, I: 21
   and alimentary toxic aleukia, I: 113
   and trichothecenes, I: 16, 61; II: 37
Botrytis cinera, II: 54, 92, 106
Botulism, see also *Clostridium botulinum*, I: 242, 243
Bradykinin, and cytochalasins, II: 67
Brain
   and aflatoxins, I: 133
   and cytochalasins, II: 54, 67, 78, 79, 83, 85, 94
   and N-nitroso compounds, I: 196, 197, 205
   and patulin, II: 126, 127
   and penicillic acid, II: 127
   and trichothecenes, I: 61; II: 38
Bread, and N-nitroso compounds, I: 230
Breast milk, I: 10
   and aflatoxins, I: 134
Brine shrimp
   and aflatoxins, II: 4
   and ochratoxins, II: 130
   and penicillic acid, II: 124
   and sterigmatocystins, II: 138
   and trichothecenes, II: 37
   and zearalenone, II: 147
Broncus, and N-nitroso compounds, I: 190
Bulgaria, I: 115
Butylnitrosamines, I: 192
Byssochlamys, II: 122

C

CA, see Cytochalasin A
C-acetate incorporation, and aflatoxins, II: 23
Calcium
   and aflatoxins, II: 26
   and cytochalasins, II: 71, 78, 79, 81
   and ochratoxins, II: 133
Calcium uptake, and cytochalasins, II: 68, 88
Calometria nivalis, II: 36
Calonectrin, II: 32, 40, 45
Canada, I: 116, 117
Capillaries
   and aflatoxins, II: 24
   and cytochalasins, II: 54
Capping, II: 81-83
Carbaryl, I: 176
Carbohydrate alterations, and cyclochlorotine, II: 102
Carbohydrate metabolism, II: 156
   and cyclochlorotine, II: 102
   and ochratoxins, II: 133-134
Carbohydrates, II: 156, 157
Carboxypeptidase, and ochratoxins, II: 131, 134
Carcinogenicity, I: 4-6, 142-150; II: 156, 157
   and aflatoxins, I: 8, 10-12, 19, 52, 55-58, 68, 120-130; II: 5-8, 10, 11, 13, 14, 16, 17, 19, 21, 24, 27
   and cyclochlorotine, II: 102
   and fish, II: 10
   and fusarenon-X, I: 69
   and luteoskyrin, II: 96
   and N-nitroso compounds, I: 164, 168, 174-176, 186-209, 221-225, 228-248
   and ochratoxins, I: 60; II: 130, 131
   and patulin, II: 123, 125
   and penicillic acid, II: 124, 125
   and psoralens, II: 108-109
   and sterigmatocystin, I: 63, 64, 70
   and T-2 toxin, I: 69
   and trichothecenes, I: 15, 62-63, 69; II: 37
   and trout, II: 10
Carcinomas
   hepatocellular, II: 5
   human epidermal, II: 31
   luteoskyrin and, II: 96
   mouse mammary, II: 5, 6, 56, 96, 123, 124, 139
   squamous cell, II: 109, 139
   stomach, II: 5
Carpathian mountains, I: 115
Cassava, I: 8
   and aflatoxins, I: 121, 122, 124, 130
Castration, II: 13
Catenarin, I: 114
Cathepsin, II: 43, 142
   and zearalenone, II: 151
Cats
   and cytochalasins, II: 71, 72
   and T-2 toxins, I: 97
   and trichothecenes, I: 15, 60, 61; II: 30
Cattle, see Cows
CB, see Cytochalasin B
CC, see Cytochalasin C
CD, see Cytochalasin D
C-4-deacetylation, II: 38
CE, see Cytochalasin E
Celery, see also Pink rot disease, I: 7
   and N-nitroso compounds, I: 174
   and pink rot disease, I: 118, 119
Celery rot, see Pink rot disease
Cell cycle, and cytochalasins, II: 87-89
Cell division, and cytochalasins, II: 57
Cell-mediated immunity, and aflatoxins, II: 26
Cell nuclei, see Nuclei
Cell sorting, and cytochalasins, II: 59-60, 75
Cell spreading, II: 59
Cellular macromolecule synthesis inhibition, and psoralens, II: 115-116
Cellular respiration, II: 127
Cellulase, and cytochalasins, II: 64, 75, 94
Central America, I: 129
Cephalosporium, I: 14, II: 30
   and trichothecenes, I: 60
Cephalosporium crotocinigenum, II: 34
Cercopithecus aethiops, I: 53
Cervical ganglia, and cytochalasins, II: 66
CF, see Cytochalasin F

CG, see Cytochalasin G
CH, see Cytochalasin H
Chaetoglobosin A, II: 52, 54, 55, 58, 71, 75
Chaetoglobosin B, II: 52, 54-56, 58
Chaetoglobosin C, II: 52, 54, 55, 58
Chaetoglobosin D, II: 52, 54, 55, 58
Chaetoglobosin E, II: 52, 54, 55, 58
Chaetoglobosin F, II: 52, 54, 55
Chaetoglobosin G, II: 53, 55
Chaetoglobosin J, II: 53, 55, 58
Chaetoglobosins, II: 48, 54
   and cytochalasins, II: 91
Chaetomium, II: 48
Chaetomium globosum, II: 52
Chang liver cells
   and aflatoxins, II: 16
   and penicillic acid, II: 125
Chara, II: 58
Cheese, and N-nitroso compounds, I: 168, 172, 173, 230
Chemical confirmatory tests, I: 35
Chemical detection methods, I: 33
Chemical detection of mycotoxins, I: 31
Chemotaxis, and cytochalasins, II: 69
Chicken embryos
   and aflatoxins, II: 8, 10
   and cytochalasins, II: 56, 58-62, 71-73, 75, 81, 82, 84-86, 92
   and ochratoxins, II: 130, 131
   and patulin, II: 122, 125
   and sterigmatocystins, II: 139
   and trichothecenes, II: 45
   and zearalenone, II: 147
Chickens, I: 97
   and aflatoxins, I: 20-22, 52; II: 4, 5, 8, 10, 11, 16, 24-26
   and N-nitroso compounds, I: 224
   and ochratoxins, I: 23, 59, 60; II: 130, 136
   and sterigmatocystins, I: 63; II: 138, 141, 143
   and T-2 toxins, I: 98, 112
   and trichothecenes, I: 16; II: 30, 45
   and zearalenone, II: 147
Childhood cirrhosis, I: 134
Chile, I: 222
China, I: 58, 119, 223, 224
Chinese hamsters
   and cytochalasins, II: 87
   and psoralens, II: 108, 114-116
Chloramphenicol, and ochratoxins, II: 135
Chlorinated cyclic polypeptides, I: 114
Chlorine, and cyclochlorotine, II: 102
p-Chloromercuribenzoate (PCMB), II: 86, 89, 90
p-Chloromercuribenzoate (PCMB) sulfonate, II: 68
2-Chloro-4(N-nitroso-N-ethyl amino)-6-isopropylaminotriazine, I: 176
Cholangiocarcinomas, I: 129; II: 5
Cholesterol, and aflatoxins, II: 23, 25
Cholesterol esters, and aflatoxins, II: 23
Cholesterol uptake, and cytochalasins, II: 76
Choline, II: 89
Choline transport, II: 89

Choline acetylesterase, II: 83
Chondrogenesis, and cytochalasins, II: 62
Chromatin, II: 18, 21, 155
   and cytochalasins, II: 76
   and psoralens, II: 113, 115
   and zearalenone, II: 150
Chromosomal abnormalities, II: 37, 54
   and aflatoxins, II: 5
   and luteoskyrin, II: 98
   and patulin, II: 122
   and penicillic acid, II: 123
   and psoralens, II: 108
   and sterigmatocystins, II: 138
Chymotrypsinogen, and cytochalasins, II: 65
Ciliary movement, II: 58
Cirrhosis, I: 10
   childhood, I: 134
   and cyclochlorotine, II: 102
Citreoviridin, I: 5, 114, 115
Citrinin, I: 5
   and ochratoxins, II: 130
   and penicillic acid, II: 124
   screening for, I: 36
Claviceps paspali, I: 112
Claviceps purpurea, I: 112
Clavine alkaloids, I: 109-111
Cleavage furrow, II: 57
C-leucine, II: 19, 41, 98, 103, 150
Clostridium botulinum, see also Botulism, I: 168-170, 242, 243
Clostridium cachlioades, II: 52
Clostridium globosum, II: 53
Clostridium perfringen, II: 113
C-methionine, and zearalenone, II: 150
CO_2
   and aflatoxins, II: 14
   and cytochalasins, II: 72
   and penicillic acid, II: 124
Cocarboxylase, and penicillic acid, II: 126
Cod, and N-nitroso compounds, I: 171, 176
Coffee, and N-nitroso compounds, I: 176
Cointoxication, and aflatoxins, I: 129-130
Colchicine, II: 77
   and cytochalasins, II: 78
Collagen, and cytochalasins, II: 66
Collagenase, II: 65
Collagen synthesis, and cytochalasins, II: 75
Colon
   adenocarcinomas of, II: 5
   and aflatoxins, I: 128
   and ochratoxins, I: 90, 91
   and risk assessment, I: 149
Colorado, I: 225
Competing microorganisms, I: 2
Complement
   and aflatoxins, II: 26
   and cytochalasins, II: 63, 67, 69, 92
   and ochratoxins, II: 136
   and sterigmatocystins, II: 142
Con A, see Concanavalin A
Concanavalin A (Con A), II: 68, 81, 82

Confirmatory tests, and aflatoxins, I: 28
Contractile proteins, II: 77
  and cytochalasins, II: 81
Contractile ring, II: 57
Contractile systems, and cytochalasins, II: 77-80
Contraction, II: 48
and cytochalasins, II: 59, 77-80, 88, 94
Copper
  and aflatoxins, II: 26
  and psoralens, II: 108
Corn, see also Maize, I: 2, 7, 8; II: 30
  and aflatoxins, I: 21
  and N-nitroso compounds, I: 222
  and ochratoxins, I: 23
  and trichothecenes, I: 13-16, 60, 62
  and zearalenone, I: 17, 18, 65, 71
Corneal epithelium, and cytochalasins, II: 59, 71
C-orotic acid, II: 17, 126, 136
Corticosterone, and aflatoxins, II: 22
Cosmetics, and N-nitroso compounds, I: 175, 176
Costa Rica, I: 11
Cotton, I: 2
Cottonseeds, I: 2
Cows, I: 5
  and aflatoxins, I: 19-20, 52; II: 8, 10, 15
  and cytochalasins, II: 63, 66, 71, 89, 90
  and luteoskyrin, II: 97
  and ochratoxins, II: 131, 134
  and psoralens, II: 111, 113
  and risk assessment, I: 143
  and trichothecenes, I: 13, 15, 61-62; II: 30
  and zearalenone, I: 17; II: 147-150
C-phenylalanine, II: 117
Cranial bone, and cytochalasins, II: 75
Creatin, I: 8
Creatine phosphokinase, II: 43
Croaker, and N-nitroso compounds, I: 171
Crotocin, II: 31, 34, 40, 42, 45, 157
Crotocol, II: 40
Croton oil, II: 37
Crustaceans, and patulin, II: 122
C-trichodermin, and trichothecenes, II: 43
Cu, see Copper
C-urine, and luteoskyrin, II: 98
Cyclic AMP
  and cytochalasins, II: 58, 64, 66, 70-74, 76
  and ochratoxins, II: 133, 134
  and zearalenone, II: 146
Cyclic nitrosamides, I: 203, 205
Cyclic nitrosamines, I: 192-196, 199-206
Cyclochlorotine, I: 5, 6; II: 2, 102, 103, 114, 127, 156
Cycloheximide, II: 39, 42
Cyclops fuscus, II: 37
  and aflatoxins, II: 4
Cylindrocarpon, II: 30, 35, 36
Cysteine, II: 68, 86, 94
  and cytochalasins, II: 86
  and luteoskyrin, II: 99
  and patulin, II: 122, 124, 125
  and penicillic acid, II: 124, 126
Cysteine adducts
  and patulin, II: 124-126
  and penicillic acid, II: 125
Cytochalasin A (CA), II: 48-50, 54-56, 59, 63, 67-69, 74, 75, 78, 80, 83, 86, 88-94
Cytochalasin B (CB), II: 48-50, 54-60, 63-94
  dihydro-, II: 57, 58,60, 68, 71, 86, 87, 90-93
  γ-lactone, II: 68, 86, 90-93
Cytochalasin B detection method, I: 26, 27
Cytochalasin B-diacetate, II: 90
Cytochalasin C (CC), II: 48-50, 54, 55, 74, 91-93
Cytochalasin congeners, potency of, II: 92-93
Cytochalasin D (CD), II: 48-50, 54-57, 59, 60, 63, 67-71, 73, 74, 78, 79, 81-83, 86, 89-93
Cytochalasin G (CG), II: 50
Cytochalasin H (CH), II: 49, 51, 55, 56, 59, 60
Cytochalasin E (CE), I: 5, 123, 130; II: 48-50, 55-57, 59, 63, 69, 71, 74, 86, 88, 90-93
  and aflatoxins, I: 131
Cytochalasin F (CF), II: 50
Cytochalasins, II: 1, 2, 155, 156
  and acetylcholine, II: 66, 71
  and actin, II: 72, 77-80, 90, 91, 93, 94
  and G-actin, II: 78, 94
  and actin-binding protein, II: 79
  and actomyosin, II: 71, 78, 79
  and actomyosin contraction, II: 68
  and acylation, II: 91
  and adenyl cyclase, II: 76
  and adhesion, II: 59, 60
  and adipocytes, II: 73, 84, 85, 89, 90
  and adrenal cells, II: 76
  and adrenocorticotropin, II: 76
  and aflatoxins, II: 48
  and agglutination, II: 73, 75, 82
  and agglutinin, II: 82
  and aggregation, II: 59-60, 79, 81
  and albumin, II: 65
  and alloantigens, II: 81
  and alveolar macrophages, II: 60, 69, 84, 92
  and amino acid incorporation, II: 74
  and amino acids, II: 88-89
  and amino acid transport, II: 75, 88
  and α-aminobutyric acid (GABA), II: 67
  and amino sugar transport, II: 74
  and amphibians, II: 72
  and antiactin, II: 80
  and antibodies, II: 64, 68
  and antigens, II: 64, 67-69, 81
  and antisera, II: 82
  and ascites, II: 54, 75
  and ATP, II: 58, 73, 76, 78, 80, 81, 84, 91
  and ATPase, II: 78, 79, 83
  and attachment, II: 59
  and axoplasmic transport, II: 72
  and bacteria, II: 54, 60, 69
  and BHK cells, II: 65
  and bladder, II: 71, 76, 88
  and blastogenesis, II: 68
  and blood, II: 68, 71, 85
  and blood clot retraction, II: 71, 92
  and blood platelets, II: 66
  and bradykinin, II: 67
  and brain, II: 54, 67, 78, 79, 83, 85, 94

and calcium, II: 71, 78, 79, 81
and calcium uptake, II: 68, 88
and capillaries, II: 54
and cats, II: 71, 72
and cell cycle, II: 87-89
and cell division, II: 57
and cell locomotion, II: 58, 59, 69, 73, 79, 81, 82, 85, 94
and cell sorting, II: 59-60, 75
and cellulase, II: 64, 75, 94
and cervical ganglia, II: 66
and chaetoglobosins, II: 91
and chemotaxis, II: 69
and chicken embryos, II: 56, 58-62, 71-73, 75, 81, 82, 84-86, 92
and Chinese hamsters, II: 87
and cholesterol uptake, II: 76
and chondrogenesis, II: 62
and chromatin, II: 76
and chymotrypsinogen, II: 65
and CO_2, II: 72
and colchicine, II: 78
and collagen, II: 66
and collagen synthesis, II: 75
and complement, II: 63, 67, 69, 92
and contractile proteins, II: 81
and contractile systems, II: 77-80
and contraction, II: 59, 77-80, 88, 94
and corneal epithelium, II: 59, 71
and cows, II: 63, 66, 71, 89, 90
and cranial bone, II: 75
and cyclic AMP, II: 58, 64, 66, 70-74, 76
and cytochrome, II: 74
and cytokinesis, II: 59, 74, 75, 79, 91, 94
and cytolysis, II: 73, 86
and cytophysiological effects, II: 56-72
and cytoplasmic gelation, II: 79
and cytoplasmic receptor protein, II: 76, 77
and cytoplasmic streaming, II: 58, 75, 94
and cytoskeleton, II: 71
and cytotoxicity, II: 48-55, 91, 92
and deoxyadenosine, II: 87
and deoxyglucose, II: 84, 88
and deoxyglucose transport, II: 86
and deoxyglucose uptake, II: 86
and deoxyribonucleosides, II: 86
and deoxyribonucleoside transport, II: 87
and desmosomes, II: 59, 83
and diffusion, II: 84, 85, 87
and disaggregation, II: 59-60, 83
and DNA, II: 68, 71, 74-75
and DNA damage, II: 56
and DNA repair, II: 56
and DNA synthesis, II: 58, 68, 87
and DNA virus, II: 70
and dogs, II: 66, 85
and dopamine, II: 67
and duodenal explants, II: 62
and edema, II: 54
and Ehrlich ascites cells, II: 59, 73, 80, 86, 92
and epinephrine, II: 64
electrophysiological effects of, II: 71-72

and endocytosis, II: 81
and endodermal cells, II: 59
and endothelial cells, II: 60
and enucleation, II: 91
and epithelial cells, II: 57, 60
and erythrocyte ghosts, II: 85, 89, 90, 93
and erythrocytes, II: 65, 69, 71, 81, 85, 86, 89, 92
and *Escherichia coli*, II: 74
and estrogen, II: 61
and F-actin, II: 78, 79
and fertilization, II: 60, 61-62
and fibrin, II: 71
and fibroblasts, II: 54, 56-58, 63, 65, 73-75, 80, 82-88, 90-93
and fibronectin, II: 82, 83
and fish, II: 72
and fluid transport, II: 71
and frog embryos, II: 59-61
and frog gastrulae, II: 60
and frogs, II: 66, 71, 72, 84, 88
and fructose transport, II: 84
and fucose, II: 76
and fucose uptake, II: 70
and fungi, II: 50-54
and gall bladder, II: 71
and ganglion, II: 84
and glial cells, II: 56, 58, 73, 76, 77, 88
and glioma cells, II: 84, 85
and glucosamine uptake, II: 70, 75, 76
and glucose, II: 56, 63, 65, 68, 71, 75, 76, 88
and D-glucose, II: 89
and glucose carrier, II: 80, 86, 88-90, 93
and glucose transport, II: 68, 69, 73, 75, 84-86, 89-91, 93
and glucose uptake, II: 60, 62
and β-glucuronidase, II: 64
and glutamine, II: 71
and glutathione, II: 68
and glycine transport, II: 89
and glycogen, II: 73
and glycolysis, II: 58, 72-73, 84
and glycoproteins, II: 70, 80
and glycoprotein synthesis, II: 84, 88
and Golgi complex, II: 81
and GPDG, II: 77
and growth hormone, II: 66, 83, 93
and guinea pigs, II: 57, 60, 64-66, 93
and hamsters, II: 54, 82, 84, 87
and heart cells, II: 58, 59
and heart fibroblasts, II: 75
and HeLa cells, II: 54, 55, 58, 59, 70, 73, 75, 79, 84, 85, 87-89, 92
and HEp-2 cells, II: 59
and hepatoma cells, II: 63
and hexokinase, II: 73
and hexose monophosphate, II: 74
and hexose monophosphate shunt, II: 73
and hexose transport, II: 73, 74, 80, 90
and hexose uptake, II: 83
and histamine, II: 63, 67, 68
and hormonal mechanisms, II: 76-77

and hormones, II: 63, 66, 76-77, 80, 83, 93
and horses, II: 58, 84
and human embryos, II: 70
and humans, II: 49, 58, 63-67, 69-71, 73, 74, 76, 79-86, 88-90, 92, 93
and hydrocortisone, II: 77
and 5-hydroxytryptamine, II: 66
and hypogastric nerves, II: 72
 immunobiological functions of, II: 68-69
and immunoglobulin G, II: 74
and immunoglobulins, II: 63, 65, 69, 74, 81
and initiation, II: 75
and insulin, II: 63, 65, 68, 76, 83, 86, 88, 92, 93
and interferon, II: 65, 68, 69, 75
and intestine, II: 54
and ion transport, II: 88
and islet cells, II: 84, 88, 92
and K cells, II: 69
and kidney, II: 54, 70, 92
and lactic acid, II: 72, 84
and L-cells, II: 70
and lectins, II: 68, 73, 75, 82
and lens cell, II: 59
and leucine, II: 65, 83, 88
and leukocytes, II: 57, 58, 63, 64, 67, 69, 73, 74, 82, 84, 89, 90, 93
and limb bud cells, II: 59, 62
and lipase, II: 65
and liver, II: 54, 59, 65, 83, 85, 90, 91
and lung, II: 60, 63, 67, 70, 76, 92
and lymph node cells, II: 65
and lymphoblastoid cells, II: 80
and lymphocytes, II: 68, 69, 81, 83-86, 88-93
and lymphocyte transformation, II: 68
and lymphotoxins, II: 65, 69
and lysosomal enzymes, II: 63, 64
and lysozyme, II: 64
and macromolecular synthesis, II: 72-77
and macrophages, II: 60, 63, 68
and macropinocytosis, II: 60
and magnesium, II: 73, 78, 79, 83
and mammary tumor cells, II: 92
and meiosis, II: 62
and melanaphore stimulating hormone, II: 66, 72
and melanin, II: 72
and melanocyte dispersion, II: 72
and melanocytes, II: 72
and membrane, II: 57
and membrane binding, II: 89-91
and membrane carriers, II: 84, 87, 89-91
and membrane receptors, II: 82, 83, 93
and membrane topology, II: 82, 83
and membrane transport, II: 80, 83-91
and mesodermal cells, II: 62
and metabolic cooperation, II: 72
and metabolism, II: 72-77
and metamorphosis, II: 61
and mice, II: 54, 56-58, 60, 62-65, 67, 68, 70, 73-76, 80, 82, 84-88, 90-93
 and mice embryos, II: 58, 62
and microfilaments, II: 57, 58, 60-63, 72, 76, 77, 79-80, 82, 85, 88, 90, 91

and micropinocytosis, II: 60
and microtubules, II: 63, 71, 77, 78, 80, 82
and microvilli, II: 80
and mitochondria, II: 73, 76
and mitogen, II: 68, 91, 92
and monkeys, II: 70
and morphogenesis, II: 60-63, 73
and motility, II: 91
and mucopolysaccharide synthesis, II: 58, 74, 88
and mucopolysaccharides, II: 75-76
and mucopolysaccharide transport, II: 72
and multinucleation, II: 74
and muscle, II: 59, 62, 72, 77-79, 84, 91, 93
and mutagenicity, II: 54-56
and myoblasts, II: 73, 88
and myofibrils, II: 62
and myosin, II: 58, 62, 77-79, 91
and necrosis, II: 54
and nerve cells, II: 56
and nerves, II: 84
and neural epithelium, II: 59
and neural retina, II: 59
and neural retina cells, II: 81
and neural tube, II: 56
and neuraminidase, II: 80
and neuroblastoma, II: 84
and neuroblastoma cells, II: 85
and neuroepithelial cells, II: 62
and neurotransmitters, II: 63, 71, 72, 81
and noradrenaline, II: 67
and norepinephrine, II: 66, 76
and Novikoff hepatoma cells, II: 85, 87-90
and nuclear division, II: 74
and nuclei, II: 57, 76, 77
and nucleic acid, II: 86-88
and nucleoside incorporation, II: 74
and nucleosides, II: 86-88
and nucleoside transport, II: 74, 87
and oocytes, II: 62
and osmotic balance, II: 71, 76, 88
and ovary, II: 82, 87
and oviduct, II: 61
and ovulation, II: 72
and oxidative phosphorylation, II: 69
and oxytocin, II: 71
and pancreas, II: 54, 65, 84, 88, 92
and pancreatic β cells, II: 63
and pancreatic islets, II: 68
and parathyroid, II: 66
and parathyroid hormone, II: 66
and peptide hormone, II: 76, 88
and peritoneal macrophages, II: 60, 63, 64
and peritoneal mast cells, II: 63, 92
and phagocytosis, II: 60, 69, 73
and phalloidin, II: 78
and phosphate, II: 73, 88
and phosphate uptake, II: 88
and phosphodiesterase, II: 73
and phospholipid, II: 80, 84
and phosphorylation, II: 73, 76, 84, 87
and pinocytosis, II: 81
and pituitary gland, II: 66

and plasma, II: 54, 85, 92
and plasma cells, II: 63
and plasmacytoma cells, II: 65
and plasma membrane, II: 48, 56-60, 63, 68, 69, 71, 72, 76, 77, 79-81, 83, 89-91, 94
and plasma protein, II: 65
and platelet activation, II: 60
and platelets, II: 64, 66, 71, 79, 89, 92
and poliovirus, II: 92
and polymorphonuclear leukocytes, II: 60, 82
and polysaccharide synthesis, II: 60
and polysome disaggregation, II: 70, 75
and potassium, II: 88
and prolactin, II: 66
and prostaglandins, II: 63, 67, 72
and protease, II: 82
and protein, II: 63, 65, 72, 75-79, 81-84, 90, 91
and protein synthesis, II: 58, 60, 62, 63, 68, 70, 74-76, 83
and protein transport, II: 60, 71, 72
and protozoa, II: 54
and pulmonary macrophage, II: 79
and pyruvate, II: 71
and rabbits, II: 58, 60, 64, 65, 69, 71, 75, 78, 79, 81, 84, 89, 90, 92, 93
and rainbow trout, II: 72
and rats, II: 54, 63, 65-68, 70, 73, 75-79, 81, 83-92
and reptiles, II: 72
and reticulocytes, II: 75
and retinal cells, II: 59
and Reuber hepatoma cells, II: 76, 77
and ribonucleosides, II: 86
and RNA, II: 58, 68, 75
and RNA synthesis, II: 68, 70, 75
and ruffling, II: 73, 85, 92
and salivary ducts, II: 88
and salivary gland, II: 62, 73, 75
and sarcoma, II: 81
and sciatic nerve, II: 72
and sea urchin eggs, II: 57, 92
and secretory granules, II: 63
and serotonin, II: 72
and serum, II: 63, 67, 71, 74, 85, 86, 88
and sheep, II: 65
and skin, II: 67, 71, 83, 88
and small intestine, II: 54
and sodium, II: 71, 88, 89
and sodium transport, II: 71, 76
and spectrin, II: 90
and spermatocyte, II: 54
and spermatozoa, II: 62, 84, 85, 89
and spinal cord, II: 61, 71, 72
and spleen, II: 54, 68, 81, 82, 92
and spleen cells, II: 65
and steroid hormones, II: 76-77
and steroid synthesis, II: 76
structure, II: 50-53
structure-activity relationships in, II: 91-94
and sucrose uptake, II: 63
and sugar transport, II: 94
and sulfate, II: 88

and sulfate incorporation, II: 88
and sulfhydryl groups, II: 86, 94
and sulfhydryl reagents, II: 78, 86, 89, 90
and superoxide dismutase, II: 74
and superoxide generation, II: 74
and sympathetic ganglia, II: 71
and synaptosomes, II: 67, 72, 78
and T cells, II: 69
and teratogenicity, II: 54-56, 60, 92
and tetraploidy, II: 62
and thrombin, II: 67
and thrombosthenin, II: 71
and thymidine, II: 68
and thymidine transport, II: 74
and thymocytes, II: 68, 69
and thymus, II: 54, 81, 82
and thyroid, II: 65, 66
and thyroid stimulating hormone, II: 65, 66
and toads, II: 71, 76, 88
and toxicity, II: 48-55, 77, 91, 92
and trout, II: 72
and trypsin, II: 82
and tryptophan, II: 75
and tubuline, II: 94
and tyrosine aminotransferase (TAT), II: 76, 77
and uracil, II: 87
and uridine, II: 75
and urinary bladder preparations, II: 76
and vasopressin, II: 71, 76
and vero cells, II: 70, 92
and viral attachment, II: 70
and viral infectivity, II: 69-71, 92
and viral replication, II: 70
and virions, II: 70
and viropexis, II: 69
and virus, II: 73, 76
and virus transformed cells, II: 56, 74, 82, 84, 87
and white fat cells, II: 84, 86
and yeast glyceraldehyde-3-phosphate, II: 94
and zygosporins, II: 91
Cytochrome c oxidase, II: 97
Cytochrome P-448, II: 13
Cytochrome P-450, II: 13
Cytochromes
  and aflatoxins, II: 23
  and cytochalasins, II: 74
Cytokinesis, II: 48, 54, 57-58, 83, 92
  and cytochalasins, II: 59, 74, 75, 79, 91, 94
Cytolysis, and cytochalasins, II: 73, 86
Cytophysiological effects of cytochalasins, II: 56-72
Cytoplasmic gelation, and cytochalasins, II: 79
Cytoplasmic receptor proteins, II: 21
  and cytochalasins, II: 76, 77
  and zearalenone, II: 150, 152
Cytoplasmic streaming, II: 72, 92
  and cytochalasins, II: 58, 75, 94
Cytosine, II: 87
and psoralens, II: 111
Cytosine uptake, II: 87
Cytoskeleton, II: 48, 57, 59, 71, 77, 80

Cytosol proteins, and aflatoxins, II: 12
Cytotoxicity
 and aflatoxin B$_1$, II: 4
 and cyclochlorotine, II: 102
 and cytochalasins, II: 48-55, 91, 92
 and luteoskyrin, II: 96
 and N-nitroso compounds, I: 241
 and ochratoxins, II: 136
 and penicillic acid, II: 123
 and psoralens, II: 107-108, 115
 and sterigmatocystins, II: 138, 139
 and trichothecenes, II: 31, 44, 45
 and zearalenone, II: 147
Czechoslovakia, I: 8

D

Dairy products, and aflatoxins, I: 27-28
Danube, I: 115
C-4-Deacetylation, II: 38
15-Deacetylcalonectrin, II: 32, 40
Degranulation, and aflatoxins, II: 22
Dehydroepiandrosterone, II: 17
Dehydrophomin, II: 50
2'-Dehydroverrucarin A, II: 34
7'-Dehydrozearalenone, II: 146
DeMe-DiAcSTG, see
 Demethyldiacetylsterigmatocystin
DeMeSTG, see 6-Demethylsterigmatocystin
O-Demethylation, and aflatoxins, II: 14
Demethyldiacetylsterigmatocystin (DeMe-DiAcSTG), II: 141, 143
6-Demethylsterigmatocystin (DeMeSTG), II: 138, 139
Denmark, I: 22, 23
Deoxaphomin, II: 51
Deoxyadenosine, and cytochalasins, II: 87
Deoxyglucose, and cytochalasins, II: 84, 88
2-Deoxyglucose, II: 84, 85
D-Deoxyglucose, II: 84
Deoxyglucose transport, and cytochalasins, II: 86
Deoxyglucose uptake, and cytochalasins, II: 86
Deoxyguanosine, and aflatoxins, II: 12
Deoxynivalenol, I: 14-16; II: 30, 33, 37, 40
Deoxyribonucleosides, and cytochalasins, II: 86
Deoxyribonucleoside transport, and cytochalasins, II: 87
Deoxyuridine, II: 87
Department of Agriculture, I: 168, 242
Dermatitis, and psoralens, II: 106
Dermis, and psoralens, II: 109
DES, see Diethylstilbestrol
Desmosomes, and cytochalasins, II: 59, 83
Detection
 of aflatoxins, I: 24-29
 of mycotoxins, I:24-37
 of ochratoxins, I: 33-36
 of trichothecenes, I: 31-33
 of zearalenone, I: 29-31
Detoxification, II: 6
Detoxification pathway, II: 13

Dexamethasone, II: 22
DH-DeMeSTG, see Dihydro-demethylsterigmatocystin
DH-OMe-STG, see Dihydro-O-methylsterigmatocystin
DH-STG, see Dihydrosterigmatocystin
Diacetate, II: 86
Diacetoxyscirpenol, I: 6, 16; II: 30-32, 37-40, 42-45
 screening for, I: 36
3,15-Diacetyldeoxynivalenol, II: 33
Diacetylnivalenol, II: 40
Diacetylverrucarol, II: 32
Dictyostelium discoideum, II: 89
Dicumarol, see Bishydroxycoumarin
Diepoxyroridin H, II: 36
Diet, see also Nutrition, II: 5
 and aflatoxins, II: 13, 14
 and nitrosamines, I: 230
Diethylnitrosamine, I: 161, 162, 168-173, 175, 177, 207, 224, 232, 236, 237
Diethylstilbestrol (DES), I: 239
 and zearalenone, II: 148, 149, 152
Diffusion, and cytochalasins, II: 84, 85, 87
Dihydro-cytochalasin B, II: 57, 58, 60, 68, 71, 86, 87, 90-93
Dihydro-demethylsterigmatocystin (DH-DeMeSTG), II: 138, 139, 143
Dihydrodiol, II: 44
Dihydro-O-methylsterigmatocystin (DH-0Me-STG), II: 138, 139
Dihydrosterigmatocystin (DH-STG), II: 138, 139, 142, 143
7,8-Dihydroxydiacetoxyscirpenol, II: 33, 40
4,8-Dihydroxytricholthecene, II: 33
6',8'-Dihydroxyzearalene, II: 146
Dimethylamine, I: 176
Dimethylnitrosamine, I: 160-162, 164, 166, 167, 169, 171-173, 175-177, 186, 206, 208, 220, 221, 224, 225, 228-234, 236, 237, 241, 242
Dinitrophenol(DNP), II: 23, 97
 and ochratoxins, II: 132
 and penicillic acid, II: 127
Diplodia maydis, I: 5
Diplodiatoxin, I: 5
Di-*n*-propylnitrosamine, I: 162
Disaggregation, II: 42
 and cytochalasins, II: 59-60, 83
Distribution
 of aflatoxins, II: 6-16
 of luteoskyrin, II: 96
 of ochratoxins, II: 131-132
 of patulin, II: 124
 of penicillic acid, II: 124
 of psoralens, II: 109
 of trichothecenes, II: 37-38
DNA, II: 155-157
 and aflatoxins, II: 6, 11-14, 17, 18, 26, 27
 attacking ability of aflatoxins, II: 6
 binding of, see DNA binding
 and cytochalasins, II: 68, 71, 74-75
 and luteoskyrin, II: 98, 99

and N-nitroso compounds, I: 239, 240
and ochratoxins, II: 134
and patulin, II: 123
and penicillic acid, II: 124
and psoralens, II: 107, 108, 111, 113, 114, 116, 118
repair of, see DNA repair
and sterigmatocystins, II: 140
synthesis of, see DNA synthesis
and trichothecenes, II: 38
and zearalenone, II: 149, 150
DNA binding
and aflatoxins, II: 6, 7, 9, 11, 12, 14, 19, 27
and N-nitroso compounds, I: 208, 241
and ochratoxins, II: 134
and psoralens, II: 110, 111, 113, 116, 118
and risk assessment, I: 149
and sterigmatocystins, II: 139
DNA complex formation, and luteoskyrin, II: 97-98
DNA damage
and aflatoxins, II: 6, 7
and cytochalasins, II: 56
and luteoskyrin, II: 96, 99
and N-nitroso compounds, I: 240, 241
and patulin, II: 123, 124
and psoralens, II: 112
and sterigmatocystins, II: 139, 142, 3
and trichothecenes, II: 37
DNA polymerase
and psoralens, II: 117
and trichothecenes, II: 38
DNA repair, II: 16-17
and aflatoxins, II: 7, 13, 17
assay, II: 6
and cytochalasins, II: 56
and luteoskyrin, II: 98
and N-nitroso compounds, I: 238, 240, 241
and psoralens, II: 108, 114, 116-117
and risk assessment, I: 149, 150
and sterigmatocystins, II: 138, 140-141
and trichothecenes, II: 39
and zearalenone, II: 147
DNAse
and psoralens, II: 115
and sterigmatocystins, II: 141
DNAse activity, and aflatoxins, II: 24
DNA synthesis, II: 16-17, 156, 157
and aflatoxins, I: 52, 53, 57; II: 7, 12, 16, 17, 19, 20, 22, 23
and cytochalasins, II: 58, 68, 87
inhibition of, II: 140
and N-nitroso compounds, I: 209
and ochratoxins, II: 135
and patulin, II: 126
and penicillic acid, II: 125
and psoralens, II: 114-116, 118, 119
and sterigmatocystins, II: 139-143
and trichothecenes, I: 60; II: 38, 39
unscheduled, II: 17, 140
and zearalenone, II: 150
DNA template

activity, II: 17
and luteoskyrin, II: 99
and N-nitroso compounds, I: 208
DNA virus, and cytochalasins, II: 70
Dogs
and aflatoxins, I: 52, 53
and cyclochlorotine, II: 103
and cytochalasins, II: 66, 85
and N-nitroso compounds, I: 186, 190, 196, 198, 221, 229
and ochratoxins, I: 59, 79-82
and trichothecenes, I: 14-16, 60, 61
Dopamine, and cytochalasins, II: 67
Dorsal root ganglia, II: 92
Dose-response relationships
of aflatoxins, I: 144-145, 148
of N-nitroso compounds, I: 232-238, 241
Drosophila, II: 37
and aflatoxins, II: 4
Drosophila melanogaster, II: 115
Drugs
nitrosation of, I: 177
and N-nitroso compounds, I: 175
Ducks
and aflatoxins, I: 20-22, 52, 53, 58, 66, 71, 72; II: 4, 5, 7, 8, 17, 25
and ochratoxins, I: 59, 73, 74; II: 130
and risk assessment, I: 151
and sterigmatocystins, I: 63; II: 139
and trichothecenes, I: 14, 16; II: 41
Duodenal explants, and cytochalasins, II: 62

E

Ear
and N-nitroso compounds, I: 197, 206
and psoralens, II: 108, 109
and zearalenone, II: 148
Edema, and cytochalasins, II: 54
EFDV, I: 7
Eggs, I: 13; II: 155
and aflatoxins, I: 21, 22; II: 16
and ochratoxins, I: 24
and patulin, II: 122
and sterigmatocystins, II: 139
Egypt, I: 225
Ehrlich ascites cells, II: 38
and cytochalasins, II: 59, 73, 80, 86, 92
and luteoskyrin, II: 98
and psoralens, II: 114, 115, 118, 119
and trichothecenes, II: 38
Elastase, II: 64
Electron transport
and aflatoxins, II: 23
and ochratoxins, II: 135
Electrophysiological effects of cytochalasins, II: 71-72
Elodea, II: 58, 92
Elongation, II: 20
and luteoskyrin, II: 99
and penicillic acid, II: 125

and trichothecenes, II: 39, 41, 42, 45
Embryotoxicity, see also Teratagenicity, II: 158
Emetic factor, I: 13, 14
Encephaly, II: 5
Edemic nephropathy, see Balkan nephropathy
Endocytosis, II: 60-63
  and cytochalasins, II: 81
Endodermal cells, II: 60
  and cytochalasins, II: 59
Endoplasmic reticulum, I: 208, 209; II: 6, 22, 43
Endothelial cells, and cytochalasins, II: 60
England, I: 19-21, 222
Enucleation, II: 57, 92
  and cyctochalasins, II: 91
Environmental Protection Agency (EPA), I: 165, 166, 228, 230
Enzyme activity
  and ochratoxins, II: 134-135
  and patulin, II: 126-127
  and penicillic acid, II: 126-127
Enzyme transport, and penicillic acid, II: 126-127
EPA, see Environmental Protection Agency
Epidemiology of aflatoxins, I: 10-11
Epidermis, and psoralens, II: 109, 116
Epinephrine, and cytochalasins, II: 64
Epithelial cells
  and cytochalasins, II: 57, 60
  and ochratoxins, II: 136
  and psoralens, II: 117
  and sterigmatocystins, II: 138, 140, 142, 143
Epithelial mucosa, and trichothecenes, II: 37
Epoxidation, II: 44
  and aflatoxins, II: 11-13, 17, 27
  and psoralens, II: 109
  and sterigmatocystins, II: 140
Epoxide, II: 43
Epoxide hydrase, II: 44
Epoxyisororidin E, II: 35
Epoxyroridin H, II: 36
12,13-Epoxytrichothecenes, I: 112
12,13-Epoxy-δ-trichothecenes, I: 6
Eppley detection method, I: 30
Ergot alkaloids, I: 108-112
  and amino acid alkaloids, I: 109
  and muscle, I: 109
  and risk assessment, I: 142
  and uterus, I: 109
Ergot poisoning, I: 108-112
Ergotamine, I: 109, 111
Erwinia aroideae, I: 119
Erythema, II: 108
  and psoralens, II: 106, 117
Erythrocyte ghosts, and cytochalasins, II: 85, 89, 90, 93
Erythrocytes
  and aflatoxins, II: 26
  and cytochalasins, II: 65, 69, 71, 81, 85, 86, 89, 92
  and patulin, II: 124, 125
  and penicillic acid, II: 125, 127
Escherichia coli, II: 54, 60, 92, 107
  and aflatoxins, II: 7

and cytochalasins, II: 74
and luteoskyrin, II: 97, 99
and ochratoxins, II: 135
and patulin, II: 125
and penicillic acid, II: 125, 126
and psoralens, II: 113-118
Esophagus
  and aflatoxins, I: 11
  and N-nitroso compounds, I: 170, 187-198, 200, 207, 221-224, 232
Esterase, II: 38
Estradiol, II: 22
17β-Estradiol, II: 17
  and zearalenone, II: 149, 150
Estradiol-receptor complex, and zearalenone, II: 150
Estrogenic syndrome, see also Hyperestrogenism, I: 16
Estrogen receptors, and zearalenone, II: 150
Estrogens, II: 157
  and cytochalasins, II: 61
  and zearalenone, II: 146, 148, 149, 151, 152
Estrus, and zearalenone, II: 148
Ethyl ester, and ochratoxins, II: 131
N-Ethylmalemide, II: 86
Ethylnitrosamines, I: 190-191
Euglena gracilis, II: 54
Europe, I: 20, 60, 108, 142
European Common Market Commission, I: 20
Excretion
  of aflatoxin B₁, II: 14-16
  of aflatoxins, I: 57; II: 6-16
  of luteoskyrin, II: 96
  of N-nitroso compounds, I: 240
  of ochratoxins, I: 59; II: 131-132
  of patulin, II: 124
  of penicillic acid, II: 124
  of psoralens, II: 109
  and risk assessment, I: 149
Exencephaly, II: 56
  and ochratoxins, II: 130
  and patulin, II: 122
Exocytosis, II: 63-68
Exposure estimates, and N-nitroso compounds, I: 228-232
Extraction of ochratoxins, I: 35
Eye, and ochratoxins, II: 130

F

F-actin, and cytochalasins, II: 78, 79
Fat, and ochratoxins, II: 130
Fat pads, and aflatoxins, II: 24
Fatty acids, I: 208
  and aflatoxins, II: 23
  and ochratoxins, II: 134
FDA, see Food and Drug Administration
Fe, see Iron
Feces
  and aflatoxins, I: 133; II: 8, 14-16
  and luteoskyrin, II: 96

and N-nitroso compounds, I: 159, 223
and ochratoxins, II: 131, 132
and penicillic acid, II: 124
and psoralens, II: 109
and trichothecenes, I: 15; II: 38
and zearalenone, II: 148
Feedstuff and zearalenone, I: 71
Fertilization, and cytochalasins, II: 60-62
Fibrin, and cytochalasins, II: 71
Fibrinogen, II: 20
Fibroblasts
  and aflatoxins, II: 7
  and cytochalasins, II: 54, 56-58, 63, 65, 73-75, 80, 82-88, 90-93
  and N-nitroso compounds, I: 241
  and psoralens, II: 108, 114-116
  and sterigmatocystins, II: 138, 140, 141
  and trichothecenes, II: 31, 37, 38
  and zearalenone, II: 147
Fibronectin, and cytochalasins, II: 82, 83
Fibrosarcomas
  and patulin, II: 122
  and penicillic acid, II: 124
Fibrosis, and cyclochlorotine, II: 102
Fire plague, see Ergot poisoning
Fish, I: 5, 190
  and aflatoxins, I: 57-58, 124; II: 4, 5, 7, 10
  and carcinogenicity, II: 10
  and cytochalasins, II: 72
  and N-nitroso compounds, I: 168, 171-173, 175, 176, 187, 230, 231
  and ochratoxins, I: 60; II: 130, 131
  and patulin, II: 122
  and sterigmatocystins, II: 138, 143
  and trichothecenes, II: 37
Fish meal, and N-nitroso compounds, I: 160
Flagellar movement, II: 58
Fluid transport, and cytochalasins, II: 71
Fluordensitometry, I: 30
Follicle-stimulating hormone (FSH), and zearalenone, II: 149
Food and Drug Administration (FDA), I: 52, 55, 159, 242
Food animals, see also specific animals, I: 11-24
Foods, see also specific foods
  and aflatoxins, I: 124
  and nitrosamines, I: 231
  and N-nitroso compounds, I: 166-175
Formation of N-nitroso compounds, I: 159-164
5-Formylzearlenone, II: 146
Fowl, see Birds; specific birds
France, I: 108-109
Frankfurters, and N-nitroso compounds, I: 170, 171, 176, 230, 231
Frog embryos, and cytochalasins, II: 59-61
Frog gastrulae, and cytochalasins, II: 60
Frogs, and cytochalasins, II: 66, 71, 72, 84, 88
Fructose transport, and cytochalasins, II: 84
FSH, see Follicle-stimulating hormone
F-2 toxin, I: 5
Fucose, and cytochalasins, II: 76
Fucose uptake, and cytochalasins, II: 70

Fungi, II: 6, 30
  "blue-eye", I: 3
  and cytochalasins, II: 54
  as cytochalasin sources, II: 50-53
  and patulin, II: 122
  and psoralens, II: 106-108
  as trichothecene sources, II: 32-36
  and zearalenone, II: 146, 149
Furocoumarins, II: 107
  and psoralens, I: 117; II: 106, 108, 110, 112-118
Fusarenon-X, I: 5, 15, 16, 62; II: 30, 31, 33, 37-40, 42-44
  carcinogenicity of, I: 69
Fusarium, I: 2, 3, 13-17, 31; II: 30, 33
  and aflatoxins, I: 123
  and alimentary toxic aleukia, I: 112
  and trichothecenes, I: 60, 61
  and zearalenone, II: 146
Fusarium concolor, II: 36
Fusarium culmorum, I: 14, 18; II: 32, 33, 36
Fusarium dimerum, II: 36
Fusarium episphaeria, II: 33, 36
Fusarium equiseti, II: 36
Fusarium graminearum, I: 5, 6, 13, 14, 62; II: 36
  and trichothecenes, I: 69
Fusarium lateritium, II: 32, 146
Fusarium merismoides, II: 36
Fusarium moniliforme, I: 2, 3, 13, 14, 16; II: 146
Fusarium nivale, I: 5, 14, 62, 69; II: 30, 33
Fusarium oxysporum, I: 16; II: 32, 33, 146
Fusarium poae, I: 6, 14
  and trichothecenes, I: 63; II: 36
Fusarium rigiduscufum, II: 32
Fusarium roseum, I: 14, 16-18; II: 32, 36
  and zearalenone, I: 65; II: 146, 152
Fusarium sambucinum, II: 36
Fusarium solani, II: 32
Fusarium sporotrichioides, I: 6, 63; II: 36, 146
Fusarium sulphureum, II: 33
Fusarium tricinctum, I: 3, 6, 15, 16, 18; II: 32, 33, 36, 146

G

G-actin, and cytochalasins, II: 78, 94
β-Galactosidase, and ochratoxins, II: 135
Gall bladder
  and aflatoxins, I: 53
  and cytochalasins, II: 71
  and N-nitroso compounds, I: 191, 206, 223
  and ochratoxins, I: 80, 81
Ganglion, II: 88
  and cytochalasins, II: 84
Garlic, and aflatoxins, I: 124
Gas chromatography, I: 157
Gastroenteritis, I: 113; II: 37
Genetic template, I: 241
Gene transcription, II: 21
Geneva, I: 120
Geotrichum candidum, I: 224
Germany, I: 8, 22, 134, 135, 164, 168, 225
Gibberella intricans, II: 36

Gibberella saubinetti, I: 13
Gibberella zeae, I: 5, 13; II: 36
Glial cells, and cytochalasins, II: 56, 58, 73, 76, 77, 88
Glioma cells, II: 87
  and cytochalasins, II: 84, 85
Globulin
  and aflatoxins, II: 26
  and sterigmatocystins, II: 142
β-Globulin, and ochratoxins, II: 136
γ-Globulin, and ochratoxins, II: 134
Glomeruli, I: 8
Glucocorticoid-cytosol receptor complex, II: 22
Glucosamine, II: 84
Glucosamine uptake, and cytochalasins, II: 70, 75, 76
Glucose
  and cyclochlorotine, II: 102
  and cytochalasin, II: 56, 63, 65, 68, 71, 75, 76, 88
  and luteoskyrin, II: 97
  and ochratoxins, II: 134
  and zearalenone, II: 149, 150
D-Glucose, and cytochalasins, II: 89
Glucose carrier, and cytochalasins, II: 80, 86, 88-90, 93
Glucose incorporation, and aflatoxins, II: 25
Glucose-6-phosphatase, II: 99
D-Glucose-6-phosphate, II: 102
Glucose transport
  and cytochalasins, II: 68, 69, 73, 75, 84-86, 89-91, 93
  and ochratoxins, II: 134
Glucose uptake, and cytochalasins, II: 60, 62
β-Glucuronidase, II: 24
  and cytochalasins, II: 64
Glucuronide conjugates
  and aflatoxins, II: 6-8, 15
  and sterigmatocystins, II: 140
  and zearalenone, II: 148
Glutamate dehydrogenase activity, II: 25
Glutamate-oxaloacetate transaminase, and ochratoxins, II: 133
Glutamate-pyruvate transaminase, and ochratoxins, II: 133
Glutamic-oxaloacetic-transaminase, II: 25
Glutamine, and cytochalasins, II: 71
Glutathione, II: 68, 94
  and aflatoxins, II: 11, 13
  and cytochalasins, II: 68
  and N-nitroso compounds, I: 164
  and patulin, II: 122, 125
  and penicillic acid, II: 127
  and risk assessment, I: 149
  and trichothecenes, II: 44
Glutathione adducts
  and patulin, II: 125
  and penicillic acid, II: 125
Glutathione transferase, II: 13, 44
Glycerol phosphate dehydrogenase (GPDH), and cytochalasins, II: 77

C-Glycine incorporation, and cyclochlorotine, II: 103
Glycine methyltransferase, and aflatoxins, II: 24
Glycine transport
  and cyclochlorotine, II: 103
  and cytochalasins, II: 89
  and luteoskyrin, II: 99
  and ochratoxins, II: 136
  and penicillic acid, II: 127
  and sterigmatocystins, II: 142
  and zearalenone, II: 151
Glycogen
  and aflatoxins, II: 25
  and cytochalasins, II: 73
  and ochratoxins, II: 133, 134
Glycogenesis, and ochratoxins, II: 134
Glycogenolysis, and ochratoxins, II: 134
Glycogen synthetase, II: 25
  and ochratoxins, II: 133
Glycolysis
  and cyclochlorotine, II: 102
  and cytochalasins, II: 58, 72-73, 84
  and ochratoxins, II: 133
Glycoproteins, II: 70
  and cytochalasins, II: 80
Glycoprotein synthesis
  and cytochalasins, II: 84, 88
  and trichothecenes, II: 43
Goats, and aflatoxins, I: 52; II: 8
Golgi complex, and cytochalasins, II: 81
Gonadotropin, and zearalenone, II: 149
Greece, I: 108
Griseofulvin, I: 6
Growth hormone
  and cytochalasins, II: 66, 83, 93
  and zearalenone, II: 149
GTP, II: 42
GTPase, II: 42
Guanine, II: 87
Guanine-hypoxanthine, II: 87
Guanosine, II: 87
Guinea pigs, II: 109
  and aflatoxins, I: 53, 66; II: 8, 25, 26
  and cytochalasins, II: 57, 60, 64-66, 93
  and N-nitroso compounds, I: 189-191, 196, 197, 221
  and ochratoxins, I: 82, 83; II: 136
  and patulin, II: 127
  and penicillic acid, II: 127
  and psoralens, II: 108, 110, 114, 117, 118
  and sterigmatocystins, II: 142
  and T-2 toxins, I: 94-96
  and trichothecenes, I: 60, 61
Guppies, and aflatoxins, I: 58

H

Haddock, and N-nitroso compounds, I: 171
Hake, and N-nitroso compounds, I: 171
Ham, and N-nitroso compounds, I: 170, 171, 176

Hamsters
    and aflatoxins, I: 66; II: 5, 7, 8, 11, 13, 17
    and cytochalasins, II: 54, 82, 84, 87
    and N-nitroso compounds, I: 187-196, 198, 236, 237
    and ochratoxins, II: 130
    and psoralens, II: 108, 114-116
    and trichothecenes, II: 31, 38, 39, 41, 43, 44
Harvard Medical School, I: 121
Hay
    and stachybotryotoxicosis, I: 6
    and zearalenone, I: 17
Heart
    and aflatoxins, I: 19, 52, 131, 133; II: 23
    and cytochalasins, II: 75
    and luteoskyrin, II: 97
    and N-nitroso compounds, I: 206
    and ochratoxins, I: 78; II: 130, 133
Heart cells, and cytochalasins, II: 58, 59
Hedgehog, and N-nitroso compounds, I: 190
HeLa cells, II: 6
    and aflatoxins, II: 17, 18, 26
    and cytochalasins, II: 54, 55, 58, 59, 70, 73, 75, 79, 84, 85, 87-89, 92
    and luteoskyrin, II: 96
    and patulin, II: 122-124, 126
    and penicillic acid, II: 125
    and sterigmatocystins, II: 138, 140
    and trichothecenes, II: 31, 37-39, 42
Helical polysomes, and aflatoxins, II: 7, 21
Helminthosporium, II: 48
Helminthosporium dematoideum, II: 50
Hemorrhagic syndrome, I: 15
HEp-2 cells, II: 59, 89, 91
Hepatic disease, I: 134; II: 4
    and aflatoxins, I: 52
Hepatic necrosis, II: 4
Hepatitis, and aflatoxins, I: 134
Hepatitis-B virus, I: 11
Hepatocellular carcinomas, II: 5
Hepatocytes, II: 5
    and aflatoxins, II: 8
    and sterigmatocystins, II: 138, 141, 143
Hepatoma, I: 9; II: 4
    and cyclochlorotine, II: 102
    and ochratoxins, II: 130
    and sterigmatocystins, II: 139
Hepatoma cells, II: 84
    and cytochalasins, II: 63
Herpes simplex virus (HSV)
    and cytochalasin B, II: 70, 71
    and cytochalasins, II: 70, 71, 92
Herring, and N-nitroso compounds, I: 161, 171
Hexokinase, II: 84, 97
    and cytochalasins, II: 73
Hexose monophosphate (HMP), and cytochalasins, II: 74
Hexose monophosphate (HMP) shunt, II: 25
    and cytochalasins, II: 73
Hexoses, II: 84-86
Hexose transport, II: 86, 87, 89, 156, 157
    and cytochalasins, II: 73, 74, 80, 90

Hexose uptake, and cytochalasins, II: 83
High-pressure liquid chromatography (HPLC), I: 26, 31
Histamine, II: 92
    and cytochalasins, II: 63, 67
Histamine release, and cytochalasins, II: 68
Histidine, II: 125
Histone, II: 12, 21
    and luteoskyrin, II: 97
    and psoralens, II: 115
Histone deacetylase, II: 21
Histopathology, of ochratoxins, I: 22, 23
HMP, see Hexose monophosphate
Hormiscium, II: 50
Hormonal mechanisms, and cytochalasins, II: 76-77
Hormonal status, II: 5
    and aflatoxins, II: 13
Hormones
    and cytochalasins, II: 63, 66, 76-77, 80, 83, 93
    and zearalenone, II: 148, 149
Horses, I: 5, 6
    and cytochalasins, II: 58, 84
    and trichothecenes, I: 13, 60; II: 30
"Hot spots", I: 2
HPLC, see High-pressure liquid chromatography
HSV, see Herpes simplex virus
HT-2 toxin, II: 32, 40
H-thymidine, II: 16
    and aflatoxins, II: 17
    and sterigmatocystins, II: 140
Human embryos, and cytochalasins, II: 70
Human epidermal carcinoma, II: 31
Humans, II: 155
    and aflatoxins, I: 7-8, 57, 120-130; II: 5, 7, 8, 10, 24
    and animal data, I: 237-242
    and bishydroxycoumarin, I: 117
    and cytochalasins, II: 49, 58, 63-67, 69-71, 73, 74, 76, 79-86, 88-90, 92, 93
    important mycotoxins for, I: 4-11
    and N-nitroso compounds, I: 220-225, 241
    and patulin, II: 122
    and penicillic acid, II: 127
    and pink rot disease, I: 118
    and psoralens, II: 108, 109, 116-118
    and risk assessment, I: 151
    and sterigmatocystins, II: 138, 140, 141
    and trichothecenes, II: 30, 38, 39, 42, 43
    and zearalenone, II: 146, 148
H-uridine, II: 18, 150
Hydrocortisone, II: 21
    and cytochalasins, II: 77
    and ochratoxins, II: 133
Hydrophobic interactions, II: 12
7-Hydroxydiacetoxyscirpenol, II: 33, 40
Hydroxylation, II: 10
    of aflatoxin B_1, II: 8, 11
    of aflatoxins, II: 6, 8, 11, 13, 27
    of psoralens, II: 109
    of sterigmatocystins, II: 140
4-Hydroxyochratoxin A, and ochratoxins, II: 130, 131

Hydroxyproline, II: 75
5-Hydroxytryptamine, and cytochalasins, II: 66
8'-Hydroxyzearalenone, II: 146, 150
Hyperestrogenism, I: 17
  and zearalenone, I: 65, 70; II: 146
Hyperglycemia, and cyclochlorotine, II: 102
Hyperplastic nodules, and sterigmatocystins, II: 141
Hypogastric nerves, and cytochalasins, II: 72
Hypoglycemia, and cyclochlorotine, II: 102
Hypophysectomy, II: 13
Hypoxanthine, II: 87
Hypoxanthine-guanine phosphoribosyltransferase (HGPRT), II: 72
Hypoxanthine uptake, II: 87

I

IARC, see International Agency for Research on Cancer
Ig, see Immunoglobulins
IgA, see Immunoglobulin A
IgG, see Immunoglobulin G
Ignis plaga, see Ergot poisoning
Immune system, and aflatoxins, I: 20
Immunobiological functions, of cytochalasins, II: 68-69
Immunoglobulin A (IgA), and aflatoxins, II: 26
Immunoglobulin G (IgG)
  and aflatoxins, II: 26
  and cytochalasins, II: 74
Immunoglobulins, (Ig), II: 82
  and cytochalasins, II: 63, 65, 69, 81
India, I: 7, 8, 120, 134
Infection, and aflatoxins, I: 129
Ingested nitrate, and N-nitroso compounds, I: 229
Inhibition of macromolecular synthesis, II: 16-21
Initiation
  and aflatoxins, II: 21
  and cytochalasins, II: 75
  and luteoskyrin, II: 99
  and trichothecenes, II: 39, 41, 42, 45
Inosine, II: 87
Insects, I: 2, 4
Insulin, I: 8
  and cytochalasins, II: 63, 65, 68, 76, 83, 86, 88, 92, 93
  and zearalenone, II: 149
Interferon, and cytochalasins, II: 65, 68, 69, 75
International Agency for Research on Cancer (IARC), I: 52, 156, 157, 159, 223, 225; II: 131
Interstitium, I: 8
Intestinal mucosa, and trichothecenes, II: 37
Intestine
  and aflatoxins, II: 24
  and cytochalasins, I: 61; II: 54
  and N-nitroso compounds, I: 188, 196-198, 206
  and ochratoxins, II: 131
  and T-2 toxins, I: 97
  and trichothecenes, I: 16, 62; II: 37

and zearalenone, II: 148
Ion transport, and cytochalasins, II: 88
Iran, I: 11, 223, 224
Ireland, I: 22
Iridoskyrin, I: 114
Iron, and aflatoxins, II: 26
Islandicin, I: 114
Islanditoxin, I: 114
Islet cells, and cytochalasins, II: 84, 88, 92
Isororidin E, II: 35
Israel, I: 225

J

Japan, I: 4, 13, 14, 60, 114, 120
Japanese killifish, and aflatoxins, II: 5

K

Kampala, I: 121
Karyokinesis, II: 57, 58
Karyorrhexis, II: 37
K cells, and cytochalasins, II: 69
Kenya, I: 9, 12, 120, 121, 125-126, 146
Kidney, I: 6, 202
  and aflatoxins, I: 19, 20, 52, 56, 131, 133; II: 6-8, 13, 15-17, 19-21, 23, 24, 26
  and Balkan (endemic) nephropathy, I: 115, 116
  and cyclochlorotine, II: 103
  and cytochalasins, II: 54, 70, 92
  and luteoskyrin, II: 96, 97
  and N-nitroso compounds, I: 170, 186-188, 190, 191, 193, 194, 196-199, 206, 208, 236, 240
  and ochratoxins, I: 22, 23, 58, 59, 90, 92, 93; II: 130, 132, 136
  and patulin, II: 126
  and penicillic acid, II: 124, 127
  and psoralens, II: 109
  and risk assessment, I: 149, 150
  and sterigmatocystins, I: 63, 64, 99; II: 138, 140-143
  and trichothecenes, I: 61; II: 30, 31, 37, 38, 41, 44
  and yellowed rice syndrome, I: 114
  and zearalenone, II: 148, 149, 151
Kinetics of formation of N-nitroso compounds, I: 160-161
Kirsten murine sarcoma-leukemia virus, II: 70
Kodo-cytochalasins, II: 49, 51, 55
Kwashiorkor, I: 10

L

Lactate, and ochratoxins, II: 134
Lactate dehydrogenase (LDH), II: 43, 76, 94
  and ochatoxins, II: 133
  and penicillic acid, II: 126
Lactic acid, and cytochalasins, II: 72, 84

γ-Lactone cytochalasin B, II: 68, 86, 90, 91, 93
Lambs, and zearalenone, II: 148
Large intestine
    and N-nitroso compounds, I: 206
    and ochratoxins, II: 131
    and trichothecenes, I: 62
Larynx, and N-nitroso compounds, I: 170, 206, 234, 236
L-cells, and cytochalasins, II: 70
LDH, see Lactate dehydrogenase
Lectin, II: 80
    and cytochalasins, II: 68, 73, 75, 82
Leguminicola, I: 5
Lens cell, and cytochalasins, II: 59
Leptostromiformis, I: 5
LETS protein, II: 83
Lettuce, and N-nitroso compounds, I: 163, 174
Leucine, II: 75
    and aflatoxins, II: 7, 26
    and cytochalasins, II: 65, 83, 88
    and ochratoxins, II: 136
    and psoralens, II: 117
C-Leucine
    and aflatoxins, II: 19
    and luteoskyrin, II: 98
    and trichothecene, II: 41
    and zearalenone, II: 150
C-Leucine uptake, and cyclochlorotine, II: 103
Leukemia, II: 37
Leukocytes, and cytochalasins, II: 57, 58, 63, 64, 67, 69, 73, 74, 82, 84, 89, 90, 93
LH, see Luteinizing hormone
Limb bud cells, and cytochalasins, II: 59, 62
Linear cancer risk model, I: 242
Lipase, and cytochalasins, II: 65
Lipid
    and ochratoxins, II: 133
    and zearalenone, II: 149
Lipid metabolism, and cyclochlorotine, II: 102
Lipid synthesis
    and aflatoxins, II: 23-24
    and trichothecenes, II: 44
Liver
    and aflatoxin B_1, I: 6, 10
    and aflatoxins, I: 6, 7, 10, 18-21, 52-58, 71, 72, 120-130, 133, 134; II: 5, 7, 8, 10-13, 15-23, 25, 26
    and bishydroxycoumarin, I: 117
    and cyclochlorotine, II: 102, 103
    and cytochalasins, II: 54, 59, 65, 83, 85, 90, 91
    and hepatitis B virus, I: 11
    and luteoskyrin, II: 96, 98, 99
    and N-nitroso compounds, I: 164, 170, 171, 173, 175, 186-196, 198, 199, 206, 208, 220, 221, 228, 232, 234, 236, 241, 242
    and ochratoxins, I: 22, 23, 58, 73-77, 79, 80, 82, 84, 86-89, 91; II: 131-134, 136
    and patulin, II: 122, 123, 126
    and penicillic acid, II: 124, 125
    and psoralens, II: 107-109, 115
    and risk assessment, I: 142-146, 149
    and sterigmatocystins, I: 63, 64, 100; II: 138, 140-143
    and trichothecenes, I: 61, 62; II: 31, 38, 42-44
    and yellowed rice syndrome, I: 114
    and zearalenone, II: 148, 149, 151
Liver fluke, and aflatoxins, I: 129
Liver necrosis, and luteoskyrin, II: 99
Long wave UV (PUVA), see also UV light, II: 106
Lucilia, II: 37
Luncheon meat, and N-nitroso compounds, I: 171
Lung, I: 10
    and aflatoxins, I: 131; II: 6, 17
    and cytochalasins, II: 60, 63, 67, 70, 76, 92
    and luteoskyrin, II: 96
    and N-nitroso compounds, I: 170, 186, 187, 189, 190-193, 195, 196, 201, 206, 207, 234, 236
    and ochratoxins, I: 89
    and patulin, II: 122
    and penicillic acid, II: 124
    and psoralens, II: 108
    and risk assessment, I: 143
    and sterigmatocystins, I: 64; II: 138, 139, 141
    and trichothecenes, I: 61
Luteinizing hormone (LH), and zearalenone, II: 149
Luteoskyrin, I: 5, 6, 114; II: 2, 156
    and adenomas, II: 96
    and Ames assay, II: 96
    and ATPase, II: 97
    and blood, II: 96
    and carcinogenicity, II: 96
    and chromosomal abnormalities, II: 98
    and cows, II: 97
    and cysteine, II: 99
    and cytotoxicity, II: 96
    distribution of, II: 96
    and DNA, II: 98, 99
    and DNA complex formation, II: 97-98
    and DNA damage, II: 96, 99
    and DNA repair, II: 98
    and DNA template, II: 99
    and Ehrlich ascites cells, II: 98
    and elongation, II: 99
    and *Escherichia coli*, II: 97, 99
    excretion of, II: 96
    and feces, II: 96
    and glucose, II: 97
    and glycine transport, II: 99
    and heart, II: 97
    and HeLa cells, II: 96
    and histone, II: 97
    and initiation, II: 99
    and kidney, II: 96, 97
    and C-leucine, II: 98
    and liver, II: 96, 98, 99
    and liver necrosis, II: 99
    and lung, II: 96
    and macromolecular synthesis inhibition, II: 98-99

and magnesium, II: 97-99
and malic dehydrogenase, II: 99
and mice, II: 96, 98, 99
and microsomal enzymes, II: 99
and microsomes, II: 96, 99
and mitochondria, II: 96
and mitochondrial function, II: 97
and mutagenicity, II: 96
and nucleus, II: 96
and oxidative phosphorylation, II: 97
and oxygen uptake, II: 97
pharmacokinetics of, II: 96
and phosphate, II: 97
and protein, II: 98, 99
and protein synthesis, II: 98
and purine, II: 98
and rabbits, II: 98, 99
and rats, II: 96
and reticulocytes, II: 98, 99
and reticuloendotheliomas, II: 96
and RNA, II: 98
and RNA polymerase, II: 98, 99
and RNA synthesis, II: 98, 99
and SGOT, II: 99
and sodium, II: 99
and thymidine, II: 98
and thymus, II: 97
and toxicity, II: 96, 97
and transcription, II: 97-99
and C-uridine, II: 98
and urine, II: 96
and zearalenone, II: 147
Lymphatic leukemia, II: 37
Lymph node cells, and cytochalasins, II: 65
Lymph nodes
  and aflatoxins, I: 131
  and trichothecenes, II: 37
Lymphoblastoid cells, and cytochalasins, II: 80
Lymphocytes
  and cytochalasins, II: 68, 69, 81, 83-86, 88-93
  and psoralens, II: 108
Lymphocyte transformation, and cytochalasins, II: 68
Lymphokine production, II: 69
Lymphotoxins, and cytochalasins, II: 65, 69
Lymph system
  and N-nitroso compounds, I: 193, 206
  and ochratoxins, I: 82, 83
Lysergic acid, I: 109, 110
Lysine, II: 39
  and patulin, II: 125
  and penicillic acid, II: 125
Lysosomal enzymes
  and aflatoxins, II: 24-25
  and cytochalasins, II: 63, 64
Lysosomes, II: 156, 157
  and aflatoxins, II: 23, 24, 27
  and ochratoxins, I: 59; II: 136
  and sterigmatocystins, II: 142
  and trichothecenes, II: 44
  and zearalenone, II: 151
Lysozyme, II: 64

and cytochalasins, II: 64

M

Macaca irus, I: 53
Macaca mullatta, I: 53
Macara fascicularis, I: 66
Macromolecular binding
  of aflatoxins, II: 11-13
  of ochratoxins, II: 134
Macromolecular complexes of aflatoxin B_1, II: 12
Macromolecular synthesis
  and cytochalasins, II: 72-77
  inhibition of, II: 16-21
  and patulin, II: 125-126
  and penicillic acid, II: 126
  and psoralens, II: 115-117
Macromolecular synthesis inhibition
  and luteoskyrin, II: 98-99
  and ochratoxins, II: 135-136
Macromolecules, and ochratoxins, II: 134-135
Macrophages, and cytochalasins, II: 60, 63, 68
Macropinocytosis, and cytochalasins, II: 60
Magnesium
  and aflatoxins, II: 26
  and cytochalasins, II: 73, 78, 79, 83
  and luteoskyrin, II: 97-99
  and penicillic acid, II: 127
  and psoralens, II: 113
Maize, see also Corn
  and aflatoxins, I: 121-123
  and risk assessment, I: 151
  and zearalenone, I: 71
Malate dehydrogenase, and aflatoxins, II: 25
Malawi, I: 221
Malic dehydrogenase, and luteoskyrin, II: 99
Maltoryzine, I: 5
Mammals, and ochratoxins, II: 130
Mammary
  and N-nitroso compounds, I: 197, 206
  and zearalenone, I: 17; II: 146, 150
Marmosets, and aflatoxins, II: 58
Massachusetts Institute of Technology, I: 121
Mass spectrometer, I: 157-158
Mathematical cancer risk models, I: 237-242
MC, see Moisture content
Meats, and N-nitroso compounds, I: 164, 168-170, 171, 173-175, 222, 230, 231
Meiosis, and cytochalasins, II: 62
Melanin, and cytochalasins, II: 72
Melanocytes
  and cytochalasins, II: 72
  and psoralens, II: 106
Melanocyte-stimulating hormone (MSH), and cytochalasins, II: 66, 72
Membrane
  and aflatoxins, II: 22
  and cytochalasins, II: 57
  and trichothecenes, II: 44
Membrane binding, and cytochalasins, II: 89-91

Membrane carriers, and cytochalasins, II: 84, 87, 89-91
Membrane flexibility, II: 81
Membrane marker proteins, II: 82-83
Membrane receptors, and cytochalasins, II: 82, 83, 93
Membrane structure alterations, II: 80-91
Membrane topology, II: 80-81, 156, 157
   and cytochalasins, II: 82, 83
Membrane transport, II: 156, 157
   and aflatoxins, II: 17, 26
   and cytochalasins, II: 80, 83-91
   and ochratoxins, II: 136
   and penicillic acid, II: 127
   and trichothecenes, II: 39
Mesenchymal cells, and sterigmatocystins, II: 141, 143
Mesodermal cells, and cytochalasins, II: 62
O-MeSTG, see O-Methylsterigmatocystin
Metabolic activation
   and aflatoxins, II: 6, 11, 12, 25, 27
   and N-nitroso compounds, I: 207-209
   and psoralens, II: 109
   and trichothecenes, II: 38
Metabolic cooperation, and cytochalasins, II: 72
Metabolism
   and aflatoxins, I: 54, 57; II: 6, 12-14, 16, 18
   and carbohydrates, II: 102
   and cytochalasins, II: 72-77
   factors affecting, II: 13-14
   and lipid, II: 102
   and N-nitroso compounds, I: 208-209, 238
   and ochratoxins, II: 131-132
   and patulin, II: 124
   and penicillic acid, II: 124
   and psoralens, II: 109
   and sterigmatocystins, II: 140
   and trichothecenes, II: 37-38
Metamorphosis, and cytochalasins, II: 61
Metarrhizium, II: 48
Metarrhizium anisopliae, II: 50
Methionine, II: 42, 75
C-Methionine, and zearalenone, II: 150
Methonium ion, I: 241
5-Methoxypsoralen (5-MOP), II: 106-108, 110, 113, 115, 117-119, 156
8-Methoxypsoralen (8-MOP), I: 7; II: 106-113, 115-118, 156
   and pink rot disease, I: 117-119
5-Methoxysterigmatocystin, II: 139
N-Methyl-4-aminoazobenzene, I: 239
Methylbenzylnitrosamine, I: 186, 224, 225
3-Methylcholanthrene, II: 13
Methyl esters, and ochratoxins, II: 130, 131
3-O-Methylglucose, II: 84, 85, 89
   and zearalenone, II: 150
3-O-Methylglucose uptake, II: 86
Methylnitrosamines, I: 187-189
O-Methylsterigmatocystin (O-MeSTG), II: 138, 139, 143
Mg, see Magnesium
Mice
   and aflatoxins, I: 55, 66; II: 5, 6, 8, 10, 13, 16, 17, 19-21, 24
   and cyclochlorotine, II: 102
   and cytochalasins, II: 54, 56-58, 60, 62-65, 67, 68, 70, 73-76, 80, 82, 84-88, 90-93
   and luteoskyrin, II: 96, 98, 99
   and mammary carcinoma, see Mouse mammary carcinoma
   and N-nitroso compounds, I: 175, 186-198, 221, 234, 236, 237
   and ochratoxins, I: 59, 60, 85; II: 130, 136
   and patulin, II: 122, 123, 125, 126
   and penicillic acid, II: 124, 127
   and pink rot disease, I: 117
   and psoralens, II: 108, 109, 115-117
   and risk assessment, I: 149
   and sterigmatocystins, I: 63, 64, 70, 100; II: 139, 141
   and trichothecenes, I: 15, 16, 62, 63, 69; II: 31, 37, 39, 43, 44
   and zearalenone, II: 146, 148, 149, 151
Mice embryo cells, and cytochalasins, II: 58
Mice embryos, and cytochalasins, II: 62
Michigan, I: 117-119
Microfilament-plasma membrane, II: 56
Microfilaments, II: 48, 155, 156
   and cytochalasins, II: 57, 58, 60-63, 72, 76, 77, 82, 85, 88, 90, 91
Microfilament structure alterations, and cytochalasins, II: 79-80
Micropinocytosis, and cytochalasins, II: 60
Microsomal activation, and sterigmatocystins, II: 139, 141
Microsomal activity, and aflatoxins, II: 14
Microsomal enzymes
   activity of, II: 5
   and aflatoxins, II: 12
   and luteoskyrin, II: 99
Microsomes
   and aflatoxins, II: 8, 10, 11, 13, 19, 22, 25
   and luteoskyrin, II: 96, 99
   and N-nitroso compounds, I: 208
   and ochratoxins, II: 131
   and patulin, II: 123
   and psoralens, II: 109
   and sterigmatocystins, II: 139, 141
   and trichothecenes, II: 37, 43
   and zearalenone, II: 147
Microspikes, and cytochalasins, II: 59
Microtubules, and cytochalasins, II: 63, 71, 77, 78, 80, 82
Microvilli, and cytochalasins, II: 56, 80
Milk, I: 13
   and aflatoxins, I: 20, 27-28; II: 15
   and N-nitroso compounds, I: 164, 176
Mink
   and aflatoxins, II: 16, 24
   and N-nitroso compounds, I: 187
Mitochondria
   and aflatoxins, II: 7, 23
   and cytochalasins, II: 73, 76
   and luteoskyrin, II: 96, 156

and ochratoxins, I: 59; II: 132-133, 157
and patulin, II: 127
and penicillic acid, II: 127
Mitochondrial function
   and aflatoxins, II: 23, 27
   and luteoskyrin, II: 97
Mitochondrial transport
   and luteoskyrin, II: 156
   and ochratoxins, II: 157
Mitogen
   and cytochalasins, II: 68, 91, 92
   and psoralens, II: 108
Mitosis
   and aflatoxins, I: 52
   and penicillic acid, II: 123
   and sterigmatocystins, II: 138, 142
Mitotic index, II: 37
   and zearalenone, II: 149
Mixed function oxidase, II: 6, 16, 44
   and aflatoxins, II: 20
   and psoralens, II: 109
   and sterigmatocystins, II: 140
Moisture content (MC), I: 2, 3
Moldy corn toxicosis, I: 60; II: 30
Molybdenum, I: 224
Monkeys, I: 10
   and aflatoxins, I: 52-54, 58, 66, 132, 133; II: 8, 10, 15-17, 19, 20, 25
   and cytochalasins, II: 70
   and N-nitroso compounds, I: 190, 193
   and ochratoxins, I: 84; II: 136
   and risk assessment, I: 150
   and sterigmatocystins, I: 63, 64, 70, 100; II: 138-140, 142, 143
   and zearalenone, II: 149
Monoacetoxyscirpenol, II: 37, 40
Monosomes, and aflatoxins, II: 20
5-MOP, see 5-Methoxypsoralen
8-MOP, see 8-Methoxypsoralen
Morphogenesis, and cytochalasins, II: 60-63, 73
Morpholine, I: 176
Motility, II: 81
   and cytochalasins, II: 91
Mouse mammary carcinoma, II: 5, 6, 56, 96
   and patulin, II: 123
   and penicillic acid, II: 124
   and sterigmatocystins, II: 139
Mouse mammary tumor cells, II: 57
   and cytochalasins, II: 92
Mouth, and trichothecenes, I: 15, 16
Mozambique, I: 11, 12, 120, 121, 126, 130, 146, 221
mRNA
   and aflatoxins, II: 20, 21
   and psoralens, II: 116
   and trichothecenes, II: 42, 43
mRNA synthesis, and aflatoxins, II: 18
Muconomycin A, II: 34
Muconomycin B, II: 34
Mucopeptide synthesis, and ochratoxins, II: 135
Mucopolysaccharides, and cytochalasins, II: 75-76
Mucopolysaccharide synthesis, and cytochalasins, II: 58, 74, 88
Mucopolysaccharide transport, and cytochalasins, II: 72
Mucor mucedo, II: 30
Mucous membranes, and alimentary toxic aleukia, I: 113
Mules, and trichothecenes, I: 13
Multinucleation
   and cytochalasin B, II: 58
   and cytochalasins, II: 58, 74
Multiple screening detection method, I: 30
Muscle
   and aflatoxins, I: 19, 21; II: 8, 15, 16, 24
   and cytochalasins, II: 59, 62, 72, 77-79, 84, 91, 93
   and ergot alkaloids, I: 109
   and ochratoxins, I: 23, 58; II: 130, 132, 134
   and penicillic acid, II: 126
   and zearalenone, I: 65; II: 148, 149
Mutagenicity, II: 156, 157
   and aflatoxins, I: 54; II: 5-9, 11, 12, 27, 37
   and cyclochlorotine, II: 102
   and cytochalasins, II: 54-56
   and luteoskyrin, II: 96
   and N-Nitroso compounds, I: 207, 209, 241
   and ochratoxins, II: 131
   and patulin, II: 124
   and psoralens, II: 107-108, 115
   and sterigmatocystins, II: 139, 143
   and zearalenone, II: 147
Myeblobastoma cells, and cytochalasins, II: 86
Myoblasts, and cytochalasins, II: 73, 88
Myofibrils, and cytochalasins, II: 62
Myosin, and cytochalasins, II: 58, 62, 77-79, 91
Myrothecium, I: 60; II: 30, 32
Myrothecium roridium, II: 32, 34, 35
Myrothecium verrucaria, II: 34, 35

N

Na⁺, see Sodium
NAD, II: 25
NADH, and aflatoxins, II: 25
NADP, II: 10
   and aflatoxins, II: 25
   and cyclochlorotine, II: 102
   and ochratoxins, II: 133
NADPH
   and aflatoxins, II: 8, 25
   and cyclochlorotine, II: 102
NADPH-dependent enzymes, II: 6, 10
Necrosis
   and cytochalasins, II: 54
   and sterigmatocystins, II: 139
   and trichothecenes, II: 37
Neosolaniol, I: 6, 15, 16; II: 30, 32, 40, 43, 44
Neosolaniol monoacetate, II: 33
Nephropathy, see also Balkan (endemic) nephropathy
   and ochratoxins, I: 22, 23

Nerve cells, and cytochalasins, II: 56
Nerves
   and cytochalasins, II: 84
   and N-nitroso compounds, I: 197, 204
Nervous system, and N-nitroso compounds, I: 193
Netherlands, I: 135
Neural epithelium, and cytochalasins, II: 59
Neural retina, and cytochalasins, II: 59
Neural retina cells, and cytochalasins, II: 81
Neural tube
   and cytochalasins, II: 56
   disorganization of, II: 5
Neuraminidase, and cytochalasins, II: 80
Neuroblastoma, and cytochalasins, II: 64
Neuroblastoma cells, and cytochalasins, II: 85
Neuroepithelial cells, and cytochalasins, II: 62
Neurospora crassa, and aflatoxins, II: 7
Neurotransmitters, and cytochalasins, II: 63, 71, 72, 81
New York City, I: 165, 166
New Zealand, I: 8, 132
Nigrosabulum, II: 50
Nitella, II: 58
Nitrosamides, I: 173, 186, 196, 197, 199-207, 209
N-Nitrosamides, I; 158
Nitrosamines, I: 156-159, 164, 166, 168-170, 173, 175, 176, 186, 198, 203, 207-209, 220, 223-225, 228, 229, 232, 234, 237
   blood levels of, I: 230-232
   dietary sources of, I: 230
   in foods, I: 231
   formation of, I: 163-164
N-Nitrosamines, I: 222
Nitrosation
   of amines, I: 160, 163, 174
   of drugs, I: 177
   of pesticides, I: 178
   rate of, I: 163
N-Nitrosocarbaryl, I: 176
N-Nitroso compounds
   and adduct formation, I: 207
   and adrenal gland, I: 197, 206
   and aflatoxins, I: 223
   in air, I: 164-166
   and air analysis, I: 158
   and alcohol, I: 223
   and anchovies, I: 171
   and ascorbic acid, I: 164
   and bacon, I: 169, 170, 171, 173, 230, 231
   and bass, I: 176
   and beer, I: 173, 176, 230
   and beets, I: 174
   and bladder, I: 170, 188-192, 204, 206, 225
   and blood, I: 159, 171, 197, 198, 204, 223, 228, 230, 232
   and blood vessels, I: 194
   and bone, I: 197, 206
   and brain, I: 196, 197, 205
   and bread, I: 230
   and bronchus, I: 190
   and carcinogenicity, I: 164, 168, 174-176, 186-209, 221-225, 228
   and celery, I: 174
   and cheese, I: 168, 172, 173, 230
   and chickens, I: 224
   and cod, I: 171, 176
   and coffee, I: 176
   and corn, I: 222
   and cosmetics, I: 175, 176
   and croaker, I: 171
   and cytotoxicity, I: 241
   and DNA, I: 239, 240
   and DNA binding, I: 208, 241
   and DNA damage, I: 240, 241
   and DNA repair, I: 238, 240, 241
   and DNA synthesis, I: 209
   and DNA template, I: 208
   and dogs, I: 186, 190, 196, 198, 221, 229
   dose-response relationships in, I: 232-238, 241
   and drugs, I: 175
   and ear, I: 197, 206
   and esophagus, I: 170, 187, 189-198, 200, 207, 221-224, 232
   excretion of, I: 240
   and exposure estimates, I: 228-232
   and feces, I: 159, 223
   and fibroblasts, I: 241
   and fish, I: 168, 171-173, 175, 176, 187, 190, 230, 231
   and fish meal, I: 160
   in foodstuffs, I: 166-175
   formation of, I: 159-164
   and frankfurters, I: 170, 171, 176, 230, 231
   and gall bladder, I: 191, 205, 223
   and glutathione, I: 164
   and guinea pigs, I: 189-191, 196, 197, 221
   and haddock, I: 171
   and hake, I: 171
   and ham, I: 170, 171, 176
   and hamsters, I: 187-196, 198, 236, 237
   and heart, I: 206
   and hedgehog, I: 190
   and herring, I: 161, 171
   and humans, I: 220-225, 241
   and ingested nitrate, I: 229
   and intestine, I: 188, 196-198, 206
   and kidney, I: 170, 186-188, 190-194, 196-199, 202, 206, 208, 236, 240
   kinetics of formation of, I: 160-161
   and large intestine, I: 206
   and larynx, I: 170, 206, 234, 236
   and lettuce, I: 163, 174
   and liver, I: 164, 170, 171, 173, 175, 186-196, 198, 206, 208, 220, 221, 228, 232, 234, 236, 241, 242
   and luncheon meat, I: 171
   and lung, I: 170, 186-193, 195, 196, 201, 206, 207, 234, 236
   and lymph system, I: 193, 206
   and mammary, I: 197, 206

and meats, I: 164, 168-171, 173-175, 222, 230, 231
and metabolic activation, I: 207-209
and metabolism, I: 208, 238
and mice, I: 175, 186-198, 221, 234, 236, 237
and microsomes, I: 208
and milk, I: 164, 176
and mink, I: 187
and monkey, I: 190, 193
and mutagenicity, I: 207, 209, 241
and nerves, I: 197, 204
and nervous system, I: 193
and nose, I: 187, 188, 190-196, 200, 205, 234, 236
and nucleic acids, I: 208
and opium, I: 223
and ovary, I: 194, 197, 205
and pancreas, I: 190, 191, 195, 204
and parakeet, I: 190
and perch, I: 176
persistence of, I: 159-164
and pharynx, I: 188-190, 193, 194, 196, 198, 203
and pigs, I: 173, 190, 207
and pork, I: 171
and potatoes, I: 163
and protein, I: 208, 242
and quantitative analysis, I: 156-159
and rabbits, I: 186, 190, 196, 221
and radishes, I: 174
and rats, I: 186-199, 206-208, 221, 224, 232-234, 236, 237, 239, 241, 242
and risk assessment, I: 242-243
and RNA, I: 239
and RNA polymerase, I: 209
and sable, I: 172
and salami, I: 171
and saliva, I: 163, 174, 228, 229
and salivary glands, I: 223
and salmon, I: 172, 176
and sausage, I: 171
and shad, I: 172
and sheep, I: 160, 166
and skin, I: 196-198, 205, 206
and small intestine, I: 198
and sodium nitrite, I: 160
in soil, I: 162
and souse, I: 171
and spinach, I: 163, 168, 173, 174, 230
and spinal cord, I: 196, 197, 205
and spleen, I: 220
and stomach, I: 174, 187, 188, 190, 193, 196-199, 201, 222, 225, 228, 229, 236
structure-activity relationships in, I: 206
and substrate, I: 208
and sugar, I: 222
and sulfhydryl compounds, I: 164
and tea, I: 176, 223
and teratogenicity, I: 207
and testis, I: 186, 192, 206
and thymus, I: 189, 206
and thyroid, I: 191, 197, 206

and tobacco, I: 223
and tomato, I: 173
and tongue, I: 171, 188, 191, 193-196, 203
and toxicity, I: 186, 187, 208, 209, 220-221
and trachea, I: 170, 188, 190-193, 196, 203, 236
and trout, I: 176, 187, 190
and tuna, I: 176
and urine, I: 159, 222, 223, 225, 232
and uterus, I: 206, 222
and vagina, I: 191, 198, 205,222
and vegetables, I: 222
in water, I: 162, 166
and water analysis, I: 158
and wine, I: 176
and zebra fish, I: 190
N-Nitroso-di-*n*-butylamine, I: 170
N-Nitrosodiethanolamine, I: 175, 176
N-Nitrosoglyphosate, I: 162
N-Nitrosoheptamethyleneimine, I: 207
N-Nitrosohexamethyleneimine, I: 177
Nitrosomorpholine, I: 164
Nitrosopiperidine, I: 170, 171, 172
N-Nitrosopiperidine, I: 230
Nitrosoproline, I: 169
Nitrosopyrrolidine, I: 169, 170-173, 232
N-Nitrosopyrrolidine, I: 169, 176, 228, 230, 231
Nivalenol, I: 5, 14-16; II: 30, 31, 33, 37-40, 44, 45
Noradrenaline, and cytochalasins, II: 67
Norepinephrine, and cytochalasins, II: 66, 76
North America, I: 142
North Carolina, I: 21
North Dakota, I: 116
Norway, I: 22, 160, 168
Nose
  and N-nitroso compounds, I: 187, 188, 190-196, 200, 205, 234, 236
  and trichothecenes, I: 15
Novikoff hepatoma, II: 84
Novikoff hepatoma cells, and cytochalasins, II: 85, 87-90
Nuclear division, and cytochalasins, II: 74
Nuclear membrane, and aflatoxins, II: 12
Nuclear RNA synthesis, and sterigmatocystins, II: 141
Nuclei, II: 21
  and aflatoxins, II: 7, 8, 11, 13, 16, 18, 19, 22, 24, 25
  and cytochalasins, II: 57, 76, 77
  and luteoskyrin, II: 96
  and penicillic acid, II: 125
  and sterigmatocystins, II: 140, 141
  and zearalenone, II: 150
Nucleic acids
  conformation of, II: 112-113
  and cytochalasins, II: 86-88
  and N-nitroso compounds, I: 208
  photoreactions with, II: 118-119
  and psoralens, II: 110-115, 118-119
  structure of, II: 112-113
Nucleic acid synthesis
  and aflatoxins, II: 17

inhibition of, II: 119
and psoralens, II: 115
and sterigmatocystins, II: 140-141
and trichothecenes, II: 38
Nucleolar polymerase, and aflatoxins, II: 18
Nucleolar RNA, and aflatoxins, II: 19
Nucleolar RNA synthesis, II: 17
Nucleoplasmic polymerase, and aflatoxins, II: 18
Nucleoside incorporation, and cytochalasins, II: 74
Nucleosides
    and cytochalasins, II: 86-88
    and psoralens, II: 111
    and sterigmatocystins, II: 140
Nucleoside transport, II: 87
    and aflatoxins, II: 27
    and cytochalasins, II: 74, 87
    and sterigmatocystins, II: 141
Nucleotides, and psoralens, II: 110, 111, 113, 114, 117
Nutrition, see also Diet
    and aflatoxins, I: 127-129

O

Occurrence of mycotoxins, I: 2-4
Ochratoxin α, II: 131, 132, 134-136
Ochratoxin β, II: 131, 134, 136
Ochratoxin A, I: 5, 22, 24, 58-60, 116, 142; II: 130, 132-136, 157
    structure of, I: 117
Ochratoxin A_1, I: 23
Ochratoxin A ethyl ester, II: 131
Ochratoxin A methyl ester, II: 131
    and penicillic acid, II: 124
Ochratoxin B, I: 24; II: 130-132, 134, 136
Ochratoxin B_1, I: 59
Ochratoxin B ethyl ester, II: 131
Ochratoxin B methyl ester, II: 131
Ochratoxin C, II: 130, 131, 134, 136
Ochratoxins, I: 4, 7, 22-24, 58-60; II: 155, 157
    and *Actinomyces*, II: 130
    and adenomas, II: 131
    and adrenal gland, I: 78
    and aflatoxins, I: 123, 130
    and albumin, II: 132, 134, 136
    and Ames assay, II: 131
    and amino acids, II: 136
    and aminoacylation, II: 135
    and antibodies, II: 136
    and antigens, II: 136
    and *Aspergillus*, I: 22, 58; II: 130
    and *Aspergillus ochraceous*, II: 130
    and ATP, II: 132, 133
    and autolysis, II: 135
    and *Bacillus cereus*, II: 130
    and *Bacillus megaterium*, II: 130
    and *Bacillus subtilis*, II: 131, 135
    and bacteria, II: 130, 135
    and Balkan (endemic) nephropathy, II: 130
    and barley, I: 59
    and birds, II: 130
    and blood, I: 22; II: 131, 133, 134
    and bone, II: 130
    and brine shrimp, II: 130
    and calcium, II: 133
    and carbohydrate metabolism, II: 133-134
    and carboxypeptidase, II: 134
    and carboxypeptidase A, II: 131
    and carcinogenicity, I: 60; II: 130, 131
    and chicken embryos, II: 130, 131
    and chickens, I: 23, 59, 60; II: 130, 136
    and chloramphenicol, II: 135
    and citrinin, II: 130
    and colon, I: 90, 91
    and complement, II: 136
    and corn, I: 23
    and cows, II: 131, 134
    and cyclic AMP, II: 133, 134
    and cytotoxicity, II: 136
    detection of, I: 33-36
    and dinitrophenol, II: 132
    distribution of, II: 131-132
    and DNA, II: 134
    and DNA binding, II: 134
    and DNA synthesis, II: 135
    and dogs, I: 59, 79 82
    and ducks, I: 59, 73, 74; II: 130
    and eggs, I: 24
    and electron transport, II: 135
    and enzyme activity, II: 134-135
    and epithelial cells, II: 136
    and *Escherichia coli*, II: 135
    and ethyl ester, II: 131
    and excretion, I: 59
    excretion of, II: 131-132
    and exencephaly, II: 130
    and extraction, I: 35
    and eye, II: 130
    and fat, II: 130
    and fatty acids, II: 134
    and feces, II: 131, 132
    and fish, I: 60; II: 130, 131
    and β-galactosidase, II: 135
    and gall bladder, I: 80, 81
    and β-globulin, II: 136
    and γ-globulin, II: 134
    and glucose, II: 134
    and C-glucose, II: 135
    and glucose transport, II: 134
    and glutamate-oxaloacetate transaminase, II: 133
    and glutamate-pyruvate transaminase, II: 133
    and glycine transport, II: 136
    and glycogen, II: 133, 134
    and glycogenesis, II: 134
    and glycogenolysis, II: 134
    and glycogen synthetase, II: 133
    and glycolysis, II: 133
    and guinea pigs, I: 82, 83; II: 136
    and hamsters, II: 130
    and heart, I: 78; II: 130, 133

and hepatoma, II: 130
histopathology of, I: 22, 23
and hydrocortisone, II: 133
and 4-hydroxyochratoxin A, II: 130, 131
and intestine, II: 131
and kidney, I: 22, 23, 58, 59, 90, 92, 93; II: 130, 132, 36
and lactate, II: 134
and lactic dehydrogenase, II: 133
and large intestine, II: 131
and leucine, II: 136
and lipid, II: 133
and liver, I: 22, 23, 58, 73-77, 79, 80, 82, 84, 86-89, 91; II: 131-134, 136
and lung, I: 89
and lymph, I: 82, 83
and lysosomes, I: 59; II: 136
and macromolecular binding, II: 134
and macromolecular synthesis inhibition, II: 135-136
and macromolecules, II: 134-135
and mammals, II: 130
and membrane transport, II: 136
metabolism of, II: 131-132
and methyl ester, II: 130, 131
and mice, I: 59, 60, 85; II: 130, 136
and microsomes, II: 131
and mitochondria, I: 59; II: 132-133
and monkeys, I: 84; II: 136
and mucopeptide synthesis, II: 135
and muscle, I: 23, 58; II: 130, 132, 134
and mutagenicity, II: 131
and NADP, II: 133
and nephropathy, I: 22, 23
and C-orotic acid, II: 136
and oxidative phosphorylation, II: 132
and pancreas, II: 131
and penicillic acid, II: 130
and *Penicillium*, I: 58; II: 130
and *Penicillium cyclopium*, II: 130
and *Penicillium viridicatum*, II: 130
and phenylalanine, II: 131, 136
and β-phenylalanine, II: 130
and phenylalanyl, II: 135
and phenylalanyl-tRNA, II: 135
and phosphate, II: 132, 133
and phosphorylase, II: 133, 134
and phosphorylase β, II: 134
and phosphorylase β kinase, II: 133
physical properties of, I: 34
and pigs, I: 23, 58, 59, 92, 93; II: 130, 132
and placenta, II: 132
and plasma membrane, II: 132
and polysomal profiles, II: 135
and poultry, I: 59-60; II: 130
and protein, II: 133, 134, 136
and protein binding, II: 134
and protein kinase, II: 134
and protein synthesis, II: 135
and protozoa, II: 130
and purification, I: 35
and pyruvate, II: 133
and quail, I: 59
and rabbits, II: 136
and rainbow trout, I: 59, 60
and rats, I: 22, 59, 60, 75-78, 86, 87, 90, 91; II: 130-134, 136
and *rec* assay, II: 131
and respiration, II: 132-135
and reticulocytes, II: 136
and RNA, II: 136
and RNA synthesis, II: 135
and *Saccharomyces cerevisiae*, II: 131
and *Salmonella typhimurium*, II: 131
and screening, I: 36
and serum, II: 132-134, 136
and small intestine, II: 131
and sodium, II: 136
and spleen, I: 81
and sterculic acid, II: 130
and *Streptococcus faecalis*, II: 134, 135
structure, I: 34
structure-activity relationships in, II: 136
and teratogenicity, I: 60; II: 130
and *Tetrahymena*, II: 130
and toxicity, I: 58-60; II: 131, 133, 136
and trachea, II: 130
and translation, II: 135
and triglycerides, II: 134
and tRNA, II: 135
and trout, I: 59, 60, 91; II: 130, 131
and trypsin, II: 131
and tubular necrosis, I: 22
and turkeys, I: 23
and urine, I: 22; II: 131, 132
and vertebrae, II: 130
and zebra fish, II: 130
Oligomycin
  and patulin, II: 126
  and penicillic acid, II: 127
Oncornaviruses, II: 70
One-hit cancer risk model, I: 238
Onions, and aflatoxins, II: 124
Oocytes, and cytochalasins, II: 62
Opisthorchis viverrini, I: 129
Opisthorciasis, I: 129
Opium, and N-nitroso compounds, I: 223
Orotic acid, and aflatoxins, II: 7
C-Orotic acid
  and aflatoxins, II: 17
  and ochratoxins, II: 136
  and patulin, II: 126
Osmotic balance, and cytochalasins, II: 71, 76, 88
Ovariectomy, and aflatoxins, II: 13
Ovary
  and cytochalasins, II: 82, 87
  and N-nitroso compounds, I: 194, 197, 205
  and psoralens, II: 108, 115
  and trichothecenes, I: 61; II: 37-39, 43
  and zearalenone, I: 17, 65
Ovary cells, and aflatoxins, II: 5
Oviduct, and cytochalasins, II: 61
Ovulation, and cytochalasins, II: 72
β-Oxidation of fatty acids, I: 208

Oxidative metabolism, II: 10
Oxidative phosphorylation, II: 23
 and cyclochlorotine, II: 102
 and cytochalasins, II: 69
 and luteoskyrin, II: 97
 and ochratoxins, II: 132
 and penicillic acid, II: 127
Oxygen consumption, II: 23
 and trichothecenes, II: 44
Oxygen uptake
 and luteoskyrin, II: 97
 and patulin, II: 127
 and penicillic acid, II: 127
Oxytocin, and cytochalasins, II: 71

P

Pancreas, II: 65
 and aflatoxins, I: 21, 52
 and cytochalasins, II: 54, 84, 88, 92
 and N-nitroso compounds, I: 190, 191, 195, 204
 and ochratoxins, II: 131
 and trichothecenes, I: 16
Pancreatic β cells, and cytochalasins, II: 63
Pancreatic islets, and cytochalasins, II: 68
Papillomas, II: 37
Parakeet, and N-nitroso compounds, I: 190
Paramecium aurelia, II: 125
Parathyroid, and cytochalasins, II: 66
Paris, I: 108
Passive diffusion, II: 73
Pasturella multocida, and aflatoxins, I: 20
Patulin, I: 5, 6
 and adduct formation, II: 124-125
 and amphibians, II: 122
 and arginine, II: 125
 and bacteria, II: 122, 125
 and brain, II: 126, 127
 and carcinogenicity, II: 123, 125
 and cellular respiration, II: 127
 and chicken embryos, II: 122, 125
 and chromosomal abnormalities, II: 122
 and crustaceans, II: 122
 and cysteine, II: 122, 124, 125
 and cysteine adducts, II: 124-126
 distribution of, II: 124
 and DNA, II: 123
 and DNA damage, II: 123, 124
 and DNA synthesis, II: 126
 and eggs, II: 122
 and enzyme activity, II: 126-127
 and erythrocytes, II: 124, 125
 and *Escherichia coli*, II: 125
 excretion of, II: 124
 and exencephaly, II: 122
 and fibrosarcomas, II: 122
 and fish, II: 122
 and fungi, II: 122
 and glutathione, II: 122, 125
 and glutathione adducts, II: 125
 and guinea pigs, II: 127
 and HeLa cells, II: 122-124, 126
 and histidine, II: 125
 and humans, II: 122
 and kidney, II: 126
 and liver, II: 122, 123, 126
 and lung, II: 122
 and lysine, II: 125
 and macromolecular binding, II: 157
 and macromolecular synthesis, II: 125-126
 and mammary carcinoma, II: 123
 and metabolism, II: 124
 and mice, II: 122, 123, 125, 126
 and microsomes, II: 123
 and mitochondria, II: 127
 and mutagenicity, II: 124
 and O_2 uptake, II: 127
 and oligomycin, II: 126
 and C-orotic acid, II: 126
 and pharmacokinetics, II: 124
 and phytotoxicity, II: 122
 and potassium, II: 126
 and protein synthesis, II: 126
 and protozoa, II: 122
 and quail, II: 125
 and rats, II: 122, 123
 and *rec* assay, II: 123
 and respiration, II: 126-127
 and risk assessment, I: 142
 and RNA polymerase, II: 125
 and RNA synthesis, II: 125, 126
 screening for, I: 36
 and sodium, II: 126
 and sulfhydryl groups, II: 122, 124
 and sulfhydryl reagents, II: 126
 and teratogenicity, II: 122, 125
 and *Tetrahymenia pyriformis*, II: 127
 and toxicity, II: 122, 124
 and trachea, II: 122
PCMB, see *p*-Chloromercuribenzoate
Peanut butter, II: 8
 and aflatoxins, I: 122, 123
 and risk assessment, I: 142
Peanuts, I: 2, 8-10, 20
 and aflatoxins, I: 18, 21, 52, 57, 58, 121, 122, 124, 127, 130, 134
 and risk assessment, I: 142-144, 151
Peas, and aflatoxins, I: 121, 122
Penicillic acid, I: 3, 5, 6; II: 157
 and adduct formation, II: 124
 and aflatoxins, I: 130
 and alcohol dehydrogenase (ADH), II: 126
 and aldolase, II: 126
 and α-amanitin, II: 125
 and amino acids, II: 125
 and α-aminoisobutyric acid, II: 125
 and antigens, II: 125
 and arginine, II: 125
 and ATPase, II: 126, 127
 and bacteria, II: 123, 125
 and bile, II: 124
 and blood, II: 124

and brain, II: 127
and brine shrimp, II: 124
and carcinogenicity, II: 124, 125
and cellular respiration, II: 127
and Chang liver cells, II: 125
and chromosomal abnormalities, II: 123
and citrinin, II: 124
and CO_2, II: 124
and cocarboxylase, II: 126
and cyclochlorotine, II: 127
and cysteine, II: 124, 126
and cysteine adducts, II: 125
and cytotoxicity, II: 123
and dinitrophenol, II: 127
distribution of, II: 124
and DNA, II: 124
and DNA synthesis, II: 125
and elongation, II: 125
and enzyme activity, II: 126-127
and erythrocytes, II: 125, 127
and *Escherichia coli*, II: 125, 126
excretion of, II: 124
and feces, II: 124
and fibrosarcomas, II: 124
and glutathione, II: 127
and glutathione adducts, II: 125
and glycine transport, II: 127
and guinea pigs, II: 127
and HeLa cells, II: 125
and histidine, II: 125
and humans, II: 127
and kidney, II: 124, 127
and lactic dehydrogenase (LDH), II: 126
and liver, II: 124, 125
and lung, II: 124
and lysine, II: 125
and macromolecular synthesis, II: 126
and magnesium, II: 127
and mammary carcinoma, II: 124
and membrane transport, II: 127
and metabolism, II: 124
and mice, II: 124, 127
and mitochondria, II: 127
and mitosis, II: 123
and muscle, II: 126
and nuclei, II: 125
and ochratoxins, II: 124, 130
and oligomycin, II: 127
and oxidative phosphorylation, II: 127
and oxygen uptake, II: 127
and pharmacokinetics, II: 124
and phenylalanine, II: 127
and potassium, II: 127
and protein, II: 125
and protein synthesis, II: 125, 126
and protozoa, II: 125
and rabbits, II: 125, 127
and rats, II: 124, 125
and *rec* assay, II: 124
and respiration, II: 126-127
and reticulocytes, II: 125, 127
and RNA, II: 125
and RNA polymerase, II: 125
and RNA synthesis, II: 125
and rRNA, II: 125
screening for, I: 36
and sodium, II: 127
and spleen, II: 124
and sulfhydryl groups, II: 123, 124
and sulfhydryl reagents, II: 126, 127
and teratogenicity, II: 125
and *Tetrahymenia pyriformis*, II: 127
and threonine, II: 127
and transcription, II: 125
and transport, II: 126-127
and urease, II: 126
and urine, II: 124
and yeast, II: 126
Penicillium, I: 14, 33, 114; II: 122, 123
  and aflatoxins, I: 123
  and Balkan (endemic) nephropathy, I: 116
  and ochratoxins, I: 58; II: 130
  and trichothecenes, I: 2
Penicillium aurantio-virens, II: 49, 52
Penicillium citreoviride, I: 4, 5, 114
Penicillium citrinum, I: 2, 4
Penicillium crustosum, I: 5
Penicillium cyclopium, I: 4; II: 123
  and ochratoxins, II: 130
Penicillium expansum, II: 106, 122
Penicillium funiculosum, I: 2
Penicillium islandicum, I: 4, 5, 114; II: 96, 102
Penicillium oxalium, I: 2, 3, 5
Penicillium palitans, I: 3, 5
Penicillium patulum, II: 122
Penicillium puberulum, II: 123
Penicillium roqueforti, I: 5
Penicillium rubrum, I: 5, 18
Penicillium rugulosum, I: 4, 5
Penicillium stoloniferum, II: 125
Penicillium tricinctum, I: 5
Penicillium urticae, I: 4, 5; II: 122
Penicillium verrucosum, I: 7
Penicillium viridicatum, I: 3, 5, 22, 33; II: 123
  and ochratoxins, II: 130
Penitrem A, I: 5
  screening for, I: 36
Peppers, and aflatoxins, I: 124
Peptide hormone, and cytochalasins, II: 76, 88
Peptidyl transferase, II: 42
  and trichothecenes, II: 43
Peptidyl transferase activity, and trichothecenes, II: 43
Peptidyl-tRNA, II: 42
  and aflatoxins, II: 20
Perch, and N-nitroso compounds, I: 176
Perithecia, and zearalenone, II: 146, 152
Peritoneal cells, and cytochalasins, II: 63
Peritoneal macrophages, and cytochalasins, II: 63, 64
Peritoneal mast cells, and cytochalasins, II: 92
Permeability, and trichothecenes, II: 45
Persistence of N-nitroso compounds, I: 159-164
Pesticide nitrosation, I: 178

Phage, and psoralens, II: 108
Phagocytosis, II: 74
   and cytochalasins, II: 60, 69, 73
Phalloidin, and cytochalasins, II: 78
Pharmacokinetics, I: 149
   and luteoskyrin, II: 96
   and patulin, II: 124
   and penicillic acid, II: 124
   and zearalenol, II: 148
Pharynx, and N-nitroso compounds, I: 188-190, 193, 194, 196, 198, 203
Phenobarbital, II: 13, 19
Phenylalanine, II: 39, 119
   and ochratoxins, II: 131, 132, 135, 136
   and penicillic acid, II: 127
β-Phenylalanine, and ochratoxins, II: 130
C-Phenylalanine, and psoralens, II: 117
Phenylalanyl, and ochratoxins, II: 135
Phenylalanyl-tRNA, II: 39
   and ochratoxins, II: 135
Philadelphia, I: 165-167
Philippines, I: 10, 11, 121-123
Phizoctonia, I: 5
Phloretin, II: 89
Phlorizin, II: 89
Phoma, II: 48, 50, 51
Phoma exigua, II: 49
Phoma sorghina, I: 5
Phomin, II: 50
Phomopsis, I: 5; II: 48
Phomopsis paspalli, II: 49, 51
Phophodiesterase, and cytochalasins, II: 73
Phosphate
   and cytochalasins, II: 73, 88
   and luteoskyrin, II: 97
   and ochratoxins, II: 132, 133
Phosphate incorporation, and aflatoxins, II: 17, 23
Phosphate uptake, II: 44
   and cytochalasins, II: 88
Phospholipids, II: 44
   and aflatoxins, II: 23
   and cytochalasins, II: 80, 84
Phosphorylase, II: 25
   and ochratoxins, II: 133, 134
Phosphorylase β, and ochratoxins, II: 134
Phosphorylase β kinase, and ochratoxins, II: 133
Phosphorylation
   and aflatoxins, II: 8
   and cytochalasins, II: 73, 76, 84, 87
Photoactivation, and pink rot disease, I: 119
Photoadducts
   formation of, II: 111-112
   identification of, II: 111-112
   and psoralens, II: 108, 110-118
Photochemical reactions, and psoralens, II: 110-115
Photoinactivation, II: 107
   and psoralens, II: 118
Photokinesis, II: 58
Photooxidation, and psoralens, II: 107
Photoreactions with nucleic acids, II: 118-119

Photosensitivity, and psoralens, II: 110, 113, 117-119
Phototoxicity
   and pink rot disease, I: 117-120
   and psoralens, II: 106, 108, 110, 117, 118
Physical properties
   of ochratoxins, I: 34
   of zearalenone, I: 30
Physiochemical properties, of aflatoxins, I: 25
Physiological effects of zearalenol, II: 148-150
Phytic acid, I: 3
Phytoalexin, I: 119; II: 107
Phytohemagglutinin, II: 68, 88
Phytotoxicity, II: 30
   and patulin, II: 122
Pigeons, and trichothecenes, I: 13, 14, 16
Pigment granule dispersion, and cytochalasins, II: 72
Pigs, see also Pork, I: 5, 7, 8
   and aflatoxins, I: 18-19, 22, 52; II: 8, 16
   and N-nitroso compounds, I: 173, 190, 207
   and ochratoxins, I: 23, 58, 59, 92, 93; II: 130, 132
   and trichothecenes, I: 13, 14, 16, 61; II: 30
   and zearalenone, I: 17, 18, 70; II: 146, 147
Pink rot disease, I: 7, 117-120; II: 106
   and celery, I: 118, 119
   and humans, I: 118
   and 8-methoxypsoralen, I: 117-119
   and mice, I: 117
   and photoactivation, I: 119
   and psoralens, I: 117-119
   and skin, I: 117, 118
   and trimethylpsoralen, I: 117, 119
Pinocytosis, and cytochalasins, II: 81
Piperidine, I: 176
Pithomyces chartarum, I: 5
Pituitary gland, and cytochalasins, II: 66
Placenta, and ochratoxins, II: 132
Plasma
   and aflatoxins, II: 25
   and cytochalasins, II: 54, 85, 92
Plasma cells, and cytochalasins, II: 63
Plasmacytoma cells, and cytochalasins, II: 65
Plasma membrane, II: 155, 156
   and cytochalasins, II: 48, 56-60, 63, 68, 69, 71, 72, 76, 77, 79-81, 83, 89-91, 94
   and ochratoxins, II: 132
Plasma proteins
   and aflatoxins, II: 20
   and cytochalasins, II: 65
Platelet activation, and cytochalasins, II: 60
Platelets, and cytochalasins, II: 64, 66, 71, 79, 89, 92
PLC, see Primary liver cancer
Poaefusarin, I: 6, 112
Poliovirus, and cytochalasins, II: 92
Polymorphonuclear leukocytes, and cytochalasins, II: 60, 82
Polynucleation, II: 48
Polynucleotides
   and aflatoxins, II: 11

and psoralens, II: 111, 113
Polypeptide, II: 75
   and trichothecenes, II: 43
Polypeptide hormone, II: 80
Polypeptide termination assay, II: 42
Polyribonucleotides, I: 209
Polyribosome profiles, II: 42
Polysaccharides, II: 12
Polysaccharide synthesis, and cytochalasins, II: 60
Polysomal disaggregation, II: 22
Polysomal profiles, II: 20, 75
   and ochratoxins, II: 135
Polysome disaggregation, II: 20, 38
   and aflatoxins, II: 21
   and cytochalasins, II: 70, 75
   and trichothecenes, II: 39
Polysomes
   and aflatoxins, II: 20-22
   and trichothecenes, II: 39, 42
Pont St. Esprit, I: 109
Pork, see also Pigs
   and N-nitroso compounds, I: 171
Potassium
   and cytochalasins, II: 88
   and patulin, II: 126
   and penicillic acid, II: 127
Potatoes, and N-nitroso compounds, I: 163
Poultry, see also specific birds
   and aflatoxins, I: 20-22
   and ochratoxins, I: 59-60; II: 130
   and trichothecenes, I: 62; II: 30
Primary liver cancer (PLC), I: 11
Probit-log-dose cancer risk model, I: 238
Progesterone, and zearalenone, II: 149
Prolactin, and cytochalasins, II: 66
Proline, II: 39, 75
Propylnitrosamines, I: 191
Prostaglandins, and cytochalasins, II: 63, 67, 72
Protease, II: 89
   and cytochalasins, II: 82
Protein, II: 156, 157
   and aflatoxins, II: 11-13, 19-21, 25, 26
   and cytochalasins, II: 63, 65, 72, 75-79, 81-84, 90, 91
   and luteoskyrin, II: 98, 99
   membrane marker, II: 82-83
   and N-nitroso compounds, I: 208, 242
   and ochratoxins, II: 133, 134, 136
   and penicillic acid, II: 125
   and psoralens, II: 115, 117
   and sterigmatocystins, II: 142
   and zearalenone, II: 146, 149-152
Protein binding
   and aflatoxins, II: 6-8, 11-13, 25
   and ochratoxins, II: 134
Protein deficiency, and aflatoxins, II: 14, 23
Protein kinase, and ochratoxins, II: 134
Protein synthesis, II: 155-157
   and aflatoxins, II: 7, 12, 16, 19, 22, 23, 25, 26
   and cyclochlorotine, II: 102-103

and cytochalasins, II: 58, 60, 62, 63, 68, 70, 74-76, 83
inhibition of, II: 38-43
and luteoskyrin, II: 98
and ochratoxins, II: 135
and patulin, II: 126
and penicillic acid, II: 125, 126
and psoralens, II: 116, 117, 119
and trichothecenes, I: 60; II: 38, 39, 42-45
Protein transport, and cytochalasins, II: 60, 71, 72
Proteinurea, I: 8
Protophomin, II: 51
Protozoa
   and cytochalasins, II: 54
   and ochratoxins, II: 130
   and patulin, II: 122
   and penicillic acid, II: 125
   and trichothecenes, II: 31, 38
Proxiphomin, II: 51
Psoralens, I: 5, 7; II: 1, 2, 108, 109, 111, 112, 114-117, 119, 155, 156
   and adduct formation, II: 113-115
   and adrenal glands, II: 108
   and amino acids, II: 113, 115, 117
   and angelicin, II: 114
   and antigens, II: 108
   and ATPase, II: 115
   and bacteria, II: 107, 108, 114, 116
   binding of, II: 113
   biological effects of, II: 107-109
   and blood, II: 109
   and carcinogenicity, II: 108-109
   and cellular macromolecule synthesis inhibition, II: 115-116
   and Chinese hamsters, II: 108, 114-116
   and chromatin, II: 113, 115
   and chromosomal abnormalities, II: 108
   and copper, II: 108
   and cows, II: 111, 113
   and cytosine, II: 111
   and cytotoxicity, II: 107-108, 115
   and dermatitis, II: 106
   and dermis, II: 109
   distribution of, II: 109
   and DNA, II: 107, 108, 111, 113, 114, 116, 118
   and DNA binding, II: 110, 111, 113, 116, 118
   and DNA damage, II: 112
   and DNA polymerase, II: 117
   and DNA repair, II: 108, 114, 116
   and DNAse, II: 115
   and DNA synthesis, II: 114-116, 118, 119
   and ear, II: 108, 109
   and Ehrlich ascites cells, II: 114, 115, 118, 119
   and epidermis, II: 109, 116
   and epithelial cells, II: 117
   and epoxidation, II: 109
   and erythema, II: 106, 117
   and *Escherichia coli*, II: 107, 113-118
   excretion of, II: 109
   and feces, II: 109
   and fibroblasts, II: 108, 114-116

fungal sources of, II: 106-107
and fungi, II: 106, 108
and furocoumarins, II: 112-118
and guinea pigs, II: 108-110, 114, 117, 118
and hamsters, II: 108, 114-116
and histone, II: 15
and humans, II: 108, 109, 116-118
and hydroxylation, II: 109
and kidney, II: 109
and leucine, II: 117
and liver, II: 107-109, 115
and lung, II: 108
and lymphocytes, II: 108
and macromolecular synthesis, II: 115-117
and magnesium, II: 113
and melanocytes, II: 106
and metabolic activation, II: 109
and metabolism, II: 109
and mice, II: 108, 109, 115-117
and microsome, II: 109
and mitogen, II: 108
and mixed function oxidase, II: 109
and mRNA, II: 116
and mutagenicity, II: 107-108, 115
and nucleic acid, II: 110-115, 118-119
and nucleic acid synthesis, II: 115
and nucleosides, II: 111
and nucleotides, II: 110, 111, 113, 114, 117
and ovary, II: 108, 115
and phage, II: 108
and C-phenylalanine, II: 117
and photoadducts, II: 108, 110-112, 113-118
and photochemical reactions, II: 110-115
and photoinactivation, II: 118
and photooxidation, II: 107
and photosensitivity, II: 110, 113, 117-119
and phototoxicity, II: 106, 108, 110, 117, 118
and pink rot disease, I: 117-119
and polynucleotides, II: 111, 113
and protein, II: 115, 117
and protein synthesis, II: 116, 117, 119
and psoriasis, II: 106, 108
and purines, II: 111
and pyrimidines, II: 110, 111
and rats, II: 108, 117
and ribosomes, II: 117, 119
and risk assessment, I: 142
and RNA, II: 107, 113
and RNA binding, II: 110, 113, 118
and RNA polymerases, II: 117, 118
and RNA synthesis, II: 115, 116, 119
and RNA virus, II: 118
and rRNA, II: 110, 113, 115, 118
and sea urchin eggs, II: 109
and serum, II: 109
and skin, II: 106, 108-110, 114, 116-119
and sodium, II: 110, 113
and spleen, II: 108
structure-activity relationships in, II: 117-119
and SV40 virus, II: 115
and template activity, II: 117, 118
and teratogenicity, II: 109

and thymidine, II: 111, 115, 116
and thymine, II: 111, 114
and thymus, II: 111, 113
and toxicity, II: 106, 108-109
and transcription, II: 116
and tRNA, II: 113, 117
and uracil, II: 111
and uridine, II: 115
and urine, II: 109
and virus transformed cells, II: 115
and xanthotoxin, II: 106
and yeast, II: 115
Psoriasis, and psoralens, II: 106, 108
Pulmonary macrophage, and cytochalasins, II: 79
Purification of ochratoxins, I: 35
Purines
and luteoskyrin, II: 98
and psoralens, II: 111
Puromycin, II: 39, 42
PUVA, see Long wave UV
Pyrimidines, and psoralens, II: 110, 111
Pyrrolidine, I: 176
Pyruvate
and cytochalasins, II: 71
and ochratoxins, II: 133

Q

Quail
and ochratoxins, I: 59
and patulin, II: 125
Quantitative analysis, of N-nitroso compounds, I: 156-159
Quantitative detection method, I: 30, 32
Quinazolone tremorgens, I: 130, 131

R

Rabbits
and aflatoxins, II: 5, 10, 17, 20, 21, 26
and cyclochlorotine, II: 103
and cytochalasins, II: 58, 60, 64, 65, 69, 71, 75, 78, 79, 81, 84, 89, 90, 92, 93
and luteoskyrin, II: 98, 99
and N-nitroso compounds, I: 186, 190, 196, 221
and ochratoxins, II: 136
and penicillic acid, II: 125, 127
and sterigmatocystins, II: 142
and trichothecenes, I: 61; II: 38-39, 41-44
and zearalenone, II: 151
Radishes, and N-nitroso compounds, I: 174
Rainbow trout
and aflatoxins, I: 55, 57, 66; II: 4, 5, 7
and cytochalasins, II: 72
and ochratoxins, I: 59, 60
Rana, II: 37
Rats, I: 7
and aflatoxins, I: 18, 52-57, 66-68, 127-129; II: 5, 7, 8, 10-26

and cyclochlorotine, II: 102
and cytochalasins, II: 54, 63, 65-68, 70, 73, 75-79, 81, 83-92
and luteoskyrin, II: 96
and N-nitroso compounds, I: 186-199, 206-208, 221, 224, 232-234, 236, 237, 239, 241, 242
and ochratoxins, I: 22, 59, 60, 75-78, 86, 87, 90, 91; II: 130-134, 136
and patulin, II: 122, 123
and penicillic acid, II: 124, 125
and psoralens, II: 108, 117
and risk assessment, I: 142, 144-5, 147-149
and sterigmatocystins, I: 64; II: 138-143
and trichothecenes, I: 15, 16, 61-63, 69; II: 31, 38, 39, 42, 43
and yellow rice syndrome, I: 114
and zearalenone, I: 17, 65; II: 146-150, 153
Rd toxin, I: 14
Rec assay, II: 96, 139
and aflatoxins, II: 6-7
and ochratoxins, II: 131
and patulin, II: 123
and penicillic acid, II: 124
and sterigmatocystins, II: 142
Rectum, I: 7
and zearalenone, I: 17
Red blood cells, see Erythrocytes
Red-mold toxicosis, I: 60; II: 30
Refusal emesis syndrome, I: 13
Relative humidity (RH), I: 2, 3
Release factor, II: 42
Renal tumors, II: 5
Reproductive hormone, and zearalenone, II: 148
Reptiles, and cytochalasins, II: 72
Resorcylic acid lactone, I: 16
β-Resorcylic acid lactone, I: 65
Respiration, II: 157
and aflatoxins, II: 23
and ochratoxins, II: 132-135
and patulin, II: 126-127
and penicillic acid, II: 126-127
Reticulocyte polysomes, and trichothecenes, II: 43
Reticulocytes, II: 26, 38, 39, 41, 42, 44, 89
and cyclochlorotine, II: 103
and cytochalasins, II: 75
and luteoskyrin, II: 98, 99
and ochratoxins, II: 136
and penicillic acid, II: 125, 127
and sterigmatocystins, II: 142
and zearalenone, II: 151
Reticuloendotheliomas
and cyclochlorotine, II: 102
and luteoskyrin, II: 96
Retina cells, and cytochalasins, II: 59
Reuber hepatoma cells, and cytochalasins, II: 76, 77
Reye's syndrome, I: 133
and aflatoxins, I: 131-133
and risk assessment, I: 151
RH, see Relative humidity
Rhizopus, I: 123

Rhizopus nigricans, II: 106
Rhodesia, I: 221
Ribonuclease, and aflatoxins, II: 18
Ribonucleosides, and cytochalasins, II: 86
Ribosomal subunits
and aflatoxins, II: 20, 21
and trichothecenes, II: 42, 43
Ribosomes
and aflatoxins, II: 20, 21
and psoralens, II: 117, 119
and trichothecenes, II: 43
Ribs, and zearalenone, II: 147
Rice, I: 8
and aflatoxins, I: 123, 124, 130, 131
and trichothecenes, I: 62
Richothecenes, and risk assessment, I: 142
Ricinus communis, II: 82
Risk assessment
and aflatoxins, I: 142, 149, 150, 151
and apple juice, I: 142
and artery, I: 149
and Balkan (endemic) nephropathy, I: 142
and colon, I: 149
and cows, I: 143
and DNA binding, I: 149
and DNA repair, I: 149, 150
and ducks, I: 151
and ergot alkaloids, I: 142
and excretion, I: 149
and glutathione, I: 149
and humans, I: 151
and kidney, I: 149, 150
and liver, I: 142-146, 149
and lung, I: 143
and maize, I: 151
and mice, I: 149
and monkeys, I: 150
and N-nitroso compounds, I: 242-243
and patulin, I: 142
and peanut butter, I: 142
and peanuts, I: 142-144, 151
and psoralens, I: 142
and rats, I: 142, 144-145, 147-149
and Reye's syndrome, I: 151
and skin, I: 150
and sterigmatocystins, I: 142
and stomach, I: 149
and substrate, I: 149
and trichothecenes, I, 142
and trout, I: 144-148
and urine, I: 149
and yellow rice toxins, I: 142
RNA, II: 156
and aflatoxins, II: 11, 18, 26, 27
and cytochalasins, II: 58, 68, 75
and luteoskyrin, II: 98
and N-nitroso compounds, I: 239
and ochratoxins, II: 136
and penicillic acid, II: 125
and psoralens, II: 107, 113
and zearalenone, II: 149
RNA binding

and aflatoxins, II: 6, 7, 11-13, 19, 27
and psoralens, II: 110, 113, 118
RNA metabolism, alterations in, II: 17-18
RNA polymerase
   and aflatoxins, II: 18
   and luteoskyrin, II: 98, 99
   and N-nitroso compounds, I: 209
   and patulin, II: 125
   and penicillic acid, II: 125
   and psoralens, II: 117, 118
   and sterigmatocystins, II: 141
RNA synthesis, II: 155-157
   and aflatoxins, II: 7, 16, 19, 20, 22, 23, 26, 27
   and cytochalasins, II: 68, 70, 75
   inhibition of, II: 141
   and luteoskyrin, II: 98, 99
   nucleolar, II: 17
   and ochratoxins, II: 135
   and patulin, II: 125, 126
   and penicillic acid, II: 125
   and psoralens, II: 115, 116, 119
   and sterigmatocystin, I: 63; II: 140-143
   and trichothecenes, II: 38
   and zearalenone, II: 150
RNA transcription, and zearalenone, II: 150
RNA virus, and psoralens, II: 118
Roridin A, II: 34, 40
Roridin C, II: 32
Roridin D, II: 35
Roridin E, II: 35
Roridin H, II: 35
Roridins, II: 30, 31, 34, 41, 44, 45
Roselinia necatrix, II: 50
Rosette formation, II: 69, 92
rRNA
   and aflatoxins, II: 12, 13, 17, 18
   and penicillic acid, II: 125
   and psoralens, II: 110, 113, 115, 118
Rubratoxin, I: 5
Rubroskyrin, I: 114
Ruffling, II: 58
   and cytochalasins, II: 73, 85, 92
Rugulosin, I: 5, 6
Rumania, I: 115
Rumen, and trichothecenes, I: 62
Russia, I: 6, 13, 60, 112, 166
Rutaceae, II: 106
Rye, I: 109
   and ergot alkaloids, I: 112

S

Sable, and N-nitroso compounds, I: 172
Saccharomyces, II: 86, 94
   and zearalenone, II: 147
Saccharomyces cerevisiae, II: 37, 54, 92, 96, 115, 123, 124
   and aflatoxins, II: 7
   and ochratoxins, II: 131
   and sterigmatocystins, II: 139
Salami, and N-nitroso compounds, I: 171

Saliva, and N-nitroso compounds, I: 163, 174, 228, 229
Salivary ducts, and cytochalasins, II: 88
Salivary gland
   and cytochalasins, II: 62, 73, 75
   and N-nitroso compounds, I: 223
Salmon
   and aflatoxins, I: 58
   and N-nitroso compounds, I: 172, 176
Salmonella, II: 102
   and trichothecenes, I: 16
   and zearalenone, II: 147
Salmonella typhimurium, II: 37, 123, 124
   and aflatoxins, II: 7
   and ochratoxins, II: 131
   and sterigmatocystins, II: 139, 143
Salmonella typhimurium assay, II: 6, 8, 12, 96
Sampling procedure, and aflatoxins, I: 24, 26
Sarcina lutea, II: 107
Sarcoma, II: 157
   and cytochalasins, II: 81
   and sterigmatocystins, II: 139
Satratoxin C, I: 7
Satratoxin D, I: 7
Satratoxin F, I: 7
Satratoxin G, I: 7
Satratoxin H, I: 7; II: 35
Satratoxins, I; 5
Sausage, and N-nitroso compounds, I: 171
Scabby grain, and trichothecenes, I: 13
Scatchard analysis, II: 89, 110, 134
Schiff base formation, II: 8, 11
Sciatic nerve, and cytochalasins, II: 72
Scientific and Technical Assessment Report (STAR), I: 166
Scirpentriol, II: 33, 40, 45
Sclerotinia, I: 5
Sclerotinia rolfsii, I: 119
Sclerotinia sclerotiorum, I: 118, 119; II: 106, 107
Scotland, I: 17
Screening
   for citrinin, I: 36
   for ochratoxins, I: 36
   for patulin, I: 36
   for penicillic acid, I: 36
   for penitrem A, I: 36
   for sterigmatocystin, I: 36
   for T-2 toxin, I: 36
   for zearalenone, I: 36
SEATO, see Southeast Asia Treaty Organization
Sea urchin eggs
   and cytochalasins, II: 57, 92
   and psoralens, II: 109
Secondary amines, I: 160, 163, 174-176
Secretory granules, and cytochalasins, II: 63
Serbia, I: 115
Serotonin, and cytochalasins, II: 72
Serum, I: 7; II: 86
   and aflatoxins, II: 26
   and cytochalasins, II: 63, 67, 71, 74, 85, 88
   and ochratoxins, II: 132-134, 136
   and psoralens, II: 109

and sterigmatocystins, II: 142
and zearalenone, II: 148, 149
Serum glutamic-oxaloactic transaminase (SGOT)
   and cyclochlorotine, II: 102
   and luteoskyrin, II: 99
Serum protein, II: 12
   and aflatoxins, II: 20
Serum triglyceride, II: 99
Sesame seeds, and aflatoxins, I: 124
Sewage effluent, I: 166
SGOT, see Serum glutamic-oxaloactic transaminase
Shad, and N-nitroso compounds, I: 172
Sheep, I: 5
   and aflatoxins, II: 8, 15
   and cytochalasins, II: 65
   and N-nitroso compounds, I: 160, 166
   and trichothecenes, II: 30
   and zearalenone, II: 150
Shoskin-kakke, see Acute cardiac beriberi
Shrews, and aflatoxins, I: 58
Shrimp, and aflatoxins, I: 124
Sialic acid, II: 80
Sialyltransferase, II: 8, 25
Siberia, I: 112
Singapore, I: 120
Sister chromatid exchanges, II: 108
Skeleton anomalies, and zearalenone, I: 17
Skin
   and aflatoxins, II: 5, 6, 17
   and alimentary toxic aleukia (ATA), I: 113
   and Balkan (endemic) nephropathy, I: 116
   and cyclochlorotine, II: 103
   and cytochalasins, II: 67, 71, 83, 88
   and N-nitroso compounds, I: 196-198, 205, 206
   and pink rot disease, I: 117, 118
   and psoralens, II: 106, 108-110, 114, 116-119
   and risk assessment, I: 150
   and sterigmatocystins, II: 139
   and trichothecenes, I: 15; II: 37
   and zearalenone, II: 151
Skyrin, I: 114
Slaframine, I: 5
Sludge, I: 158
Small intestine
   and cytochalasins, II: 54
   and N-nitroso compounds, I: 198
   and ochratoxins, II: 131
   and trichothecenes, I: 16; II: 37
Smooth microsomes, and aflatoxins, II: 22
Sodium
   and cyclochlorotine, II: 103
   and cytochalasins, II: 71, 88, 89
   and luteoskyrin, II: 99
   and ochratoxins, II: 136
   and patulin, II: 126
   and penicillic acid, II: 127
   and psoralens, II: 110, 113
   and sterigmatocystins, II: 142
   and zearalenone, II: 151
Sodium-dependent glycine transport, II: 26, 44
Sodium nitrite, and N-nitroso compounds, I: 160

Sodium transport, and cytochalasins, II: 71, 76
Solaniol, II: 32
Sorghum, I: 2
   and zearalenone, I: 71
Souse, and N-nitroso compounds, I: 171
South Africa, I: 20, 168
South America, I: 120, 129
Southeast Asia, I: 120, 127, 142
Southeast Asia Treaty Organization (SEATO), I: 131
Soviet Union, see Russia
Soybeans, and trichothecenes, I: 62
Spectrin, and cytochalasins, II: 90
Spectrophotometric method, I: 24
Spermatocyte, and cytochalasins, II: 54
Spermatozoa, II: 73, 155
   and aflatoxins, I: 19
   and cytochalasins, II: 62, 84, 85, 89
Spinach, and N-nitroso compounds, I: 163, 168, 173, 174, 230
Spinal cord
   and cytochalasins, II: 61, 71, 72
   and N-nitroso compounds, I: 196, 197, 205
Spleen
   and aflatoxins, I: 19, 21, 52, 131; II: 26
   and cytochalasins, II: 54, 65, 68, 81, 82, 92
   and N-nitroso compounds, I: 220
   and ochratoxins, I: 81
   and penicillic acid, II: 124
   and psoralens, II: 108
   and trichothecenes, I: 16, 61, 62; II: 37, 38
   and zearalenone, II: 149
Sporidesmins, I: 5
Sporofusarin, I: 6
Squamous cell carcinomas
   and 8-MOP, II: 109
   and sterigmatocystins, II: 139
Stachybotryotoxicosis, I: 6-7, 60; II: 30
   and hay, I: 6
Stachybotrys, I: 31, 60; II: 30
Stachybotrys alternans, I: 6; II: 36
Stachybotrys atra, I: 5; II: 35, 36
Stachybotrys chartarum, II: 36
Staphylococcus aureus, II: 107
STAR, see Scientific and Technical Assessment Report
Starvation, II: 20
Sterculic acid, and ochratoxins, II: 130
Sterigmatocystins (STG), I: 5, 6, 63-64; II: 138-143, 157
   and adenocarcinomas, II: 139
   and adenomas, II: 139
   and aflatoxins, I: 123, 130; II: 138
   and albumin, II: 141
   and Ames assay, II: 139, 143
   and artery, I: 100
   and *Aspergillus*, II: 138
   and aspertoxin, II: 138
   and *Bacillus subtilis*, II: 139, 142, 143
   and bile duct, II: 139
   and birds, II: 138
   and brine shrimp, II: 138

and carcinogenicity, I: 63, 64, 70
and chicken embryos, II: 139
and chickens, I: 63; II: 138, 141, 143
and chromosomal aberrations, II: 138
and complement, II: 142
and cytotoxicity, II: 138, 139
and DNA, II: 140
and DNA binding, II: 139
and DNA damage, II: 139, 142, 143
and DNA repair synthesis, II: 138, 140-141
and DNAse, II: 141
and DNA synthesis, II: 139-143
and ducks, I: 63; II: 139
and eggs, II: 139
and epithelial cells, II: 138, 140, 142, 143
and epoxidation, II: 140
and fibroblasts, II: 138, 140, 141
and fish, II: 138, 143
and globulin, II: 142
and glucuronide conjugates, II: 140
and glycine transport, II: 142
and guinea pigs, II: 142
and HeLa cells, II: 138, 140
and hepatocytes, II: 138, 141, 143
and hepatomas, II: 139
and humans, II: 138, 140, 141
and hydroxylation, II: 140
and hyperplastic nodules, II: 141
and kidney, I: 63, 64, 99; II: 138, 140-143
and liver, I: 63, 64, 99, 100; II: 138, 140-143
and lung, I: 64; II: 138, 139, 141
and lysosomes, II: 142
and mammary carcinoma, II: 139
and mesenchymal cells, II: 141, 143
and metabolism, II: 140
and mice, I: 63, 64, 70, 100; II: 139, 141
and microsomal activation, II: 139, 141
and microsomes, II: 139, 141
and mitosis, II: 138, 142
and mixed-function oxidase, II: 140
and monkeys, I: 63, 64, 70, 100; II: 138-140, 142, 143
and mutagenicity, II: 139, 143
and necrosis, II: 139
and nuclear RNA synthesis, II: 141
and nuclei, II: 140, 141
and nucleic acid synthesis, II: 140-141
and nucleosides, II: 140
and nucleoside transport, II: 141
and protein, II: 142
and rabbits, II: 142
and rats, I: 64; II: 138-143
and rec assay, II: 2
and reticulocytes, II: 142
and risk assessment, I: 142
and RNA polymerase, II: 141
and RNA synthesis, I: 63; II: 140-143
and RNA synthesis inhibition, II: 141
and *Saccharomyces cerevisiae*, II: 139
and *Salmonella typhimurium*, II: 139, 143
and sarcomas, II: 139
screening for, I: 36

and serum, II: 142
and skin, II: 139
and sodium, II: 142
and squamous cell carcinomas, II: 139
structure-activity relationships in, II: 142, 143
and sunflower, I: 64
and thymidine, II: 138, 140
and H-thymidine transport, II: 140
and thymidine uptake, II: 141
and toxicity, I: 63-64, 70; II: 139, 142, 143
and trachea, II: 138
and urease, II: 142
and uridine, II: 141
and urine, I: 63; II: 140
and *Xeroderma pigmentosum*, II: 138, 140, 141
and zebra fish, II: 138, 143
Steroid binding, and aflatoxins, II: 22
Steroid hormones
  and cytochalasins, II: 76-77
  mechanisms of, II: 21-22
Steroid synthesis, and cytochalasins, II: 76
STG, see Sterigmatocystins
Stomach
  and aflatoxin G_1, I: 6
  and N-nitroso compounds, I: 174, 187, 188, 190, 193, 196-198, 201, 222, 225, 228, 229, 236
  and risk assessment, I: 149
  and T-2 toxins, I: 94, 96
  and trichothecenes, I: 14, 61, 62; II: 37
  and zearalenone, II: 148
Stomach carcinomas, and aflatoxins, II: 5
Stool, see Feces
Streptococcus faecalis, and ochratoxins, II: 134, 135
Streptomyces griseus, II: 30
Structure
  of aflatoxins, I: 25, 121
  of amine alkaloids, I: 110
  of amino acid alkaloids, I: 110
  of bishydroxycoumarin, I: 118
  of citroviridin, I: 115
  of clavine alkaloids, I: 110
  of cytochalasins, II: 50-53
  of lysergic acid, I: 110
  of ochratoxin A, I: 117
  of ochratoxins, I: 34
  of T-2 toxins, I: 113
  of trimethylpsoralen, I: 118
Structure-activity relationships
  in aflatoxins, II: 26-27
  in cytochalasins, II: 91-94
  in N-nitroso compounds, I: 206
  in ochratoxins, II: 136
  in psoralens, II: 117-119
  in sterigmatocystins, II: 142, 143
  in trichothecenes, II: 44-45
  in zearalenone, II: 151-153
Substrate
  composition of, I: 2
  and N-nitroso compounds, I: 208
  and risk assessment, I: 149

and secondary metabolites, I: 3
Succinic dehydrogenase, II: 97
Sucrose uptake, and cytochalasins, II: 63
Sugar, see also specific sugars
    and N-nitroso compounds, I: 222
Sugar transport, and cytochalasins, II: 94
Suguinas oedipomidas, I: 58
Sulfate, and cytochalasins, II: 88
Sulfate conjugates, II: 6
    and aflatoxins, II: 8, 15
    and zearalenone, II: 8
Sulfate incorporation, and cytochalasins, II: 88
Sulfhydryl groups, II: 156, 157
    and cytochalasins, II: 86, 94
    and patulin, II: 122, 124
    and penicillic acid, II: 123, 124
Sulfhydryl reagents
    and cytochalasins, II: 68, 78, 86, 89, 90, 94
    and N-nitroso compounds, I: 164
    and patulin, II: 126
    and penicillic acid, II: 126, 127
    and trichothecenes, II: 43
Sunflower, and sterigmatocystin, I: 64
Superoxide dismutase, and cytochalasins, II: 74
Superoxide generation, II: 93
    and cytochalasins, II: 74
SV40 virus, II: 17, 58, 82, 87, 89
    and psoralens, II: 115
Swaziland, I: 9, 12, 120, 121, 127, 128
Sweden, I: 22
Sweet clover disease, I: 117
Swine, see Pig
Switzerland, I: 120
Sympathetic ganglia, and cytochalasins, II: 71
Synaptosomes, and cytochalasins, II: 67, 72, 78
Synthetic cutting fluids, I: 176

T

T cells, and cytochalasins, II: 69
3T3 cells, II: 82-84, 88, 93
T-2 toxin, I: 5, 6, 14-16, 112; II: 30-33, 37-40, 42-44
    and aflatoxins, I: 130
    carcinogenicity of, I: 69
    and cats, I: 61, 97
    and chickens, I: 97, 98, 112
    and cows, I: 62
    and guinea pigs, I: 94-96
    and intestine, I: 97
    and mice, I: 63
    screening for, I: 36
    and stomach, I: 94, 96
    structure of, I: 113
    and trichothecenes, I: 15
Taiwan, I: 130
Tannins, I: 164
TAT, see Tyrosine aminotransferase
Tea, and N-nitroso compounds, I: 176, 223
Temperature, I: 2
Template activity

    and aflatoxins, II: 18
    and psoralens, II: 117, 118
Teratogenicity, II: 37, 156-158
    and aflatoxins, I: 55
    and cytochalasins, II: 54-56, 60, 92
    and N-nitroso compounds, I: 207
    and ochratoxins, I: 60; II: 130
    and patulin, II: 122, 125
    and penicillic acid, II: 125
    and psoralens, II: 109
    and trichothecenes, I: 15
    and zearalenone, I: 65; II: 147
Termination
    and aflatoxin B$_1$, II: 20
    and aflatoxins, II: 21
    and trichothecenes, II: 39, 41, 43, 45
Tertiary amines, I: 160, 174-175
Testis
    aflatoxins, II: 23
    and N-nitroso compounds, I: 186, 192, 206
    and trichothecenes, I: 61; II: 37
    and zearalenone, I: 17; II: 147
Testosterone, and aflatoxins, II: 14, 22
Tetrahymena, II: 44
    and ochratoxins, II: 130
Tetrahymena cells, II: 98
Tetrahymena lignorum, II: 36
Tetrahymena pyriformis, II: 38, 126
    and patulin, II: 127
    and penicillic acid, II: 127
Tetraploidy, and cytochalasins, II: 62
Thailand, I: 7-9, 12, 120, 121, 123-125, 129-133, 146, 151, 225
Thermal energy analyzer, I: 157
Thermionic nitrogen detector, I: 16, 31, 156
Thiocyanate, I: 163
Thiol, II: 43
Threonine, cancer risk model, I: 238
Thrombin, and cytochalasins, II: 67
Thrombosthenin, and cytochalasins, II: 71
Thymidine, II: 26, 87
    and aflatoxins, I: 53, 56; II: 7
    and cytochalasins, II: 68
    and luteoskyrin, II: 98
    and psoralens, II: 111, 115, 116
    and sterigmatocystins, II: 138, 140
    and trichothecenes, II: 38
H-Thymidine, II: 16
    and aflatoxins, II: 17
    and sterigmatocystins, II: 140
Thymidine incorporation, and trichothecenes, II: 39
Thymidine kinase, II: 38
Thymidine transport, II: 87, 88
    and aflatoxins, II: 17
    and cytochalasins, II: 74
    and sterigmatocystins, II: 0
Thymidine uptake, and sterigmatocystins, II: 141
Thymine, and psoralens, II: 111, 114
Thymocytes, and cytochalasins, II: 68, 69
Thymus
    and aflatoxins, I: 20, 131; II: 13

and cytochalasins, II: 54, 81, 82
and luteoskyrin, II: 97
and N-nitroso compounds, I: 189, 206
and psoralens, II: 111, 113
and trichothecenes, I: 61, 63; II: 37
Thyroid
   and cytochalasins, II: 65, 66
   and N-nitroso compounds, I: 191, 197, 206
   and zearalenone, II: 147
Thyroid hormones, and cytochalasins, II: 66
Thyroid stimulating hormone (TSH), and cytochalasins, II: 65, 66
Tissue, and aflatoxins, II: 25-26
Tissue distribution, of aflatoxin B_1, II: 14-16
TLC, see Thin layer chromatography
TMP, see 4,5',8-Trimethylpsoralen
Toads, and cytochalasins, II: 71, 76, 88
Tobacco, and N-nitroso compounds, I: 223
Tomato, and N-nitroso compounds, I: 173
Tongue, and N-nitroso compounds, I: 171, 188, 191, 193-196, 203
Tonsil, and trichothecenes, II: 39, 42
Toxicity, I: 4, 5, 150; II: 5
   and aflatoxins, I: 19, 52-58, 66, 67, 130-135; II: 6-9, 11-13, 15-17, 19, 21, 23, 24, 27
   and cyclochlorotine, II: 102
   and cytochalasins, II: 48-55, 77, 91, 92
   factors in, II: 2
   and luteoskyrin, II: 96, 97
   and N-nitroso compounds, I: 186, 187, 208, 209, 220-221
   and ochratoxins, I: 58-60; II: 131, 133, 136
   and patulin, II: 122, 124
   and psoralens, II: 106, 108-109
   and sterigmatocystins, I: 63-64, 70; II: 139, 142, 143
   and trichothecenes, I: 14, 15, 60-63, 69; II: 37, 44, 45
   and zearalenone, I: 65; II: 146, 147
Toxicology, I: 114; II: 155
Toxin production, I: 2
Trachea, I: 6
   and aflatoxins, II: 4
   and N-nitroso compounds, I: 170, 188, 190-193, 196, 203, 236
   and ochratoxins, II: 13
   and patulin, II: 122
   and sterigmatocystins, II: 138
Transcription
   and aflatoxins, II: 27
   and luteoskyrin, II: 97-99
   and penicillic acid, II: 125
   and psoralens, II: 116
   and zearalenone, II: 150
Transkei, South Africa, I: 168
Translation
   and aflatoxins, II: 27
   and ochratoxins, II: 135
Translocation, II: 42
Transpeptidation, II: 42
Triacetoxyscirpendiol, II: 32, 40
Trichoderma, I: 31, 60; II: 30

Trichoderma roseum, II: 34
Trichoderma viride, II: 32, 36
Trichodermin, II: 31, 32, 38, 40, 42, 43, 45
C-Trichodermin, and trichothecenes, II: 43
Trichodermol, II: 32, 40, 42, 45
Trichothecenes, I: 7, 13-16, 60-63; II: 1, 2, 20, 33, 156
   and acetate incorporation, II: 44
   and adrenal gland, I: 61
   and aflatoxins, I: 15
   and bacteria, II: 31, 39
   biochemical effects of, II: 40-41
   biological activity of, II: 30-37
   biological effects of, II: 40-41
   and bladder, I: 62; II: 30
   and bone marrow, I: 16, 61; II: 37
   and brain, I: 61; II: 38
   and brine shrimp, II: 37
   and carcinogenicity, I: 15, 62-63, 69; II: 37
   and cats, I: 15, 60, 61; II: 30
   and *Cephalosporium*, I: 60
   and chicken embryos, II: 45
   and chickens, I: 16; II: 30, 45
   and corn, I: 13-16, 60, 62
   and cows, I: 13, 15, 61-62; II: 30
   and cytotoxicity, II: 31, 44, 45
   detection of, I: 31-33
   distribution of, II: 37-38
   and DNA, II: 38
   and DNA damage, II: 37
   and DNA polymerase, II: 38
   and DNA repair, II: 39
   and DNA synthesis, I: 60; II: 38, 39
   and dogs, I: 14-16, 60, 61
   and ducks, I: 14, 16; II: 41
   and Ehrlich ascite cells, II: 38
   and elongation, II: 39, 41, 42, 45
   and epithelial mucosa, II: 37
   and feces, I: 15; II: 38
   and fibroblasts, II: 31, 37-39
   and fish, II: 37
   fungal sources of, II: 32-36
   and *Fusarium*, I: 60, 61
   and *Fusarium graminearum*, I: 69
   and *Fusarium poae*, I: 63
   and glutathione, II: 44
   and glycoprotein synthesis, II: 43
   and guinea pigs, I: 60, 61
   and hamsters, II: 31, 38, 39, 41, 43, 44
   and HeLa cells, II: 31, 37-39, 42
   and horses, I: 13, 60; II: 30
   and humans, II: 30, 38, 39, 42, 43
   and initiation, II: 39, 41, 42, 45
   and initiation complex, II: 42
   and intestinal mucosa, II: 37
   and intestine, I: 16, 61, 62; II: 37
   and kidney, I: 61; II: 30, 31, 37, 38, 41, 44
   and large intestine, I: 62
   and C-leucine, II: 41
   and lipid synthesis, II: 44
   and liver, I: 61, 62; II: 31, 38, 42-44
   and lung, I: 61

and lymph node, II: 37
and lysosome, II: 44
and membrane, II: 44
and membrane transport, II: 39
and metabolic activation, II: 38
and metabolism, II: 37-38
and mice, I: 15, 16, 62, 63, 69; II: 31, 37, 39, 43, 44
and microsomes, II: 37, 43
and mouth, I: 15, 16
and mRNA, II: 42, 43
and mules, I: 13
and necrosis, II: 37
and nose, I: 15
and nucleic acid synthesis, II: 38
occurrence of, II: 30
and ovary, I: 61; II: 37-39, 43
and oxygen consumption, II: 44
and pancreas, I: 16
and *Penicillia*, I: 2
and peptidyl transferase, II: 43
and peptidyl transferase activity, II: 43
and permeability, II: 45
and pigeons, I: 13, 14, 16
and pigs, I: 13, 14, 16, 61; II: 30
and polypeptide, II: 43
and polysome disaggregation, II: 39
and polysomes, II: 39, 42
and poultry, I: 62; II: 30
and protein synthesis, I: 60; II: 38-45
and protozoa, II: 31, 38
and rabbits, I: 61; II: 38-39, 41-44
and rats, I: 15, 16, 61-63, 69; II: 31, 38, 39, 42, 43
and reticulocyte polysomes, II: 43
and ribosomal subunits, II: 42
and ribosomes, II: 43
and ribosome subunits, II: 43
and rice, I: 62
and RNA synthesis, II: 38
and rumen, I: 62
and *Salmonella*, I: 16
and scabby grain, I: 13
and sheep, II: 30
and skin, I: 15; II: 37
and small intestine, I: 16; II: 37
and soybeans, I: 62
and spleen, I: 16, 61, 62; II: 37, 38
and stomach, I: 14, 61, 62; II: 37
structure, II: 30, 32-36
structure-activity relationships in, II: 44-45
and teratogenicity, I: 15
and termination, II: 39, 41, 43, 45
and testis, I: 61; II: 37
and thymidine, II: 38
and thymidine incorporation, II: 39
and thymus, I: 61, 63; II: 37
and tonsil, II: 39, 42
and toxicity, I: 14, 15, 60-63, 69; II: 37, 44, 45
and C-trichodermin, II: 43
and turkeys, I: 16
and uracil, II: 38
and uridine, II: 38
and urine, II: 38
and uterus, I: 63
and yeast, II: 39, 43
and yeast polyribosomes, II: 43
and yeast spheroplasts, II: 38, 39, 42
and zearalenone, I: 15, 17
and zebra fish, II: 37
Trichothecin, II: 31, 40
Trichothecium, I: 31; II: 30
Trichothecium roseum, II: 33
Trichothecolone, II: 33, 37, 40
Triethanolamine, I: 176
Triglycerides, II: 23, 44
and aflatoxins, II: 25
and ochratoxins, II: 134
Trimethylpsoralen
and pink rot disease, I: 117, 119
structure of, I: 118
4,5′,8-Trimethylpsoralen (TMP), I: 7; II: 106-119, 156
Triturus, II: 37
tRNA
and aflatoxins, II: 12, 19, 20, 25
and ochratoxins, II: 135
and psoralens, II: 113, 117
tRNA methylase, and aflatoxins, II: 24
tRNA synthetase, and ochratoxins, II: 135
Tropomyosin, II: 77
Troponin, II: 78
Trout
and aflatoxins, I: 55, 57, 58, 66, 148
and carcinogenicity, II: 10
and cytochalasins, II: 72
and N-nitroso compounds, I: 176, 187, 190
and ochratoxins, I: 59, 60, 91; II: 130, 131
and risk assessment, I: 144-148
Trypsin
and cytochalasins, II: 82
and ochratoxins, II: 131
Tryptophan, and cytochalasins, II: 75
Tryptophan pyrrolase, II: 21
Tryptoquivaline, I: 123
Tryptoquivalone, I: 123
TSH, see Thyroid stimulating hormone
T-2 tetraol, I: 6; II: 33, 45
Tubular necrosis, and ochratoxins, I: 22
Tubules, I: 8
Tubulin, and cytochalasins, II: 94
Tuna, and N-nitroso compounds, I: 176
Tupaia glis, I: 58
Turkeys
and aflatoxins, I: 52, 53; II: 4
and ochratoxins, I: 23
and trichothecenes, I: 16
and zearalenone, I: 65; II: 147
Turkey X disease, II: 4
and aflatoxins, I: 20
Tyrosine aminotransferase (TAT), and cytochalasins, II: 76, 77
Tyrosine transaminase, II: 21

U

Uganda, I: 8, 9, 120-122, 130-131, 151
Umbelliferae, II: 106
United States, I: 8, 13, 21, 60, 132, 176
Unscheduled DNA synthesis, II: 17, 140
Unsymmetrical dimethylhydrazine, I: 164
Uracil
   and cytochalasins, II: 87
   and psoralens, II: 111
   and trichothecenes, II: 38
Urea, and aflatoxins, II: 26
Urea nitrogen, and zearalenone, II: 149
Urease, II: 25, 43
   and penicillic acid, II: 126
   and sterigmatocystins, II: 2
Uridine, II: 26
   and cytochalasins, II: 75
   and penicillic acid, II: 125
   and psoralens, II: 115
   and sterigmatocystins, II: 141
   and trichothecenes, II: 38
   and zearalenone, II: 150
H-Uridine, II: 18
   and zearalenone, II: 150
Uridine-5-pyrophosphate, and cyclochlorotine, II: 102
Uridine transport, and aflatoxins, II: 18
Urinary bladder preparations, and cytochalasins, II: 76
Urine, I: 7, 10; II: 6
   and aflatoxins, I: 57, 123, 134; II: 8, 14-16
   and Balkan (endemic) nephropathy, I: 115
   and luteoskyrin, II: 96
   and N-nitroso compounds, I: 159, 222, 223, 225, 232
   and ochratoxins, I: 22; II: 131, 132
   and penicillic acid, II: 124
   and psoralens, II: 109
   and risk assessment, I: 149
   and sterigmatocystins, I: 63; II: 140
   and trichothecenes, II: 38
   and zearalenone, II: 148
C-Urine, and luteoskyrin, II: 98
U.S. National Academy of Sciences, I: 174, 228-230
Uterotropic activity, and zearalenone, I: 17; II: 148-153
Uterus
   and aflatoxins, II: 22
   and ergot alkaloids, I: 109
   and N-nitroso compounds, I: 206, 222
   and trichothecenes, I: 63
   and zearalenone, I: 17, 65; II: 149-151, 153
UVA, see UV light
UV light (UVA), see also Long wave UV, II: 107-112, 115-117, 157

V

Vaccinia viruses, II: 70
Vagina, and N-nitroso compounds, I: 191, 198, 205, 222
Vasopressin, and cytochalasins, II: 71, 76
Vegetables, see also specific vegetables
   and N-nitroso compounds, I: 222
Vero cells, and cytochalasins, II: 70, 92
Verrucarin A, II: 30, 31, 34, 38-40, 42, 43
Verrucarin B, II: 34, 40
Verrucarin J, II: 34, 40
Verrucarins, II: 30, 31, 34, 41, 44, 45
Verrucarol, II: 40, 45
Vertebrae, and ochratoxins, II: 130
Verticimonosporium, I: 60; II: 30
Verticimonosporium diffractum, II: 35
Vertisporin, II: 35
Vinblastine, II: 77, 82
Vinca, II: 77
Vincristine, II: 77, 82
Vinyl chloride, I: 239
Viral attachment, and cytochalasins, II: 70
Viral hepatitis, I: 129
Viral infectivity, and cytochalasins, II: 69-71, 92
Viral replication, II: 69
   and cytochalasins, II: 70
Virions, and cytochalasins, II: 70
Viropexis, and cytochalasins, II: 69
Viruses
   and cytochalasins, II: 73, 76
   DNA, II: 70, 107
   hepatitis-B, I: 11
   herpes simplex (HSV), II: 70, 71, 92
   kirsten murine sarcoma-leukemia, II: 70
   oncornavirus, II: 70
   poliovirus, II: 92
   RNA, II: 107, 118
   SV40, II: 17, 58, 82, 87, 89, 115
   vaccinia, II: 70
Virus transformed cells, II: 17, 58
   and cytochalasins, II: 56, 74, 82, 84, 87, 89
   and psoralens, II: 115
Vitamin A
   and aflatoxins, II: 26
   deficiency of, II: 5
Vitiligo, II: 106
Vomitoxin, I: 5, 6, 14, 16; II: 33
Vulva, and zearalenone, I: 65

W

Waltham, Massachusetts, I: 167
Water analysis, and N-nitroso compounds, I: 158
Wheat, I: 2, 168
   and Balkan (endemic) nephropathy, I: 116
White fat cells, and cytochalasins, II: 84, 86
WHO, see World Health Organization
Wilmington, Delaware, I: 165, 167
Wine, and N-nitroso compounds, I: 176
World Health Organization (WHO), I: 52, 223

X

Xanthotoxin, and psoralens, II: 106
Xanthotoxol, II: 115, 118
Xeroderma, and sterigmatocystins, II: 141
Xeroderma pigmentosum, II: 17, 116
   and sterigmatocystins, II: 138, 140

Y

Yeast, II: 6
   and penicillic acid, II: 126
   and psoralens, II: 115
   and trichothecenes, II: 39, 43
Yeast glyceraldehyde-3-phosphate, and cytochalasins, II: 94
Yeast polyribosomes, and trichothecenes, II: 43
Yeast spheroplasts, and trichothecenes, II: 38, 39, 42
Yellow rice disease, I: 4, 114-115
Yellow rice toxins, I: 5
   and risk assessment, I: 142
Yugoslavia, I: 115, 116, 142

Z

Zaire, I: 221
Zambia, I: 168, 221
Zearalane, II: 153
Zearalanol, II: 150-153
Zearalanone, II: 151, 153
Zearalenol, II: 146, 147
   pharmacokinetics of, II: 148
   physiological effects of, II: 148-150
cis-Zearalenol, II: 151
trans-Zearalenol, II: 151
Zearalenone, I: 6, 16-18, 65-66, 112; II: 2, 157
   and aflatoxins, I: 65
   and α-aminoisobutyric acid, II: 150
   and anabolic activity, II: 148-149
   and anabolic properties, I: 17
   and anabolism, II: 148, 149
   and antiestrogens, II: 149
   and *Bacillus subtilis*, II: 147
   and barley, I: 18
   and bile, II: 148
   and blood, II: 149
   and brine shrimp, II: 147
   and cathepsin, II: 151
   and chicken embryos, II: 147
   and chickens, II: 147
   and chromatin, II: 150
   and corn, I: 17, 18, 65, 71
   and cows, I: 17; II: 147-150
   and cyclic AMP, II: 146
   and cytoplasmic receptor proteins, II: 150, 152
   and cytotoxicity, II: 147
   detection of, I: 29-31
   and diethylstilbestrol (DES), II: 148, 149, 152
   and DNA, II: 149, 150
   and DNA repair, II: 147
   and DNA synthesis, II: 150
   and ear, II: 148
   and 17β-estradiol, II: 149, 150
   and estradiol-receptor complex, II: 150
   and estrogen, II: 146, 148, 149, 151
   estrogenic potency of, II: 152
   and estrogen receptors, II: 150
   and estrus, II: 148
   and feces, II: 148
   and feedstuff, I: 71
   and fibroblasts, II: 147
   and follicle-stimulating hormone, II: 149
   and fungus, II: 146, 149
   and *Fusarium*, II: 146
   and *Fusarium roseum*, I: 65; II: 146, 152
   and glucose, II: 149, 150
   and glucuronide conjugates, II: 148
   and glycine transport, II: 151
   and gonadotropin, II: 149
   and growth hormone, II: 149
   and hay, I: 17
   and hormones, II: 148, 149
   and humans, II: 146, 148
   and hyperestrogenism, I: 65, 70; II: 146
   and insulin, II: 149
   and intestine, II: 148
   and kidney, II: 148, 149, 151
   and lambs, II: 148
   and C-leucine, II: 150
   and lipids, II: 149
   and liver, II: 148, 149, 151
   and luteinizing hormone, II: 149
   and luteoskyrin, II: 147
   and lysomes, II: 151
   and maize, I: 71
   and mammary, I: 17; II: 146, 150
   and C-methionine, II: 150
   and 3-0-methylglucose, II: 150
   and mice, II: 146, 148, 149, 151
   and mitotic index, II: 9
   and monkeys, II: 149
   and muscle, I: 65; II: 148, 149
   and mutagenicity, II: 147
   and nuclei, II: 150
   and ovary, I: 17, 65
   and perithecia, II: 146, 152
   physical properties of, I: 30
   and pigs, I: 17, 18, 70; II: 146, 147
   and progesterone, II: 149
   and protein, II: 146, 149-152
   and rabbits, II: 151
   and rats, I: 17, 65; II: 146-150, 153
   and rectum, I: 17
   and reproductive hormone, II: 148
   and reticulocytes, II: 151
   and RNA, II: 149
   and RNA synthesis, II: 150
   and RNA transcription, II: 150
   and *Saccharomyces*, II: 147
   and *Salmonella*, II: 147

screening for, I: 36
and serum, II: 148, 149
and sheep, II: 150
and skeleton anomalies, I: 17
and skin, II: 151
and sodium, II: 151
and sorghum, I: 71
and spleen, II: 149
and stomach, II: 148
structure-activity relationships in, II: 151-153
and sulfate conjugates, II: 148
synthetic analogs of, II: 152
and teratogenicity, I: 65; II: 147
and testis, I: 17; II: 147
and thyroid, II: 147
and toxicity, I; 65; II: 146, 147
and transcription, II: 150
and trichothecenes, I: 15, 17
and turkeys, I: 65; II: 147
and urea nitrogen, II: 149
and uridine, II: 150
and H-uridine, II: 150
and urine, II: 148
and uterotropic activity, I: 17; II: 148-153
and uterus, I: 17, 65; II: 149-151, 153
and vulva, I: 65
cis-Zearalenone, II: 150
trans-Zearalenone, II: 150
Zebra fish
 and aflatoxins, II: 4, 5, 7
 and N-nitroso compounds, I: 190
 and ochratoxins, II: 130
 and sterigmatocystins, II: 138, 143
 and trichothecenes, II: 37
Zeiosis, II: 56, 57, 59, 92, 94
Zinc, I: 3
Ziram, I: 176
Zoxazolamine hydroxylase, II: 21
Zygosporin D, II: 51, 54, 55
Zygosporin E, II: 51, 54, 55
Zygosporin F, II: 54, 55
Zygosporin G, II: 51, 54, 55
Zygosporins, II: 48, 54
 and cytochalasins, II: 91
Zygosporium, II: 48
Zygosporium masonii, II: 50, 51